T0296625

Cognitive Computing for Human-Robot Interaction

Principles and Practices

Cognitive Data Science in Sustainable Computing

Cognitive Computing for Human-Robot Interaction

Principles and Practices

Edited by

Mamta Mittal

Department of Computer Science and Engineering, G. B. Pant Government Engineering College, New Delhi, India

Rajiv Ratn Shah

Department of Computer Science and Engineering (joint appointment with the Department of Human-centered Design), IIIT-Delhi, New Delhi, India

Sudipta Roy

PRTTL, Washington University in Saint Louis, Saint Louis, MO, United States

Series Editor

Arun Kumar Sangaiah

School of Computing Science and Engineering, Vellore Institute of Technology (VIT), Vellore, India

Academic Press is an imprint of Elsevier
125 London Wall, London EC2Y 5AS, United Kingdom
525 B Street, Suite 1650, San Diego, CA 92101, United States
50 Hampshire Street, 5th Floor, Cambridge, MA 02139, United States
The Boulevard, Langford Lane, Kidlington, Oxford OX5 1GB, United Kingdom

British Library Cataloguing-in-Publication Data
A catalogue record for this book is available from the British Library

Library of Congress Cataloging-in-Publication Data
A catalog record for this book is available from the Library of Congress

ISBN: 978-0-323-85769-7

For Information on all Academic Press publications
visit our website at https://www.elsevier.com/books-and-journals

Publisher: Mara Conner
Acquisitions Editor: Sonnini R. Yura
Editorial Project Manager: Rachel Pomery
Production Project Manager: Niranjan Bhaskaran
Cover Designer: Matthew Limbert

Typeset by MPS Limited, Chennai, India

Contents

List of contributors

Shivani Agarwal KIET School of Management, KIET Group of Institutions, Delhi-NCR, Ghaziabad, India

Sushila Aghav-Palwe School of CET, MIT World Peace University, Pune, India

Shaik Vaseem Akram School of Electronics & Electrical Engineering, Lovely Professional University, Phagwara, India

A. Anny Leema Vellore Institute of Technology, Vellore, India

Gopi Battineni Telemedicine and Tele Pharmacy Centre, School of Medical Products and Sciences, University of Camerino, Camerino, Italy

S.R. Boselin Prabhu Surya Engineering College, Mettukadai, Erode, India

Debatri Chatterjee TCS Research and Innovation, Tata Consultancy Services, Kolkata, India

Sushabhan Choudhury University of Petroleum and Energy Studies, Dehradun, India

Da Peng School of Mechanical Engineering, Ningxia University, Yinchuan, China

Prabin Kumar Das School of Electronics & Electrical Engineering, Lovely Professional University, Phagwara, India

Suman Deb Department of Computer Science and Engineering, National Institute of Technology Agartala, Agartala, India

D. Devi Sri Krishna College of Engineering and Technology, Coimbatore, India

Leena N. Fukey Christ University, Bengaluru, India

Rahul Gavas TCS Research and Innovation, Tata Consultancy Services, Kolkata, India

Anita Gehlot School of Electronics & Electrical Engineering, Lovely Professional University, Phagwara, India

Lalit Mohan Goyal Department of CE, J. C. Bose University of Science & Technology, YMCA, Faridabad, India

Anita Gunjal School of CET, MIT World Peace University, Pune, India

H.M.K.K.M.B. Herath Department of Mechanical Engineering, Faculty of Engineering Technology, The Open University of Sri Lanka, Nugegoda, Sri Lanka

K.K.L. Herath Department of Mechanical Engineering, Faculty of Engineering Technology, The Open University of Sri Lanka, Nugegoda, Sri Lanka

Divneet Singh Kapoor Embedded Systems & Robotics Research Group, Chandigarh University, Mohali, India

Kalpana Katiyar Department of Biotechnology, Dr. Ambedkar Institute of Technology for Handicapped, Kanpur, India

Sarthak Katiyar Department of Computer Science & Engineering, Rajkiya Engineering College, Sonbhadra, India

Ravinder Khanna Department of ECE, MM University, Sadopur, Ambala, India

Rajender Kumar Department of Computer Science & Engineering, IKG Punjab Technical University, Jalandhar, India

Surender Kumar Department of Computer Science & Engineering, Guru Teg Bahadur College, Bhawanigarh, Sangrur, India

Quanjin Ma School of Mechanical Engineering, Ningxia University, Yinchuan, China; Structural Performance Materials Engineering (SUPREME) Focus Group, Faculty of Mechanical & Automotive Engineering Technology, Universiti Malaysia Pahang, Pekan, Malaysia

B.G.D.A. Madhusanka Department of Mechanical Engineering, Faculty of Engineering Technology, The Open University of Sri Lanka, Nugegoda, Sri Lanka

S.S. Manjunath ATME, Mysore, India

Mamta Mittal Department of CSE, G. B. Pant Government Engineering College, New Delhi, India

Tutan Nama Department of Computer Science and Engineering, Indian Institute of Technology Kharagpur, Kharagpur, India

Krishna Sai Narayana Renault Nissan Technology & Business Centre Pvt. Ltd.

Anubha Parashar Department of CSE, Manipal University Jaipur, Jaipur, India

Apoorva Parashar Department of CSE, Maharshi Dayanand University, Rohtak, India

H.D.N.S. Priyankara Department of Mechanical Engineering, Faculty of Engineering Technology, The Open University of Sri Lanka, Nugegoda, Sri Lanka

S. Ravi Shankar Department of CSE, Vidya Academy of Science and Technology, Thrissur, India

M.R.M. Rejab School of Mechanical Engineering, Ningxia University, Yinchuan, China; Structural Performance Materials Engineering (SUPREME) Focus Group, Faculty of Mechanical & Automotive Engineering Technology, Universiti Malaysia Pahang, Pekan, Malaysia

Sanjoy Kumar Saha Department of Computer Science and Engineering, Jadavpur University, Kolkata, India

Roopkatha Samanta Department of Computer Science and Engineering, Jadavpur University, Kolkata, India

Atharva Sandeep Vidwans Pune, India

E.D.G. Sanjeewa Department of Mechanical Engineering, Faculty of Engineering Technology, The Open University of Sri Lanka, Nugegoda, Sri Lanka

K. Seemanthini DSATM, Bangalore, India

Subramani Sellamani Renault Nissan Technology & Business Centre Pvt. Ltd.

Kiran Jot Singh ECE Department, Chandigarh University, Mohali, India

Rajesh Singh School of Electronics & Electrical Engineering, Lovely Professional University, Phagwara, India

Ashutosh Sinha Robonomics AI India Pvt Ltd., Bangalore, India

Mudita Sinha Christ University, Bengaluru, India

Balwinder Singh Sohi Embedded Systems & Robotics Research Group, Chandigarh University, Mohali, India

S. Sophia Sri Krishna College of Engineering and Technology, Coimbatore, India

Bo Sun School of Mechanical Engineering, Ningxia University, Yinchuan, China

Zidong Yang School of Mechanical Engineering, Ningxia University, Yinchuan, China

Hao Yao School of Mechanical Engineering, Ningxia University, Yinchuan, China

Guangxu Zhu School of Mechanical Engineering, Ningxia University, Yinchuan, China

About the editors

Mamta Mittal graduated in Computer Science & Engineering from Kurukshetra University Kurukshetra in 2001 and received a Master's degree (Honors) in Computer Science & Engineering from YMCA, Faridabad. She completed her PhD at Thapar University Patiala in Computer Science and Engineering. Her research areas include data mining, big data, and machine learning. She has been teaching for the last 18 years with an emphasis on data mining, machine learning, soft computing, and data structure. She is a lifetime member of CSI. She has published and communicated more than 70 research papers in SCI, SCIE, and Scopus indexed Journals and attended many workshops, FDPs, and Seminars. She has filed nine patents in the area of Artificial Intelligence and Deep Learning, out of which two are granted and many are published online. Presently, she is working at G.B. PANT Government Engineering College, Okhla, New Delhi (under Government of NCT Delhi) and supervising PhD candidates of Guru Gobind Singh Indraprastha University, Dwarka, New Delhi. She is the main editor of *Data Intensive Computing Applications for Big Data* published by IOS press, the Netherlands and also the editor of *Big Data Processing Using Spark in Cloud* by Springer and many more. She is the managing editor of the International Journal of Sensors, Wireless Communications and Control Published by Bentham Science. She is working on DST approved Project "Development of IoT based hybrid navigation module for mid-sized autonomous vehicles" with a research grant of 25 Lakhs and Handing Pradhan Mantri YUVA project for Entrepreneur Cell Activity as Faculty Coordinator/ Representative from G. B. Pant Govt. Engineering College. She is the reviewer of many reputed journals (Including Wiley, IEEE Access, Elsevier, etc.) and has chaired several conferences.

Rajiv Ratn Shah currently works as an Assistant Professor in the Department of Computer Science and Engineering (joint appointment with the Department of Human-centered Design) at IIIT-Delhi. He is also the director of the MIDAS Lab at IIIT-Delhi. He received his PhD in Computer Science from the National University of Singapore, Singapore. Before joining IIIT-Delhi, he worked as a research fellow in Living Analytics Research Center at the Singapore Management University, Singapore. Before completing his PhD, he received his MTech and MCA degrees in Computer

Applications from the Delhi Technological University, Delhi and Jawaharlal Nehru University, Delhi, respectively. He has also received his BSc in Mathematics (Honors) from the Banaras Hindu University, Varanasi. Dr. Shah is the recipient of several awards, including the prestigious Heidelberg Laureate Forum and European Research Consortium for Informatics and Mathematics fellowships. He won the best student poster award at the 33rd AAAI Conference on Artificial Intelligence in Honolulu, HI, United States and won the best poster runner-up award at the 20th IEEE International Symposium on Multimedia (ISM) conference in Taichung, Taiwan. Recently, we also won the best poster and best industry paper awards at the 5th IEEE International Conference on Multimedia Big Data (BigMM) conference. He is also the winner of the 1st ACM India Student Chapter Grand Challenge 2019. He has also received the best paper award in the IWGS workshop at the ACM SIGSPATIAL conference 2016, San Francisco, CA, United States and was the runner-up in the Grand Challenge competition of ACM International Conference on Multimedia 2015, Brisbane, QLD, Australia. He is involved in organizing and reviewing many top-tier international conferences and journals. He is TPC cochair for IEEE BigMM 2019 and BigMM 2020. He has also organized the Multimodal Representation, Retrieval, and Analysis of Multimedia Content (MR2AMC) workshop in conjunction with the first IEEE MIPR 2018 and 20th IEEE ISM conferences. His research interests include multimedia content processing, natural language processing, image processing, multimodal computing, data science, social media computing, and the Internet of Things.

Sudipta Roy received his PhD in Computer Science & Engineering from the Department of Computer Science and Engineering, University of Calcutta. He is the author of more than 40 publications in refereed national/international journals and conferences, including IEEE, Springer, and Elsevier. He is the author of one book and many book chapters. He holds a US patent in medical image processing and filed an Indian patent in smart agricultural system. He has served as a reviewer of many international journals including IEEE, Springer, Elsevier, and IET, and international conferences. He has served international advisory committee member and program committee member of AICAE-2019, INDIACom-2019, CAAI 2018, ICAITA-2018, ICSESS-2018, INDIACom-2018, ISICO-2017, AICE-2017, and many more conferences. He is serving as an associate editor of IEEE Access, IEEE, and International Journal of Computer Vision and Image Processing, IGI Global Journal. He has more than 5 years of experience in teaching and research. His fields of research interests are healthcare image analysis, image processing, steganography, artificial intelligence, big data analysis, machine learning, and big data technologies. Currently, he is working at PRTTL, Washington University in St. Louis, Saint Louis, MO, United States.

Preface

Cognitive computing is the use of computerized models to simulate the human thought process in complex situations where the answers may be ambiguous and uncertain. Using self-learning algorithms that use data mining, pattern recognition, and natural language processing, the computer can mimic the way the human brain works. So it can be incorporated with Human–Robot Interaction (HRI) for assisting humans in various forms.

The study of interactions between humans and robots is thus fundamental to ensure the development of robotics and to devise robots capable of socially interacting intuitively and easily through speech, gestures, and facial expressions. HRI is an integrative field that has provided considerable improvement in various applications, including human–computer interaction (HCI), robotics and artificial intelligence, the service robots combined with cognitive computing to design intelligent robotic services. This robotics provides novel services in terms of companionship for the elderly and disabled and does household chores in the home environment. It lays at the crossroad of many subdomains of AI and, for effect, it calls for their integration: modeling humans and human cognition; acquiring, representing, manipulating in a tractable way abstract knowledge at the human level; reasoning on this knowledge to make decisions; and eventually instantiating those decisions into physical actions both legible to and in coordination with humans. Unique features of *Cognitive Computing for Human–Robot Interaction: Principles and Practices* are as follows:

- Introduces several new contributions related to the representation and the management of humans in an autonomous robotic system.
- The main strength of this powerful and reputable collection is to collect and provide knowledge regarding cognitive computing and HRI.
- It will give a challenging approach of those several repercussions of cognitive computing and HRI in the actual global scenario.
- This covers a wide spectrum of topics that will be of interest among groups of readers consisting of professionals, educator, academics, and students.
- It will also help educators be aware of the changes and of what society needs from them.

Supported by experimental results, this book focuses on how explicit knowledge management, both symbolic and geometric, proves to be instrumental to richer and more natural HRIs by pushing for pervasive, human-level semantics within the robot's deliberative system. It has been organized into 17 chapters.

Chapter 1, *Introduction to Cognitive Computing and its Various Applications*, explains that cognitive computing is an intelligent system that converses with and mimics the human being in a natural form by learning at scale, reasoning with purpose. It falls under the third era of computing and now has attracted considerable attention in both academia and industry. Nowadays, there is explosive data growth, business conditions are also changing rapidly; so smart, hassle-free, and enhanced interactions amongst human beings and technology can be effectively addressed by cognitive systems. Some examples of personal assistants that use cognitive computing are Alexa, Siri, Google Assistant, and Cortana. Some more applications that can use cognitive computing to gain benefits from this type of technology are cognitive computing for changing business values, financial and investment firms, healthcare and veterinary medicine, travel and tourism, health and wellness, education and learning, agriculture, communication, and network technology.

Chapter 2, *Recent Trends Toward Cognitive Science: From Robots to Humanoids*, focuses on how cognitive processing is to replicate the human thought processes in a computerized model. Employing self-learning algorithms that use data mining, pattern recognition, and natural language processing concerning this specific subject in which the computer can mimic the human brain function. Following the diversity of user's interpretation of problems, the cognitive computing system presents the reorganization of types of data and sways meaning and analysis. When concluding, the cognitive computing system usually relies on considering contradictory proof and proposes a solution that is "best" more than precisely "right." The systems with cognitive computing ability create situations quantifiable. Machines with cognitive systems recognize as well as excerpt factor characteristics for example location, time, history, work, or description to portray the dataset suitable for a person or a relying implementation involved in a particular method at an exact schedule and location. These systems reframe the scenery involving the association of individuals and their progressively all-encompassing digital environment. This has been shown to play a larger role as the mentor or assistant for the user; similarly, they may behave digitally independently for resolving various problems.

Chapter 3, *Cognitive Computing in Human Activity Recognition With a Focus on Healthcare*, presents that human activity recognition is a quintessence to empower a robot to distinguish the conduct of a personal care-receiver. As opposed to outward appearances, an activity recognition can see practices of a consideration beneficiary, who might be a senior adult, a youngster, or a chronic patient. Through human activity recognition, a robot tracks the care patient activity and perceives human practices like unhealthy habits and anomalous activities. However, patient activity recognition through simple images is a highly challenging task. Several challenges such as the likeness of unmistakable human behaviors, disorder background, similarities in different human activities may significantly reduce the classification performance. Because of rapid developments in cutting-edge machine learning models, substantial solutions can arise from distinct deep learning algorithms, including Convolutional Neural Network (CNN), Generative Adversarial Network. In this chapter, the authors review several cognitive computing approaches in the advancement of HRI, especially in healthcare industries.

Chapter 4, *Deep Learning-based Cognitive State Prediction Analysis Using Brain Wave Signal*, presents the analysis of cognitive state during different learning tasks using EEG signals. Online learning tools such as videoconferencing, multimedia lessons, digital materials, and e-learning platforms with options for both real-time learning and self-paced learning provide a pleasant and immersive experience. In addition to these features, assessment of cognitive state during the learning phase has been proven to improve the efficiency of online learning. The analysis of cognitive state during different learning tasks using EEG signals that were obtained using 128 channel Emotive Epoch headset device is the main focus of this study. Artifacts prominent in raw signals were filtered by using linear filtering. To determine the exact concentration levels, the fuzzy fractal dimension measures and the Discrete Wavelet Transform were adapted to the same extracted Electroencephalogram (EEG) signals for feature extraction. The extracted parameters are then classified into concentration levels using the deep learning algorithm Enhanced Convolutional Neural Network (ECNN), which has proven to be of higher accuracy compared to other classifiers. ECNN can then be used to control cognitive states as a feedback mechanism.

Chapter 5, *EEG-Based Cognitive Performance Evaluation for Mental Arithmetic Task*, discusses an appropriate framework to assess participant's cognitive performance based on their brain activity dynamics recorded through an EEG device. To this aim, the authors have used a publicly available EEG dataset. The dataset contains EEG recording of 36 subjects before and during a mental arithmetic task. The participants were divided into two subgroups (good and bad performers) based on the accuracy of the task performed. A simple but novel approach has been proposed to summarize these window-level features and formulate the signal-level descriptor. The descriptor thus formed, captures the distribution of the feature values effectively. Experimental results suggest that the proposed descriptor, obtained after summarizing the window level EEG domain features, performs satisfactorily in discriminating between the two sets of performers. Mean classification accuracy obtained was about 85% using Gaussian naïve Bayes classifier which outperformed EEG domain feature-based classification models.

Chapter 6, *Trust or No Trust in Chatbots: A Dilemma of Millennial*, empirically investigates what dilemma millennials have regarding the perception of trust while interacting with chatbots and how the perception of trust can be measured in terms of this research context. The major findings of the study are that millennials' perception of trust has been measured into two dimensions. The first dimension of trust is cognitive-based trust which represented that "we choose whom we will trust in which respect and under what circumstances, and we base the choice on what we take to be "good reasons," constituting evidence of trust-worthiness" conversing with the chatbot. The second dimension is *affect-based trust* "which is made by emotional bonds between individuals that go beyond a regular business or professional relationship." So, it is recommended for the organizations to implement chatbots in their premises for the initial interaction with the millennials.

Chapter 7, *Cognitive Computing in Autonomous Vehicles*, discusses the neuromorphic architecture and how it is inspired by the human brain which mimics micro neurobiological architecture present in nervous systems, and the Von Neumann architecture model combines to form what we call Cognitive Computing, its applications and advantages in AVs. For this, It takes into account hardware components and mathematical models required for the design of an autonomous vehicle. It focuses on different Cognitive Artificial Intelligence techniques and Algorithms that will help us to achieve closeness to human-level performance or what we call Level 5 autonomy in AVs having unlimited Operational Design Domain. Achieving level 5 autonomy is an extremely difficult task because it requires almost perfect decision making, Object and Event Detection and Response, localization even in uncertain conditions like cloudy weather, fog, extreme darkness, and rain, which act as a forestall to vision task and localization. That is where cognitive computing comes into the picture. Cognitive computation techniques enhance the accuracy of the model to achieve human-like performance in decision making or even in object detection. Yet even above all this, there is greater scope for development, as perfect Level 5 autonomy is still not achieved. Furthermore, this chapter explores the further developments possible in this field of AVs and the effects which it will cast on future generations.

Chapter 8, *Optimized Navigation Using Deep Learning Technique for Automatic Guided Vehicle*, presents that autonomous driving has passed the point of being called the biggest step, as the smart car revolution is already taking shape around the world. Self-driving cars are relevant if not prevalent and the biggest obstacles to reach mass adoption are customer acceptance, cost, infrastructure, and the reliance on several onerous algorithms that include perception, lane marking detection, path planning, and variation in pathways. This study tackled the mentioned problems with a straightforward and cost-effective solution, using end-to-end learning and replacing the numerous sensors with a camera and commandeering just the forward, backward, left, and right controls. In this research, the authors have used the most popular method of deep learning that is CNN to train the collected data on the VGG16 model. Later these have optimized directly by the proposed system with cropping each unnecessary image and mapping pixels from a single front-facing camera to direct steering instructions. It has been observed from the experimental work that the proposed model has given a better result than the existing work that is increasing in the accuracy from 88% (Udacity training dataset) to 98% (proposed). This model is suitable for industrial use and robust in real-time scenarios, therefore, can be applied in modern industrialized systems.

Chapter 9, *Vehicular Middleware and Heuristic Approaches for ITS Systems of Smart Cities*, discusses that a smart city comprises the intellectual use of the vehicular system. This technical revolution from wireless telephony to Vanet networks provides Intelligent transport system for Smart cities. The middleware and heuristic approaches of Vehicular system are essential for designing and analyzing research problems related to smart cities. The review of the literature played a critical part in determining numerical knowledge about the smart cities with the vehicular system, middleware and heuristic approaches, find gaps in published research, and produce new original ideas for the intelligent transport system of Smart cities.

Chapter 10, *Error Traceability and Error Prediction Using Machine Learning Techniques to Improve the Quality of Vehicle Modeling in Computer-Aided Engineering* discusses how machine learning techniques playing the role of bug prediction and effectively increases the accuracy rate in vehicle modeling. While designing a vehicle, the designer follows a step-by-step process to reduce the unintended software behavior to complete the model. Locating Errors in industry-size software systems is time-consuming and challenging. Hence an automated approach is proposed to trace the errors helps the CAD engineer to complete the designing task in less time. Error Traceability and Error Prediction are the two major tasks focused on the designing stage. Software fault prediction techniques are used to predict the faults in the software and machine learning techniques is playing an important role in detecting the software default. Bug prediction and correction of bugs improve the software quality and reduces the maintenance cost.

Chapter 11, *All About Human–Robot Interaction*, illustrates a generalized framework and metric taxonomies for robot design through a detailed review of the literature. The framework presents an end-to-end product design overview for hobbyists and researchers interested in the general steps to create a fully autonomous, consumer-focused social robots with perception and reasoning capabilities that aim to achieve high technology readiness. The latest developments in science and technology have led to the expansion and deployment of robots in various applications covering all spheres of human life. This necessitates the interaction of humans with robots on a larger scale, as past evidence suggests the robots are considered more as companions rather than machines/tools. HCI provides insights in understanding and improving interactions with computer-based technologies. However, HRI takes a cue from HCI by introducing autonomy, physical proximity, and the ability to make decisions in addition to HCI techniques for a robotic system, which makes HRI a distinct area for research. Taking into consideration the huge number of interactions between humans and robots, there is a stringent requirement to standardize and make fixed protocols to ensure the usage of robotic technology in a responsible and principled way.

Chapter 12, *Teleportation of Human Body Kinematics for a Tangible Humanoid Robot Control*, presents that People learn best when they use sensory-based perceptual learning styles. To model this "action-learning," a didactic design was used to create an instructional resource and applied through humanoid robot interaction. The quasiexperimental result of the interaction analysis has revealed higher retention of learning contents by participants. This work is a systematic approach to involve the kinetic movement of human limbs with sensory organs with teleportation of gesture movements on humanoid robots, which has made twofold coordination between human and machine in a live interaction. People with learning difficulty as well as gamification of learning in elementary classes can be addressed with this augmented approach in conventional pedagogy. The algorithmic percept action sequence has created a unified order of closed-loop interaction across different cognitive level people. The experimental results have revealed substantial evidence of learning enhancement and higher order logical understanding by using the proposed immersive extension of natural body movement identification by machines and teleportation of the same to another machine.

Chapter 13, *Recognition of Trivial Humanoid Group Event Using Clustering and Higher Order Local Auto-correlation Techniques* presents that a video surveillance system is explicitly established to manage a small human group. The proposed methodology concentrated on the trivial humanoid groups that remained in an identical location for some time and characterized the group activity. The defined methodology has widespread applications in numerous areas such as video reconnaissance systems, cluster interface, and activities classification. The video surveillance system primarily covenants with the action recognition and classifying the cluster activity by considering violent activities such as fighting. The steps in action recognition include generation of frames, segmentation using fuzzy c-mean clustering, feature extraction by completed local binary pattern and high order local autocorrelation, classification by Recurrent Neural Network. Spatial Gray Level Difference Method extracts the statistical features while bag-of-words technique creates vocabulary features set which is derived from Local Group Activity descriptors. CNN classifier employed to classify the human activities.

Chapter 14, *Understanding the Hand Gesture Command to Visual Attention Model for Mobile Robot Navigation: Service Robots in Domestic Environment*, presents that in recent years, robotic systems and techniques have acquired the unprecedented capacity for perceiving and understanding their world not just in a low-level manner but even close to humanly understandable concepts. HRI is used frequently in the care of the aged and the disabled population. Human behavior is expected to achieve natural interaction from these robots. Human activity is essential both before and after contact with a human user is initiated. Intelligent service robots in evolving fields of robotic technologies, from entertainment to healthcare, are currently being built to satisfy demand. The service robots have controlled by nonexpert users, and direct contact with them and their human users will be their support activities. With these service robotics, human-friendly social features are typically favored. Individuals tend to use voice commands, responses, and suggestions to express their peers' opinions.

Chapter 15, *Mobile Robot for Air Quality Monitoring of Landfilling Sites using Internet of Things*, discusses air quality monitoring is a vital mechanism for continuously observing the quality of air in the environment. Generally, landfilling sites are filled with mixed waste of various materials, this mixed waste generates stinky gases. Robots are the solution for monitoring the air quality of landfilling sites. In this study, an IoT and cloud server–enabled mobile robot–based model has been proposed for monitoring the air quality at the landfill sites. LoRa radio-based sensor nodes are embedding with distinct gas sensors for sensing in landfill sites and transmits to the mobile robot. A mobile robot is an integration of LoRa radio and GPRS communication, it transmits the sensory data to the cloud server via internet protocol. The sensory data of different gases in the landfilling sites are recorded in the cloud server and also a graphical representation of the sensory data is discussed briefly.

Chapter 16, *AI and IoT Readiness: Inclination for Hotels to Support a Sustainable Environment*, presents that Smart Cities has been one of the key driving factors for the urban transformation to a low carbon climate, sustainable economy and mobility in recent years because of the alarming situation of Global warming. One of the industries with swift growth is the hotel sector and hence is one of the key contributors to carbon emission and leaves environmental footprints. The new emerging concept of sustainable tourism is envisaged as an important part of the Smart Cities paradigm. Improving sustainability by saving energy is becoming a primary task today for many hotels. A great opportunity is provided by AI and IoT to assimilate different systems on a platform by encouraging and assisting hotel guests to operate through a single device and optimizing hotel operations. Current research focuses to identify the strategic positions of a hotel in terms of sustainability, AI and IoT technology. Components that will be considered by Hotels for the strategic intention of adopting AI and IoT for environmental sustainability.

Different development and modification needed to be taken if management wants high sustainability readiness and/or IoT readiness. This conceptual paper constructs the comprehensive study and systematic review of different areas where the Hotels can feasibly implement AI and IoT for improving sustainably.

Chapter 17, *Design and Fabrication of an Automatic Classifying Smart Trash Bin Based on IoT*, presents that Municipal Solid Waste is an increasing waste resource and challenging task to solve environmental pollution and waste collection problems. The smart trash bin is sufficient equipment to reduce human intervention and improve living conditions, lacking multifunctional purposes. To solve the problem that traditional trash can lacks self-identification of recyclable garbage and nonrecyclable garbage, a smart trash bin based on WIFI environment and Arduino control is proposed. The smart trash bin takes Arduino as its master controller and interacts with Arduino in the WIFI environment to realize automatic garbage classification and communication between people and the trash bin. This paper aims to describe the hardware and software design ideas of trash bins in the implementation process, and the mechanism of garbage classification and the application of the Internet of Things in garbage recycling are emphatically studied. The results indicate that the system can run stably and achieve accurate classification within 2 seconds, which reduces the waste of resources, changes the traditional garbage management mode, improves management efficiency, and realizes recycling of garbage.

The editors are very thankful to Rachel Pomery, Editorial Project Manager: Elsevier and Series Editor Professor Arun Kumar Sangaiah for providing us the opportunity to edit this book.

Mamta Mittal[1], Rajiv Ratn Shah[2] and Sudipta Roy[3]

[1]Department of CSE, G. B. Pant Government Engineering College, New Delhi, India, [2]Department of Computer Science and Engineering (joint appointment with the Department of Human-centered Design), IIIT-Delhi, New Delhi, India, [3]PRTTL, Washington University in Saint Louis, Saint Louis, MO, United States

Chapter 1

Introduction to cognitive computing and its various applications

Sushila Aghav-Palwe and Anita Gunjal
School of CET, MIT World Peace University, Pune, India

Introduction

In this era of modern computing system, cognitive computing is emerging technological mode, as process automation is expected to evolve the old systems. According to Gartner, cognitive computing will change the technology era unlike none of the technology introduced in the last two decade. In technology, cognitive computing is becoming modern buzzword, capturing the attention of many entrepreneurs and tech enthusiasts.

Cognition is the combination of brain related processes and activities used to learn, perceive, think, understand and remember, to acquire knowledge and understand, experience, and the senses. Cognitive Science is the broad scientific study of the human-mind, the brain, and human-intelligent behavior. As cognitive system stimulates the mechanism of human thinking, the use-cases and benefits of cognitive computing are much more advanced and smarter than artificial intelligence (AI) systems. As it is able to analyze and handle Big data, cognitive computing is capable to be used in today's smart and complex system to solve real-life problem. AI creates new ways to solve problems better than human but cognitive computing tries to replicate human behavior to solve problems (Megha et al., 2017). Cognitive system must adopt the cognitive skills of human like perception, decision making ability, motor skills, language skills and social skills to solve problems. AI can only be intelligent as how individual teach it, but it is not suitable for current cognitive era. Advanced cognitive computing framework uses natural language processing (NLP) with context-aware emotion intelligence, AI, machine learning, neural networks as a foundation to tackle complex routine problems as humans-being. Cognitive computing is defined as "An advanced system that learns at scale, reason with purpose and interacts with humans in a natural form" (Megha et al., 2017).

Cognitive Computing for Human-Robot Interaction. DOI: https://doi.org/10.1016/B978-0-323-85769-7.00009-4

Cognitive system learns the patterns and advises human react appropriately depending on acquired knowledge. It provides an assistance to human rather than one completing the task. The main purpose of cognitive computing is to assists people in decision making which gives them superior accuracy in analysis. Since cognitive computing has the ability to analyze data faster and more accurately, there is no need to worry about the incorrect decisions. If we see example of healthcare system using AI and using cognitive computing, treatment decisions are made without consulting with human doctor in AI system, whereas cognitive computing work one step ahead and assist doctors for disease diagnosis with the help of data repositories and data analysis which helps in smart decision making.

To provide human assistance, Chabot's are important techniques. Chabot development involves a collection of various advance techniques like machine learning, AI, also analytics all of which are involved to make Chatbot experience more conversational.

Cognitive system must analyze data accurately, also it must learn efficient business process. In cognitive systems data collection, comparison and cross-referencing must be highly-efficient to analyze a situation effectively (Megha et al., 2017). In this era of digital evolution, manual processes are identified so that it can be automated by adopting cognitive technology. IBM is becoming pioneered for the cognitive technology sphere by which it is driving many digital organizations across the globe.

Explosive data is analyzed to get insights from historical data for betterment of future. Cognitive technology understands the past and helps in providing assistant for predicting future events with more accuracy. In future, it will be immensely used by the businesses due to its robust and agile nature. Financial and healthcare domain is already leveraged by power of cognitive computing. In the future, with the use of such a technology, it will be helpful to make human-being more work-efficient.

Cognitive computing overview

Cognition is the word with origin from Latin base cognition "get to know," another meaning of the same is "thinking and awareness." Multiple mental processes are involved in these activities related to cognition such as perception, learn, think on, understand and remember. It also includes grasping knowledge and analyze through thinking and past experience.

Cognitive computing involves the methods, algorithms and approaches that supports the system to "think on," "understand" in similar way that the human brain does. In a way the cognitive computing methods works and inculcates all the mentioned activities which human mental processes does.

Cognitive computing arena includes various techniques, that helps to simulate the human cognition in computing world. Various technologies and tools such as machine learning, AI, NLP, exploratory data analytics and

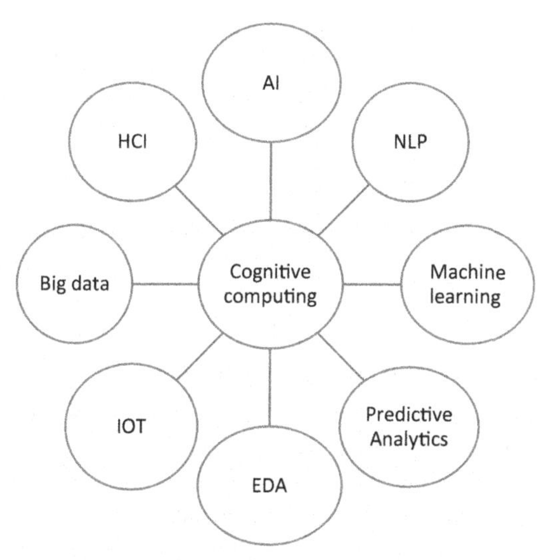

FIGURE 1.1 Cognitive computing technologies. *AI*, Artificial intelligence; *HCI*, Human-Computer-Interaction; *IoT*, Internet of things; *NLP*, Natural Language Processing.

predictive analytics are useful in simulating the human cognition like understanding, remembering, learn and perception. Along with these various technologies like internet of things (IOT), human−computer interfacing, Big data contributes to make cognitive systems reactive.

As shown in Fig. 1.1, various advance technologies make the computer system as cognitive system. Big data is crucial component of cognitive computing systems, as it allows to get insights from the huge voluminous historical data to predict the feature. Some of the other sources of the data to cognitive systems are the data collected through IOT, human−computer interaction (HCI). HCI and IOT retrieve pass the data collected through various sensors or collected through interned across physically distinct locations. Sometimes data is available in audio form or in text form. To process such data using cognitive system, it is needed to convert it in proper understandable format. NLP do the task of conversion of audio data or text data in structured form which is suitable for analysis purpose. Valuable insights from this collected data can be drawn using exploratory data analytics like charts and graphs, or using intelligent machine learning techniques like predictive analytics, prescriptive analytics or denotive analytics.

Cognitive computing techniques

Natural language processing

NLP contributes in cognitive computing by realizing, processing and simulating the human expressions in terms of language expressed in terms of speech or written. In literature (Aghav-Palwe & Mishra, 2017, 2020; Chen & Argentinis, 2016; Chen et al., 2017; Gupta et al., 2020; Orozco et al., 2010; Tian et al., 2016) various methods for NLP are proposed and implemented are bag of words, topic modeling, and lemmatization. Various models for NLP in computer science domain majorly used are state machines and automata, formal rules systems, logic and probability theory. Supervised machine learning methods like linear regression and classification proved helpful in classifying the text and mapping it to semantics.

Artificial intelligence and machine learning

AI methods contributes in cognitive computing by supporting perception, reasoning, learning and problem solving. AI methods like reinforcement learning helps in analyzing the context by learning with the help of past experience and apply/reply with the suggestion/prediction and reasoning of the context. Various intelligence for AI is ranging from musical, personal, logical, spatial and linguistic. Various AI methods and algorithms are proposed for solving the problems in each domain and context. Some of the examples of AI. Current AI application which helps in cognition simulation are driving car, game playing, Chatbots and many more.

Machine learning methods like predictive analytics and exploratory data analytics helps in analyze the hidden information and patterns from the given data. This learning is important in cognitive computing to react upon the situation for business decisions, AI-ML found useful for predicting the future possibilities.

Fig. 1.2 shows the usage of decision tree, supervised machine learning method to classify the customer in aspirant customer category. Considering the past experience as training data and <age, credit rating, income, student> features, the system is able to predict that the customer will buy the computer or not. With such decisions, the cognitive system can then further suggest/percept important strategies to motivate customers for improvement in business.

Internet of things and Big data

IOT and Big data are considered as foundations of the cognitive computing. These are the major resources of data, which is helpful in drawing valuable insights for decision making. Unstructured data which is generated from these resources need to be processed to make it suitable for analyzing and for finding meaningful insights. Cognitive systems process such collected unstructured data and consolidate it to propose the semantics from it.

IOT data gathered from the various sensors/devices is useful in predicting the need of maintenance of devices, relocation, rescheduling, resource management of devices.

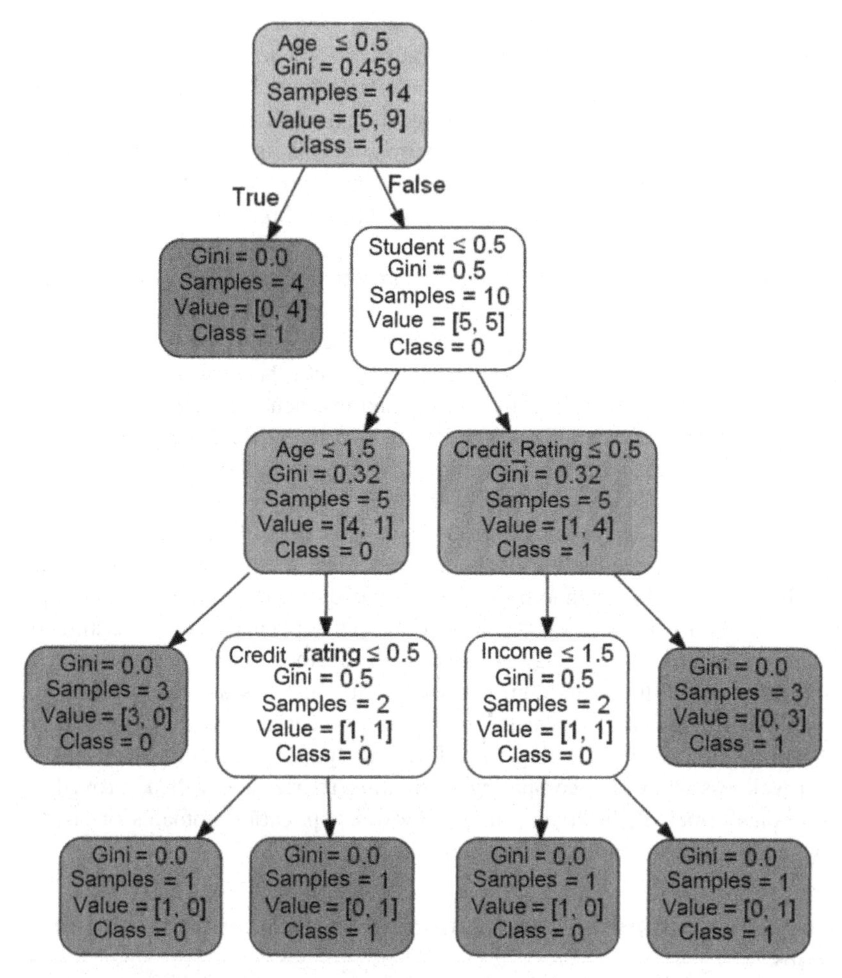

FIGURE 1.2 Decision tree to predict the aspirant customer. *Class 0: No, Class 1: Yes.*

Whether alarming systems are helpful to save and protect the health and wealth in real life. Whether related IOT data collected from various sensors after analyzing, generates the prevention alarms to save and protects the resources, similar to human rescue force.

Cognitive computing challenges

Security—Security is the major concern as digital devices manage critical information. As cognitive computing system analyze the big data, most important challenge is data security and privacy. Full-proof security mechanism must be followed in cognitive systems to detect fraudulent incidences to enhance data integrity (Makadia, 2019).

Adoption—Voluntary adoption is crucial obstacle in the progress of emerging technology. For cognitive computing to be successful, the main vision is to define the model to make the effective use of technology for betterment of business. The adoption stage is executed through collaboration with stakeholders like enterprises, government and individuals, technology developers, It is also important to include data protection policy that will help to increase adoption of cognitive computing (Makadia, 2019).

Accepting change—People are not easily ready to accept changes due to natural human tendency and also as cognitive computing is capable to mimic human-learning quality, technologist are scared as they think that someday machines would replace humans. Cognitive technology is designed to assist and associate humans in synchronization. By feeding the data into the systems, humans will nurture the technology. This is the distinct and excellent example of association amongst human and technology everyone needs to accept (Makadia, 2019).

In coming years, regardless of these challenges, cognitive technology will be proved as revolution in variety of industries. For any industry, it is unique chance to change and automate the industry processes. Understanding of the modern trends is required to make effective use of this revolutionary outreach like cognitive technology. It is engaged in creation of cutting-edge technologies for businesses. Many leading and top businesses have already adapted cognitive technology in their daily business processes. Variety of promising cognitive computing examples and use-cases shows how efficiently it is applied to make the businesses successful. Nowadays, in each aspect of human lives ranging from travel to sports, entertainment to health and wellness, cognitive computing is creating impact and will be proved as real transformer of the human lives. Various application domains of cognitive computing are explained in below section.

Cognitive computing application in financial and investment firms

The financial world is observing real changes and fluctuations now-a days. The most of the changes are observed as positive and helps in transformation of financial sectors like banking, investment, and share market from untrustworthy environment towards transparent services (5 Ways Cognitive Computing is Disrupting Finance https://disruptionhub.com/5-ways-cognitive-computing-disrupting-finance/). It is been mostly supported by the use of advance technology like cognitive computing. Data collection, analysis and usage are the most important aspects for finance sector which make cognitive computing really suitable for such industry.

Cognitive computing applications can be designed to understand the requirements in domain specific way as per the clients and assist the system to make context-aware suggestions. To provide target-oriented, real-time investment suggestions to managerial level knowledge workers, it must learn from huge historical data repositories and use it to for smart decisions in client funds investment in a highly unpredictable market. Reporting and analytics capabilities must be provided to investment companies. In order to provide better insight and decision making, large amount of data needs to be analyzed to understand the available market resources, risk factors, and financial aspects. Most of the industries are using digital tools for huge data analysis and decision making. Use of cognitive computing in finance is motivated by the technology's capability to support the companies for managing huge real-time data for various subjects. As volume of financial data continuously growing, efficient data collection, analysis, exploit is very essential. This large amount of data increases complexity in decision making for financial sector. Cognitive computing can be used to accurate analysis, identifying patterns and relationships and correlation of qualitative and quantitative data of financial sector.

By using NLP and text analytics, intelligent automation can be done for customer interaction, claims management. It will give cost saving solutions to achieve customers satisfaction in financial sector. Most of the customers are using mobile and online banking for doing their business at any time. So through cognitive systems, customers' requests can be handled quickly and effectively as per customers' needs in their own language. It will provide valuable insight to customer. Cognitive systems can easily extract the data, identify trends, generate analytics and enhance the efficiency of automation.

Adoption of cognitive computing will notably transform financial sector. Knowledge workers in finance domain would make effective use of machines to promote customer satisfaction and enhance decision-making. Similarly, various managerial level person like advisors will make use of cognitive computing to make their task more accurate and easier. This will enhance client engagement and will lead to operational excellence across the finance sector.

Choosing virtual agents over digital channels which are empowered by cognitive computing can help financial organizations see meaningful transactions in convoluted areas like brokerage and mortgages. Using cognitive computing, the financial industry can now address fraud and risk management more efficiently and effectively. Advanced predictive analytics and incorporating external and internal data is enabling the detection of fraud and emphasize on proactive prevention. Operations cost and IT are the two areas where much of the financial industry struggle, in terms of both productivity and predictability. Productivity gains are in demand and always given a substantial amount of spending from these areas. Cognitive computing is a new way of containing costs with exceptional optimization options which is possible through automation and cognitive tooling. Following are various areas in financial sector where cognitive technology will be helpful:

1. Organization: Cognitive computing helps financial sector to gather context-aware data about the personal and use it to suggest and recommend the spending-investing so as to offer personalized experience.

As an example, installation of nifty apps which associate users' calendars with dues, reminding personnel to select best wishes card for relative. Personalized finance services showcase how smartly such automated systems enhance the customer experience with financial services. Such apps can also make customer aware of the their spending in financial limits by providing real-time message-prompts on such purchases.

2. Advice: Cognitive technologies in finance sector made customer motivated to use and explore this sector by provide self-pace and personalized platforms and services which mimic the human-being as adviser. Such techniques really helpful to answer the complex queries raised by customer and involve the customer in conversation to understand their needs better. For example, Chatbots as finance advisor are the AI adapted software apps modeled to set services and to enhance customer satisfaction.

Cognitive computing application for changing business values

Cognitive computing systems must learn business processes efficiently. Data is valuable asset in every business as with this data valuable insights can be drawn with the help of analytical methods and practices. In cognitive computing, as data analytics is important component, it offers a smart decision support mechanism which helpful for betterment of businesses. In real-time, cognitive computing can evaluate emerging trends, recognize market opportunities for business and solve process centric problems. These system helps in streamlining the business stages, helps in risk management and offers the solution with respect to changing environment. Businesses can build a specific action on uncontrollable factors, also can create lean business processes using cognitive systems.

Customer experiences can be improved by designing chatbots which offers conversational interactive interface. Chatbots can provide related and required data to customers without need of human intervention.

Cognitive computing in veterinary medicine

With historical medical data, cognitive computing application offers a medical practitioners an assistant which offers consolidated information about the specific disease and medication. With complexity of case studies and big data it is difficult to get accurate solutions on health-related question. Cognitive computing provides the assistance to human being by offering ease of use and consolidated data. To address individuals' questions and to make intelligent, personalized recommendations, immediate processing of vast volumes of data must be done (Behera et al., 2019). Health insurers, providers, and similar organizations can assist the individual enrollments and patients to enhance their overall medical related data analysis.

In the case of the medical field, cognitive computing systems consolidate and analyze health-data collected from heterogeneous and diverse resources such as diagnostic tools, health-related journals, past reports and past data from the medical fraternity thereby assists medical practitioners to provide treatment suggestions are useful to practitioners and patients too. Cognitive computing will speed up the data analysis in process automation in healthcare industry. Cancer fighter patients can also be provided with intelligent phone assistant.

Helping veterinarians to detect and to handle the diseases of pets. Cognitive computing not only helps humans, but also helps veterinarians to take better care of the animals. Veterinarians ask a question in much simpler way as they communicate with their colleague. Instant support and advice must be provided, that saves time of busy vets and quality care is provided to their patients.

Cognitive computing application in health and wellness

Currently smart wearable devices generating data input that can help to get suggestions regarding healthcare program like diet, exercise etc., It also offers stress management programs which have crucial adverse impact on overall health. Such applications are the new ways that cognitive computing offers and can be scaled up with respect to future need. Here, fitness plans of individual and effective tracking system for health-progress is developed. This helps in tracking the health in accurate and hassle-free manner and offers effective support to fitness trainers and medical practitioners.

CaféWell—healthcare concierge by Welltok

Welltok designed an effective health caretaker, CaféWell (Makadia, 2019). This app records appropriate medical history records of patient through the collection of huge health-related data. This health-app effectively used by many health insurance companies to offer their customers the valuable data which in turn will enhance their health. Based on data gathering and query processing, it offers accurate and smart health-related suggestion which improve the wellness od personnel. To provide optimize health status of patient, the industry needs to be realistic, different approach need to engage that impact health of population.

Cognitive computing application in travel and tourism

Cognitive system applications plays important role in tourism and travels for suggesting the optimal itinerary. For this purpose system combine, merge and evaluate travel, destination information, for example routes, resources, availability of flight and resort, their prices etc. and consolidate it with individual likings, financial aspects, etc.

Case study: personal travel planner to simplifying travel planning by WayBlazer

WayBlazer has created personal travel planer by using power of cognitive technology to make it easy for travelers to plan for trips (Makadia, 2019). Traveler can ask questions in natural language. By gathering and analyzing trip data as well as knowledge regarding traveler likings, the tourism agent asks simple questions and offers personalized results. To finalize the travel journey, time for hotel booking, flight search can be saved by cognitive tool. Travel agents have used this technology effectively which is important to improve their sales and customer loyalty both hand in hand.

Cognitive computing application in education and learning

It is most promising application area of cognitive computing. Research challenges and practical appeal are recognized and gives broad spectrum of human learning. Cognitive computing equally applicable in education and learning with advancement in digital education.

In online tutorials, cognitive computing will provide personalized learning and assessment materials. Cognitive tutor will assess the student for their learning and accordingly prompt the new topic to learn for precious learning outcome (Irfan & Gudivada, 2016). Students may ask topic related questions to cognitive tutor based on learning material and assessment components. The answers revealed by the cognitive tutor are context-aware which makes the learning experience more paced and personalized. And more importantly, the cognitive tutor mimics the teacher's ability to teach, assess and mentor multiple students parallelly; enabling the facilitator to offer exactly similar support to large number of students enrolled for the course (Irfan & Gudivada, 2016).

Cognitive assistant will also help in searching and applying for courses as per their profile. Based on historical student's data current relevant courses will be recommended. This can be made possible using the conversation and interaction using NLP. To initiate this conversation, students can ask for cognitive assistance to make course choices. Eventually, with teaching learning experience and observing the student's performance, solving capabilities, more related courses are offered perfectly. This entire process also takes care of financial affairs like course fees, module wise fees collection, offers and discounts, rewards etc. As per shown in Fig. 1.3 course recommender systems is modeled to recommend the courses to aspirants as per their need. Speech recognition module is used for text to speech and speech to text

FIGURE 1.3 Course recommender systems.

conversions. Speech recognition facilitate the human-computer conversation in audio form.

Digital campus—apart from the course registration and recommendations, cognitive system help to keep students more connected to campus life by offering campus-news, newsletters as per the student's interest, branch and classes. This is really helpful for huge campus to make students aware of campus happening and conduct the student affairs with good collaborations. Such services use context-aware approach to facilitate the information broadcasting services to students.

Virtual librarian—learning resources plays vital role in Teaching Learning process. To make this process more intuitive and effective, the cognitive system can be adopted as virtual librarian. For students' assessments components, based on the relevant topic, Virtual Librarian will offer the books, articles which are useful for students to solve the assessment. Moreover, the student's reading and involvement can be traced here to make this process more effective.

Library book searching is tedious job in case of vast book banks. This can be made easier and more reachable by providing conversation assistance for book searching, making the learning assistance system more user friendly.

A reward service which gives vouchers or credits to students—cognitive services can be utilized by schools, colleges and universities to encourage and support positive behavior amongst their students. Institutions could use cognitive services to promote better attendance or punctuality to lessons, the timely submission of coursework by students or improved grades. Colleges or universities can use cognitive computing to offer incentives to students in order to raise enrollment rates for courses or to raise the number of early applications to courses.

A virtual essay checker which takes advantage of natural language understanding to advise students about how they could improve their work before submitting it to their tutors. This service has the potential to reduce the percentage of assignments that need to be resubmitted by students and it has the potential to raise the average grade profile for courses.

Cognitive computing in agriculture

The cognitive computing is truly revolutionary in the field of agriculture, as it is the next milestone to assist, guide and cooperate farmers for the crop production, analysis, monitoring till sales. Cognitive system with personal assistant in local language is the need in rural areas of many countries. Better opportunities must be provided to farmers in a user friendly manner to acquire the required data and understand the current market needs. Latest farming practices and trends needs to be followed. Adoption of latest technologies in farming is slow, due to this reasons people involved in the occupation of farming are comparatively very less. Traditionally, field officers visit the farmlands and provide training, advice, and support to the farmers (Jain et al., 2018). Better decisions related to the crops that they cultivate must be taken. Many of the rural villages lack the ease of accessibility, this leads to reduced crop yield, increased wastage of valuable labor, and market inefficiency. These reasons add upto severely impact a farmer's earnings, time and opportunities to increase the crop yield (Fig. 1.4).

Speech-based apps offer various functionalities to farmers and to agriculture related professionals, this makes agriculture field more prospectus and engaged. Literacy is the basic need to make such agriculture supportive technologies more usable. These technologies offer simple interfaces which requires minimal understanding of new technical concepts or interaction methods. Data repositories and corpus are easy to edit and manage by

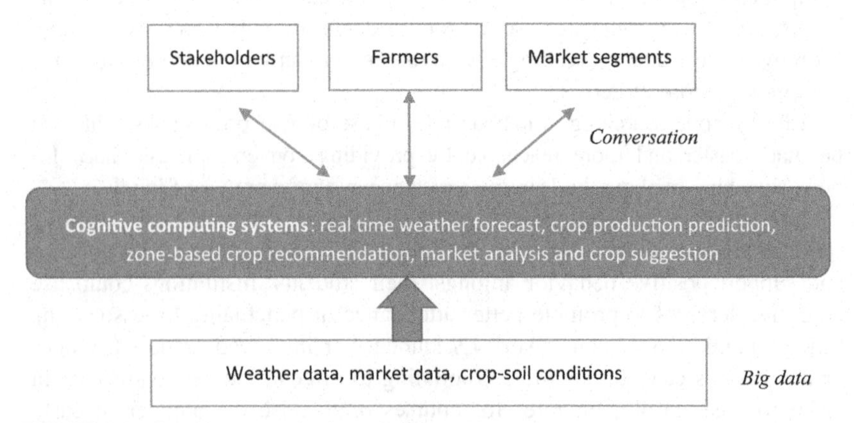

FIGURE 1.4 Cognitive computing in agriculture.

agriculture experts. However, chatbots for farmers are still seek the need of the acceptance and adaptation in rural spaces (Jain et al., 2018).

Some of the major areas wherein the information support is critically requiring are plant protection, pesticides and diseases, weather suitability for crop, and yield improvement (Bagchi, 2018). Many farmers are taking help of agri-experts for recommendations of pesticides for a particular crop disease. Farmers can interact with chatbots in their own language to get recommends medicine with dosage information. Farmers always seek the weather-related information, as adverse weather conditions can cause the losses and also understanding weather conditions may enhance crop management and profit. Information related to best practices can help increase yield in terms of the quantity or quality. Chatbots will give tips to increase yield based on data analysis and helpful on increase income of farmers. Farmers which seek crop-based recommendations from agri-experts on various agri-products they should purchase. Information related to fertilizers and seeds are collected by farmers from agri-experts.

Cognitive computing in communication and network technology

Communication and network technology are rapidly developing and demanding field of many services and applications. These technologies have high impact on society and the way people are leading their lives. To meet user-expectation, efficiency of energy and complex network environment, an optimal design-model its setup and efficient network communication are required (Sharma et al., 2020). Power of cognitive computing will be very useful to handle dynamic and complex network topology. Deep learning methodologies will give effective outcome for complex and smart decision making for optimal management of networks, optimal resource utilization and deep knowledge extraction in dynamic environment. Some of the areas of networking where cognitive computing will be effective to use are, to find and identify hot-spots with enhanced capacity, connections with less power, minimal delays and maximum reliability.

Cognitive computing enables to design new architecture and infrastructure for smart communication/network. Also, effective network management will be done by analysis and behavior prediction of data. In smart networks, cognitive computing will provide enhancement for complex-information-models, context-aware data management and deep knowledge analysis, performance evaluation, security and privacy protection (Sharma et al., 2020).

Cognitive computing surely offers solutions for planning and optimization of networks, best customer-services and leads for capital acquirements by combining together the techniques to support such tasks like analytics of networks, understanding the schematics of network topology, relationships amongst entities etc. It also conducts the survey and examine regularity compliance standards for various matrices related to finance to understand the suitable cost and prioritization in case of investment related to networks. Today, these activities are conducted by team of experts and knowledge worker for decision making and planning. With cognitive computing, it will be perfect use-case of human-machine association, as it will lead to enhance processes and methods, but human involvement is equally important to model such expert systems as per the need of context. Thus, cognitive computing will be very effective and powerful for providing advanced intelligence and interactive communication\network services and applications.

Cognitive systems is basically developed as a generic approach, so need to be modeled as per the context requirement to create the specific solution targeted to specific business.

Cognitive computing in renewable energy

With increasing demand for non-renewable energy, there is drastic need to utilize and focus on renewable energy, its sources and effective methods to make its use with ease. Wind energy, solar energy etc. are types of renewable energy. Major hurdle in using these sources as energy is that it's depends on a weather condition. More or less energies are getting generated based on the weather condition, and may not be very reliable source of energy. To find an optimal way in usage of renewable energy, one can predict the weather conditions and can tune other reliable energy sources accordingly. To make this possible, cognitive computing may play vital role. As shown in Fig. 1.5, With cognitive computer, prediction of whether condition by analyzing the past weather data is possible, with the help of predictive analytics and exploratory data analytics. Based on the prediction, the switching amongst energy sources in automated fashion is possible.

Cognitive computing in predicting the units of solar energy based on past data

To predict the units of energy generated in future with certain whether conditions is important to balance and manage and schedule various renewable and non-renewable energy sources as per the consumption requirement.

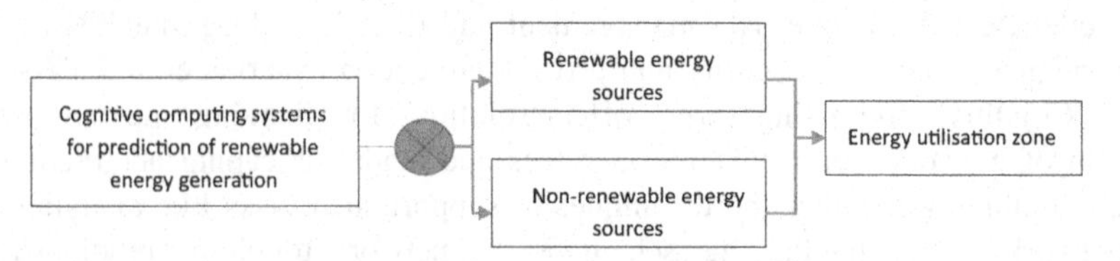

FIGURE 1.5 Cognitive computing model for renewable energy.

For this application, regression analysis is suitable machine learning algorithm for predicting the continuous solar energy value in given whether condition.

All attribute of the data is continuous value.

Features: {Cloud coverage(%), visibility (Miles), temperature (C), dew point (C), relative humidity (%), wind speed (mph), station pressure (inchHg), altimeter (inchHg)}

Target: {Solar energy (watt)}

Considering the sample dataset (Renewable energy dataset available from http://s39624.mini.alsoenergy.com/Dashboard/) described in Table 1.1, regression analysis models the relationship of all features with output values as regression line. Once this regression model is build based on past data, it would be useful to predict the units of solar energy for unseen data sample (Vinay & Mathews, 2014).

As shown in above Fig. 1.6: simple linear regression and multiple linear regression, the regression model tries to establish the relationship between dependent variable with independent variable. In simple linear regression dependent variable can be analyzed using single independent variable. Regression model created in this approach is using linear line, so called as linear regression. The linear line modeled in this approach is used to predict the value of unseen variable.

TABLE 1.1 Solar energy data attribute.

Cloud coverage	Visibility	Temperature
Dew point	Relative humidity	Wind speed
Station pressure	Altimeter	Solar energy

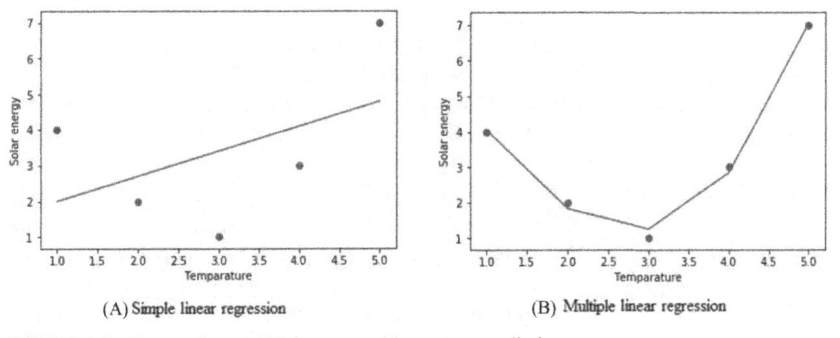

(A) Simple linear regression (B) Multiple linear regression

FIGURE 1.6 Regression model for renewable energy prediction.

Multiple linear regression is a type of linear regression, wherein multiple independence features are available to predict the dependent variable. This regression is more accurate than the simple linear regression.

With this solar energy prediction, cognitive system switch the energy sources accordingly to make system reliable and available all the time.

Conclusion

Cognitive computing framework is broad and includes a range of methods, techniques, algorithms, and approaches.

Its complex framework itself suggests the applicability of the cognitive systems in state-of-art researches. Cognitive systems are not only multidisciplinary but also transdisciplinary and hence provide capable applications for various context, domain. These applications range from architecture, medical, education, business corporates, Environment and many more. Due to its adaptability and applicability, cognitive computing is a core research area. In this chapter we discussed various fundamental techniques of Cognitive computing. AI-ML, Big data, IOT, NLP are discussed in this chapter with examples of methods. With these methods, various case studies and challenges are discussed and stated. As discussed with various case studies, the cognitive systems are the aspirant evolutions in many areas.

References

Aghav-Palwe, S., & Mishra, D. (2017). Color image retrieval using DFT phase information. In *2017 international conference on computing, communication, control and automation (ICCUBEA)*, Pune (pp. 1–5). Available from https://doi.org/10.1109/ICCUBEA.2017.8463836.

Aghav-Palwe, S., & Mishra, D. (2020). Statistical tree-based feature vector for content-based image retrieval. *International Journal of Computational Science and Engineering*, *21*(4), 556.

Bagchi, A. (2018). *Executive report – Artificial Intelligence in Agriculture*. Mindtree.

Behera, R. K., Bala, P. K., & Dhir, A. (2019). The emerging role of cognitive computing in healthcare: A systematic literature review. *International Journal of Medical Informatics*, *129*, 154–166.

Chen, M., Hao, Y., Kai, H., Wang, L., & Wang, L. (2017). Disease prediction by machine learning over big data from healthcare communities. *IEEE Access*, *5*(1), 8869–8879.

Chen, Y., & Argentinis, E. (2016). IBM Watson: How cognitive computing can be applied to big data challenges in life sciences research. *Clinical Therapeutics*, *38*(4), 688–701.

Gupta, A., Palwe, S., & Keskar, D. (2020). *Fake email and spam detection: User feedback with naives bayesian approach. Proceeding of International Conference on Computational Science and Applications. Algorithms for Intelligent Systems*. Singapore: Springer. Available from https://doi.org/10.1007/978-981-15-0790-8_5.

Irfan M.T., & Gudivada, V.N. (2016) Cognitive computing applications in education and learning, cognitive computing: Theory and applications. In V. N. Gudivada, V. V. Raghavan, V.

Govindaraju (Ed.), *Handbook of statistics*, (vol. 35, pp. 283–300). https://doi.org/10.1016/bs.host.2016.07.008.

Jain, M., Kumar, P., Bhansali, I., Vera Lio, Q., Truong, K., & Patel, S. (2018). FarmChat: A conversational agent to answer farmer queries. *Proceedings of the ACM on Interactive Mobile Wearable and Ubiquitous Technologies*, *2*(4), 1–22.

Makadia, M. (2019). What is cognitive computing? How are enterprises benefitting from cognitive technology? Towards Data Science. Available from https://towardsdatascience.com/what-is-cognitive-computing-how-are-enterprises-benefitting-from-cognitive-technology-6441d0c9067b.

Megha, C.R., Madhura, A., & Sneha, Y.S. (2017). Cognitive computing and its applications. In international conference on energy, communication, data analytics and soft computing (ICECDS-2017).

Orozco, H. et al. (2010). Making empathetic virtual humans in human computer interaction scenarios. In Proc. of the 11th Computer Graphics International (pp. 1–4).

Sharma, P., Singh A., & Jatain A. (2020) Cognitive computing for smart communication. In Krishna Kant Singh, Akansha Singh, Korhan Cengiz, Dac-Nhuong Le (Eds.), *Machine learning and cognitive computing for mobile communications and wireless networks*. https://doi.org/10.1002/9781119640554.ch4.

Tian, D., Zhou, J., Sheng, Z., & Leung, V. (2016). Robust energy-efficient mimo transmission for cognitive vehicular networks. *IEEE Transactions on Vehicular Technology*, *65*(6), 3845–3859.

Vinay P. & Mathews M.A. (2014). Modelling and analysis of artificial intelligence based MPPT techniques for PV applications. In *Advances in Green Energy International Conference* (pp. 56–65).

Further reading

Ait Khayi, N., & Franklin, S. (2018). Initiating language in LIDA: Learning the meaning of vervet alarm calls. *Biologically Inspired Cognitive Architectures*, *23*, 7–18. Available from https://doi.org/10.1016/j.bic.2018.01.003.

Kalogirou, S. & Sencan, A. (2010). Artificial intelligence techniques in solar energy applications. In: R. Manyala (Eds.), *Solar collectors and panels, theory and applications.* doi: 10.5772/10343.

Madl, T., Franklin, S., Chen, K., & Trappl, R. (2018). A computational cognitive framework of spatial memory in brains and robots. *Cognitive Systems Research*, *47*, 147–172.

Perveen, G., Rizwan, M., & Goel, N. (2019). *Comparison of intelligent modelling techniques for forecasting solar energy and its application in solar PV based energy system* (pp. 34–51). IET Energy Systems Integration.

Peterson, S., & Christiani, R. (2016). Executive Report., *Beyond bots and robots - exploring the unrealized potential of cognitive computing in the travel industry*. IBM Institute for Business Value.

Poria, S., Cambria, E., Bajpai, R., & Hussain, A. (2017). A review of affective computing: From unimodal analysis to multimodal fusion. *Information Fusion*, *37*, 98–125.

Puri, V., Jha, S., Kumar, R., Priyadarshini, I., Son, L. H., Abdel-Basset, M., ... Long, H. V. (2019). *A hybrid artificial intelligence and internet of things model for generation of renewable resource of energy* (pp. 111181–111191). IEEE Access.

Şerban, A. C., & Lytras, M. D. (2020). *Artificial intelligence for smart renewable energy sector in europe—smart energy infrastructures for next generation smart cities* (pp. 77364–77377). IEEE.

Shah, R., & Zimmermann, R. (2017). *Multimodal analysis of user-generated multimedia content.* Springer International Publishing.

Shah, R. R., Yu, Y., Verma, A., Tang, S., Shaikh, A. D., & Zimmermann, R. (2016). Leveraging multimodal information for event summarization and concept-level sentiment analysis. *Knowledge-Based Systems*, *108*, 102–109.

Snaider, J., & Franklin, S. (2014). Modular composite representation. *Cognitive Computation*, *6* (3), 510–527. Available from https://doi.org/10.1007/s12559-013-9243-y.

Chapter 2

Recent trends towards cognitive science: from robots to humanoids

Sarthak Katiyar[1] **and Kalpana Katiyar**[2]
[1]Department of Computer Science & Engineering, Rajkiya Engineering College, Sonbhadra, India, [2]Department of Biotechnology, Dr. Ambedkar Institute of Technology for Handicapped, Kanpur, India

Introduction

The concept of computing has evolved over the years, and it can be divided into three eras named as the Tabulating era (1900–49), the Programming era (1950–2010), and the Cognitive Computing (CC) era (2011-to present) in terms of timeline (Noor, 2014).

Each era of computing is characterized by significant technological development that fundamentally changes how the machine operates.

From dawn till dusk periods of the 20th century, behaviorism swerve gradually declined, but the swerve of information theory, linguistics, and data science gradually flourished. The popularization of computer technologies has commanded an impressive and thought-provoking cognitive revolution. This is given the birth of cognitive science that investigates the distribution and processing of information in the human mind.

Cognitive science emulates human beings' mental ability through a closer look at language, memory, perception, reasoning, attention, and emotion. Human beings' inbuilt cognitive conduct results from its perceptive sense organs capability to extract information as input from an ambient physical environment. In the next stage, this input is transmitted to the brain. The neurons are the functional unit of the human brain, which processes information and make a decision or output. The brain, nerves, and the spinal cord make the nervous system of humans. The processing results are conveyed through the neural network to various body parts, and then the appropriate body part makes appropriate behavior response. Thus, necessary information (input) from the environment is fed through the sense organs, the process by

Cognitive Computing for Human-Robot Interaction. DOI: https://doi.org/10.1016/B978-0-323-85769-7.00012-4

the nervous system, and the output produced by body parts creates a closed loop of human decision-making and an action part. The birth of humans starts the process of cognizing the world. The constant communication with the external environment helps in information accumulation for processing. The natural gifted human brain cognitive system is all-sufficient, but the machine-made cognitive system is very complex and employs the tools, methods from different topics to achieve the cognitive system's desirable goals. Therefore, cognitive science emerges as a multidisciplinary domain encompassing the methods and tools of psychology, linguistic, artificial intelligence (AI), philosophy, anthropology, neuroscience, and many more subjects.

Cognitive informatics (CI) is a transdisciplinary division of cognitive science, computer science, information science, and intelligence science. CC is the latest, advanced engineering application of CI in which machines have gained the capability of reasoning and learning through independent inferences and perceptive mimicking the mechanism of the brain (Wang et al., 2010). A wide range of CI and CC applications have been identified and developed in intelligence science, cognitive science, knowledge science, abstract intelligence, computational intelligence, intelligence information processing, and software engineering (Feldman, 2012).

CC uses AI and signal processing disciplines to simulate human thought processes in a computerized model. The self-learning algorithms, pattern recognition, and natural language processing (NLP) are the few foundation pillars on which CC foundation is laid (Hurwitz et al., 2015; Noor, 2014).

CC manages and computes new type of glitches. It tackles difficult instances or human problems that are characterized by ambiguity. The difficult instances or human problems are diverse, ample-data, and varying circumstances characterized by frequent change in data and confliction. CC practices tackle the fluid nature of the user's problems by reframing information sources and weighing influences, contexts, and insights. In this way, the system computes the conflicting statement and proposes the "best" rather than "right" solutions to human problems (Kelly & Hamm, 2013). To provide the best-suited solution for human problems, context remains computable in the CC system. The system recognizes and extracts features such as place, work, records, or description to generate a suitable bag of data for an individual. The cognitive system also considers the influence of a specific schedule and location to process information. The system performs computer-assisted solution by computing the enormous variable data for identifying particulars structure. In the next move, the system relates those particular structure to acknowledge to the demands of the instant (Feldman, 2012).

From the short review above, it is clear that the CC system is reframing the connotation between humans and their progressively all-encompassing digital environment. In this way, the CC system plays the role of helper or mentor for the user to assist in many problem-solving circumstances.

In Fig. 2.1, the general cognitive system architecture with closed perception action loop is shown (Bannat et al., 2010).

How cognitive computing works

The cognitive system task is to integrate information through numerous references and factors and contradictory representation to propose the most appropriate solutions. The system employs a mixture of AI, neural networks, NLP, machine learning (ML), contextual awareness, and sentimental analysis to resolve everyday human problems to achieve this goal (Feldman, 2012). The human brain inherits the capacity to resolves their problems in-situ, but for this task, the computer system requires vast quantities of unstructured and structured data nourished ML algorithms. In due course of time, the cognitive system can improve how they identify particular structure and how they process information in a manner that makes it to arrive at justified result (Ferrucci et al., 2010).

The IBM Watson CC system is a pioneer organization in the cognitive science field. It establishes "the concept of three transformational technologies" for the cognitive system working are described below (High, 2012):

- The ability to understand natural language and human communication.
- The ability to generate and evaluate evidence-based hypotheses.
- The ability to adapt and learn from its human users.

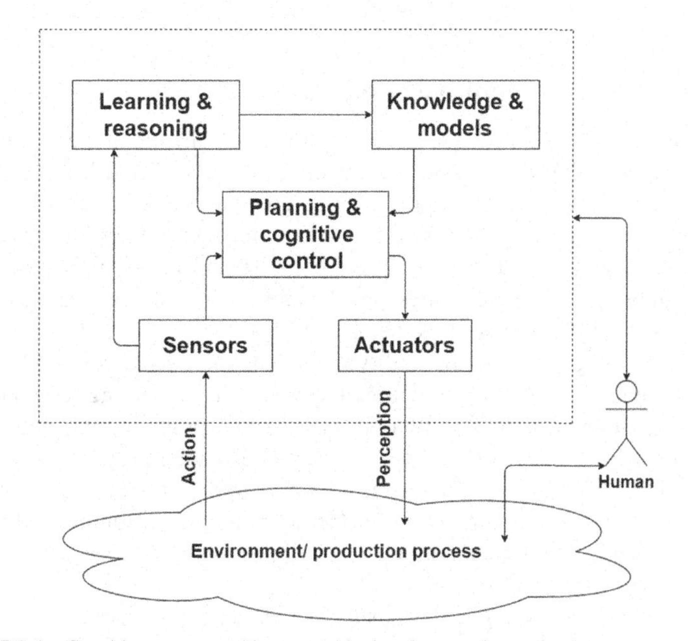

FIGURE 2.1 Cognitive system architecture with closed perception action loop.

The aim behind this concept is that the machine can solve human problems but not push human beings' cognitive abilities out of the frame. In order to attain this new height of computing, as listed by the CC Consortium, CC systems should be (Mounier, 2014; Reynolds & Feldman, 2014):

Adaptive—The CC system must learn whenever information and data vary, in addition to the development of purpose and needs. The system must address and resolve problems that are ambiguous and unpredictable in nature. The system must be configured to process the dynamic data in real-time, or near real-time.

Interactive—The CC system must communicate efficiently with users so that the system can easily identify the needs of the users and the users can describe their requirements easily. The system also communicates with other systems, processors, devices, and Cloud services, as well as with people.

Iterative and stateful—The CC system must help in identification of a problem in case the problem statement is insufficient or ambiguous by asking questions or seeking additional source input. The systems must memorize past experiences throughout the operation and give back solution that is appropriate for the particular application at that point in time.

Contextual—The CC system must recognize, define, and retrieve conditional factors like definition, pattern, schedule, place, suitable domain, prescribed guidelines, user data, method, objective and purpose. Structured and unstructured data, along with sensory inputs like gestural, auditory, visual and other sensor data, provides the system to draw information from different sources.

Functions of the cognitive computing

CC is defined in several ways. According to IBM, "the cognitive computing system is possessing three basic understanding, reasoning and learning capabilities missing in a regularly programmed computer system" (High, 2012). These capabilities can be transformed into components using technologies from research fields related to AI. These components can be applied in a reference architecture for the uses in domain-specific knowledge systems (Greenemeier, 2013; Reynolds & Feldman, 2014).

Understanding—This is the ability of the cognitive system to understand natural language and human communication. Understanding components exploits pretrained and cloud-based application programming interfaces (APIs) to extract information from unstructured data sources such as text, speech video, and sound. The extracted information, in general, is not yet domain-specific but is the baseline for the downstream, reasoning, and learning components. The preannotation capabilities enable a fast start in the implementation of a CC system.

Reasoning—Reasoning is described as the second cognitive system's second ability useful for the generation and evaluation of the evidence-based hypothesis. The reasoning component aims to deliver customized advice by interpreting a user's personality, tone, emotion, or a customer's business use cases and end-to-end processes.

Learning—The learning ability help machine to accommodate and assimilate through humans. The following two principles can apply in learning. Sieve out wrong hypothesis arriving from the reasoning components and therefore increases the precision rate. Generate a high recall hypothesis, which can then be filtered by more specific rules in a downstream reasoning component.

In a nutshell, the essential functions of CC are understanding, reasoning, and learning.

The components of a cognitive computing system

The critical component of a cognitive system as shown in Fig. 2.2 are (Hurwitz et al., 2015; Kelly & Hamm, 2013).

AI is the expression of intelligence by machines that empower machines to detect and perceive information (input) from the environment for decision-making. Algorithms are like manual books that need to be followed to solve a specific problem by machines. ML is a standard operating procedure for input data processing and making a copy of that future work.

Data mining is the process of implying knowledge and associations of data from large, pre-existing datasets.

Reasoning and decision automation are the part which encompasses learning quality to the CC system to achieve goals. The consequences of the

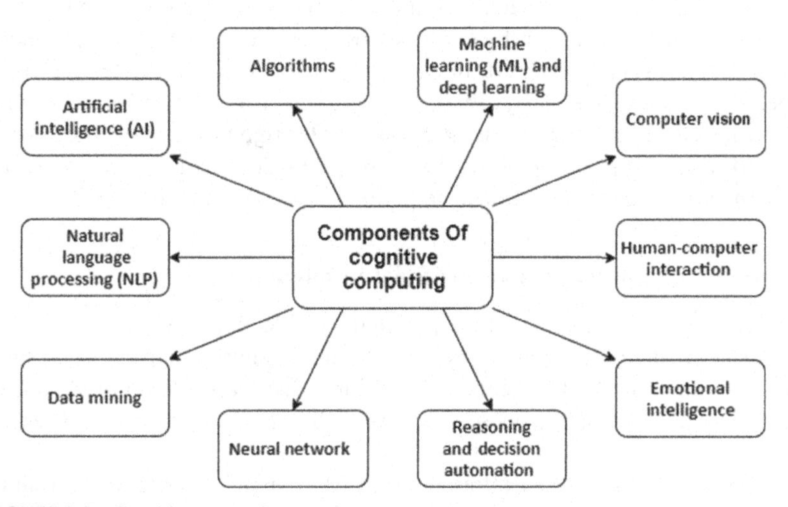

FIGURE 2.2 Cognitive computing system components.

reasoning process reflect in decision automation. That is, the software autonomously generates and implements a solution to a problem.

NLP is computing techniques deployed for response generation that are comprehensible to humans in spoken and written form. It is further subdivided into natural language understanding and natural language generation. Speech recognition authorizes machines to convert speech input into a written language format.

Human–computer interaction (HCI) and dialog and narrative generation: This work as an interface between humans and machines for purposeful and enjoyable communication. This is a subtype of NLP The inclusion of this trait makes CC system more enthralling and upgrades the quality and frequency of interaction among users and machines.

Visual recognition employs high-level algorithms and deep learning to examine individual pictures and patterns of particular images, such as face recognition.

Since the incubation of the CC concept, researchers are focusing on how to integrate the emotional intelligence aspect of humanity in the machine. In this area, a project entitled "MIT startup Affectiva" is launched for building a computing system which recognizes facial movement of humans and respond accordingly.

The neural network is analogous to the human nervous system and consists of nodes/neurons with biased associations. The deep neural network is made up of sheets of neurons. The learning process takes place by upgrading the values of nodes association—the neural network of machine work as a complex decision tree for proposing an answer.

As the practice makes a man perfect, the machine also improves their learning capacity through training, termed deep learning in cognitive system terminology. In this process, different training data set is processed over the neural network of the system, and the result is cross checked with the accurate result given by the living counterpart. In case the result is same, the system has finished its task otherwise the system has to adjust the neural network of its neural interconnections and process the data again. After rigorous training, the neural network learns to generate output that is very similar to human-generated output. The system has now "learned" to perform task as humans do.

Cognitive computing versus artificial intelligence

The AI provides a solution to a problem by selecting the best algorithm, whereas CC delivers solutions by considering a number of parameters just like the human brain. CC behaves and mimics the human brain thinking by observing patterns and proposes the best solution based on its knowledge analysis.

The processing of input information in AI is predefined in the algorithm, and the whole process remains under the control of the system. Whereas in

CC system serves as an assistant in accomplishing the task. Thus, it enables the human to have quicker and more accurate data analysis without annoying the wrong decision taken by the ML system in AI.

It is generally agreed that AI and CC are frequently operated interchangeably because these technologies' fundamental intention is to simplify tasks, but the difference lies in the way they approach tasks. AI is developed to augment human thinking from predefined statements for decision-making to solve difficult problems and deliver accurate solutions. On the other hand, CC tackle human complex problems by familiarizing them through human reasoning in the same way as human brain. An AI-supplemented system takes decision from predefined functions without consultation with human expert whereas in CC the human diagnosis is also supplement and considered with its own set understanding and reasoning analysis that assist humans in decision-making and also make a human touch to critical processes.

Ultimately, we can say that AI and CC are similar in their intent but different in their approaches.

Cognitive computing architecture

A cognitive architecture is a design framework for building a CC system to achieve decision-making ability like humans. As described by Newell, Anderson and et al., the architecture represents functional design component pillars for linguistic skills, real-time process, plastic behavior, human-like understanding, substantial learning foundation, growth and expansion, and self- conscious aspects (Anderson & Lebiere, 2003; Newell, 1994). Later on, more research in this area added more pillars to the existing framework. These new columns are ecological realism, biological reformation, cognitive realism, adaptations, modular, synergism, similarly interaction. The cognitive system organization comprise a suitable manual system for information illustration, stored data sorting, modular blocks, and I/O devices. The software-assisted outputs represent a stable model for particular computation, whereas cognitive architecture system outputs represent human brain-like analysis and in justified way apply knowledge to attain distinct tasks in spite of definite issue-resolution (Chandiok & Chaturvedi, 2018; Chen, Herrera, & Hwang, 2018). Work in this arena intends to achieve the mission to construct machines having human-like intelligence for problem-solving.

Human-centered cognitive cycle

Humans, machines, and cyberspace are the part of human-oriented cognitive cycle. Cognitive cycle, when talking in terms of cloud computing and deep learning, the machine describes the physical components for example computer networks, transmission devices, and intelligent robots. On the other hand, the cyberspace describes the virtual components for example data

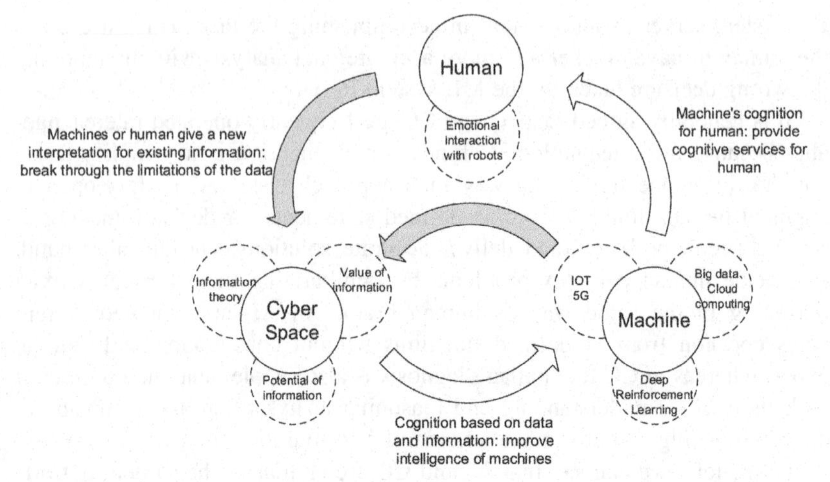

FIGURE 2.3 Human-centered cognitive cycle.

management. This cyclic system integrates human and machines with algorithms and assistance of cyberspace (Chen et al., 2018). This results in effective interaction among humans, machines and cyberspace as shown in Fig. 2.3. Therefore, a new arena opens for practical solutions to tackle the internal needs of humans and deliver intelligent cognitive services according to user requirements.

Cognitive computing layers

CC systems are capable of addressing ambiguity and a shifting set of variables. The system also constantly reevaluate information based on the changes in the user, task, goal, context and information. Before finding answers, the system must grasp the question or context. In this way, the cognitive system is able to offer multiple useful solutions that are weighted for query relatedness (Noor, 2014). CC system perform the task of converting big data into smart data and useful knowledge so that user efficiently interact with the system in an ongoing conversation. The system architecture of CC has four layers as shown in Fig. 2.4 (Chen et al., 2018).

Different types of architecture

Various cognitive framework and architecture are being developed to achieve human-like capabilities for evaluating real life issues (Kirk, Mininger, & Laird, 2016; Vernon, Metta, & Sandini, 2007). Some of the previously proposed cognitive architectures are LIDA (Duch, Oentaryo, & Pasquier, 2008), NARS, ACT-R, CLARION, SOAR, ICARUS, EPIC, MDB (Bellas, Duro, Faiña, & Souto, 2010), CogPrime (Goertzel et al., 2013), Sigma (Pynadath, Rosenbloom,

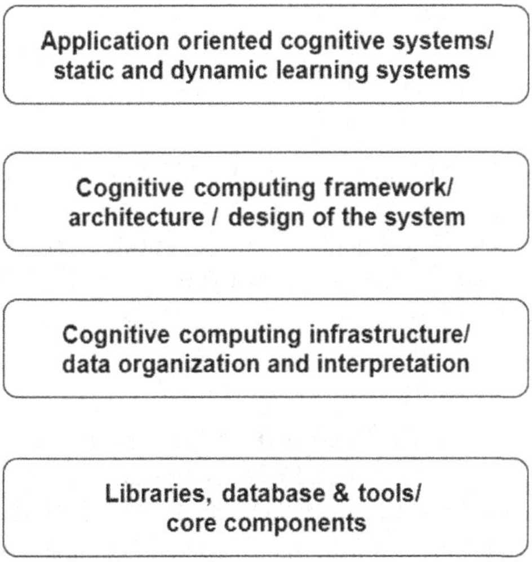

Application oriented cognitive systems/ static and dynamic learning systems

Cognitive computing framework/ architecture / design of the system

Cognitive computing infrastructure/ data organization and interpretation

Libraries, database & tools/ core components

FIGURE 2.4 Cognitive computing layers.

& Marsella, 2014), MLECOG (Starzyk & Graham, 2015), EBICA (Samsonovich, 2013), and ECA (Georgeon, Marshall, & Manzotti, 2013).

SOAR

SOAR is a cognitive system-based architecture which denotes symbolic designs and consists of various compact functional components including procedural, episodic and semantic (long-term memories), working memory (short-term) (Laird, 2008); and cognitive processing of symbolic input data (Persiani, Franchi, & Gini, 2018). Over the last 35 years, the SOAR cognitive architecture building process has centered on symbolic data handling, and SOAR9 is the latest version (Gudwin et al., 2017; Laird, Kinkade, Mohan, & Xu, 2012).

ACT-R

ACT-R architecture is based on hybrid cognitive system and consist of three main modules termed as memory and perceptual-motor module, pattern matching module, and buffers (Anderson, 2005). The latest developed version is ACT-R 7.

CLARION

Connectionist Learning with Adaptive Rule Induction Online (CLARION) is an architecture based on hybrid cognitive system (Sun, Merrill, & Peterson, 2001). CLARION obtains sound and visual data and keeps in symbolic representation by making use of different neural networks models, high-level data and deep learning. This architecture includes a specific computing structure for accomplishing perception, cognition, and action (Gray, 2007; Sun, 2009).

LIDA

Learning Intelligent Distribution Agent (LIDA) is another combined cognitive system architecture centered on concept of the global workspace theory (Franklin, Strain, McCall, & Baars, 2013). It has different types of functional memory modules for execution of cognitive cycles (Franklin, Madl, D'Mello, & Snaider, 2013).

EPIC

Executive-Process/Interactive Control (EPIC) is also mixed cognitive system architecture. The main purpose of EPIC's intend to develop, execute, as well as evaluate the learning system for the data handling same as humans that specifically centered on the "full timing process" of human brain-like perception, cognition, and action (Kieras & Meyer, 1997).

MDB

The Multilevel Darwinist Brain (MDB) is a cognitive architecture that represent an innovative approach towards cognitive system development (Bellas et al., 2010). The MDB allows for the alteration of objectives by providing a satisfaction model and enabling quick responsive actions.

To summarize, the essential requirements taken are how effectively these models meet human-like functionalities in all the cognitive architecture. The essential attributes of the cognitive system are Attention, Perception, Learning, Reasoning, Memory, Emotions, and Actions (Chandiok & Chaturvedi, 2018).

So, the essential functional parameters for evaluating the cognitive architectures are shown in Fig. 2.5.

Cognitive information technology architecture

The cognitive information technology (CIT) is a representative design for the creation and management of cognition and data handling abilities within the model to carry forward the internal movements of information and artificial general intelligence. This architecture plan efficiently follows the principle of CC and artificial agents. These fields' amalgamation establishes a modern modified method defined as "Cognitive Computing Agent Technology" (Chandiok & Chaturvedi, 2018), described in Fig. 2.6. A CC agent technology is an innovative approach possessing advanced abilities of execution and completion of the particular operation like that of humans.

The main components of the cognitive agents, as shown in Fig. 2.6, are

1. Concept model (CM), constitutes human semantics and ontologies.

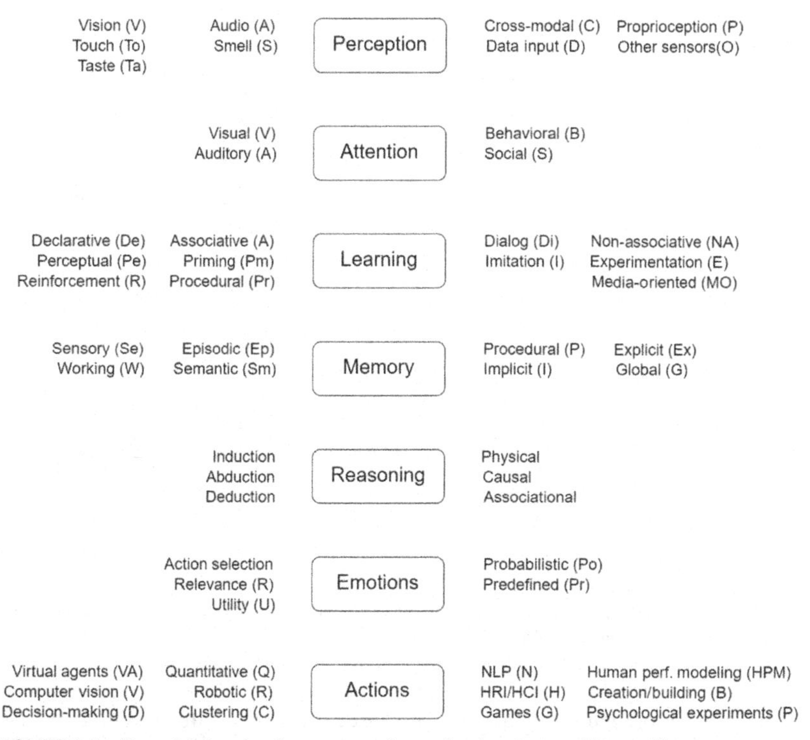

FIGURE 2.5 Essential functional parameters for evaluating the cognitive architectures.

2. Experience model (EM), constitutes human expert knowledge behavior.
3. Knowledge model (KM), constitutes human natural intelligence patterns facts.
4. Interface model (IM), constitutes natural human perception and action skills.

The CC model inculcates the human-like functionality to the system. The Cognitive Information Technology (CIT) framework design module has both CC technologies and cognitive agent blocks to create concept model, experience model, knowledge model, and interface model as given in Fig. 2.6. The cognitive agent block, subblocks are learning, perception, knowledge, memory, action, reasoning, and emotion. Robotics, deep learning, ML, NLP, Data Mining and knowledge representation are some of the technologies used in CC block. When the above-mentioned assemblies are organized in a proper model form, the machine performs cognition like human brain for example questioning-answering, conversation, robotic behavior (picture, subject, and social), identification, verbal, perception, and textual script. CIT architecture based CC agents can be expert systems (software agents) or humanoids (robotic agents).

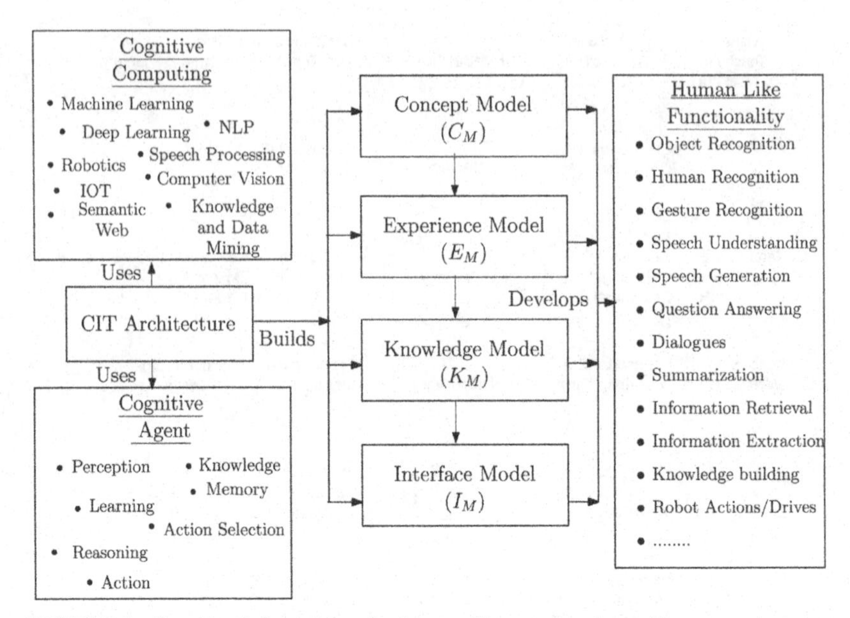

FIGURE 2.6 Cognitive information technology architecture based cognitive computing agent model for human-like functionality.

Sense, comprehend and act are the three skills that are executed inside the CIT architecture as shown in Fig. 2.7. The sensory skills receives voice, pictorial, sensor data and text script data that is focused on attention. The Comprehend Skills develops learning, knowledge and experience information from interaction. The Action skills manages plans and creates model information (Chandiok & Chaturvedi, 2018).

The CC agent technology-laden humanoids have characteristics of perception, autonomy, learning capability, knowledge, and understanding, which are essential characteristics of human beings.

Additional technologies influencing humanoids are speech activity, visual activity using computer vision, perceiving linguistic by using NLP, data processing, analyzing information with the help of semantic web and networks, reasoning with inference engine. Infusion of cognitive system and cognitive agent blocks within one system will offer a superior approach to resolve real-world problems when compared to the past developed humanoid model. The decision-making capacity in CIT based CC agent technologies are derived from experiences rather than procedures and input are fed through attention. The structural plan of Function of CIT is influenced by the LIDA cognitive architecture.

In the Fig. 2.7, CIT manages "Query information" (Question, Commands, Statements, Images, Sensor Data). Input Interface is (Computer Vision, Audio Processing, Sensor Processing) that provides "Dialog Information."

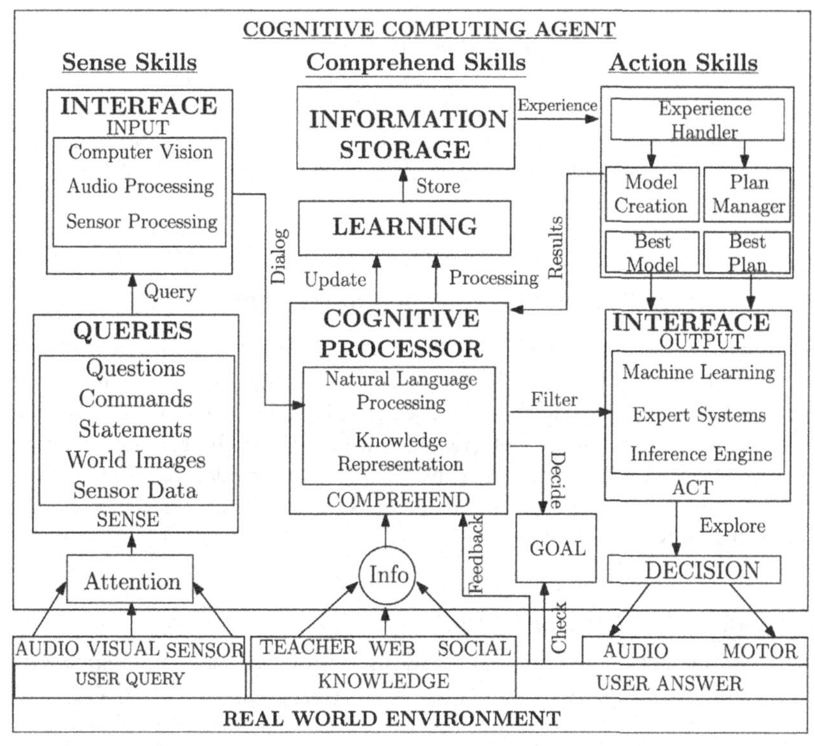

FIGURE 2.7 Cognitive information technology architecture information flow in cognitive computing agents.

Natural Language and Image Processing are applied by the Cognitive Processor and represent "Knowledge Information" accommodated in Information Storage. Experience Handler uses "Experience Information and generates Result Information." Processed results are generated by the Cognitive Processor and "Best Action Information" is given to reach the goal. CIT improves the previously gained experiences with "Feedback Information" to provide better results in future.

Human–computer interaction

The objective is to develop robotic machines that can socialize with people in a comprehensive manner. Despite enormous research, still, there are no robots that can interact and understand human needs efficiently (Toumi & Zidani, 2014). This seems due to nature gifted enormous capacity of a human to interact socially. This hurdle is somewhat overruled by integrating multiple disciplines of science in robot's development. The emotional state of human beings is endorsed from works in physiology as an analytical model; however, it is obstinate to use these indices and lacking true felt emotions (Ortony, Clore, & Collins, 1990; Trappl, Petta & Payr, 2002). There is also a research gap among the poor knowledge of stimulated emotions and emotions actually felt by humans, but despite this, an analytical model can interact emotionally with user and produce a human-like touch in the device. The OCC model is the best emotional model founded on theories of evaluations.

Different models of human–computer interaction

The most recent architecture and theory in cognitive robotics is Norman (Norman & Draper, 1986) action theory which extends an evaluation method established on circumstantial acknowledge. The other HCI models are the Instrumental interaction model (Beaudouin-Lafon, 2000), Physical interaction model of SATO (Sato & Lim, 2000), Fitts model (Carroll, 2003), Guiards model (Carroll, 2003), Model of Seeheim, Arch reference model, and Multiagents model.

Instrumental interaction model

The instrumental interaction model design could act as an intermediate between the user and the manipulated object (Beaudouin-Lafon, 2000).

Physical interaction model of SATO

SATO's physical interaction model lets mingle real-world objects (physical space) to the digital world object (media space /virtual world) (Sato & Lim, 2000).

Fitts' model

The Fitts' model represents the human motor system's information- processing capability and is a predictive model (Carroll, 2003).

Guiards' model

Guiards' model is a descriptive model for the bimanual control task (Carroll, 2003).

Model of Seeheim

The model of Seeheim is divided into two parts named as the interface part and the functional part. The interface comprises three components: the functional core, the presentation, and the controller dialog.

Arch reference model

Revised Seeheim model, also known as the arch reference model consists of five components arranged in the arch shape. These five components are Dialog controller, Presentation component, Functional core, Interaction component and Adapter domain.

Multiagents models

Multi-agent model is based on multi agent approach and made up of Model-View Controller (MVC), Presentation-Abstraction Control (PAC), and Agents Multi-faceted (AMF) model.

How software communicate with each other, is the basic criteria on which all models organized.

Norman's model

The Norman model, founded on the Norman theory, effectively integrates emotions and capabilities concepts for social interaction between robots and humans (Norman & Draper, 1986). The Norman model explains different cognitive steps of task performing in HCI as shown in Fig. 2.8.

The Norman model has seven stages as shown in Fig. 2.8

1. Formulation of Goal: Explain the main objective of the task
2. Formulation of Intention: Clarify what is required to satisfy task goal
3. Specification of the action: Describe essential action for intention completion
4. Execution of the action: Describe the selection of commands needed to carry out the task
5. Perception of the system state: describe what user noticed
6. Interpretation of the system state: Described how user knowledge is interpreted by the system
7. Evaluation of the outcome: Explain the what system has perceived and interpreted in comparison of the user's need and next required action

All these seven stages work in an iterated manner until the intention and goal are attained.

Human–robot interaction

The main objective behind this is the study of interactions between humans and robots. Different models of HRI are - visual, vocal and social.

Human–robot visual interaction

This trait enables a robot to visually interact with humans with the help of a vision system. Robots can identify different geometric shapes and colors

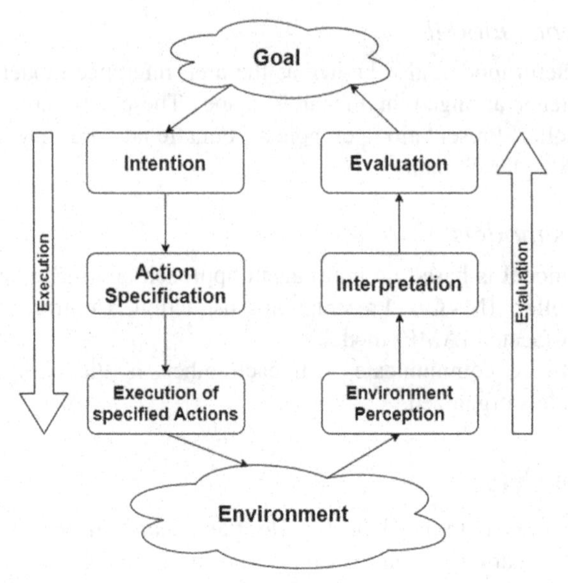

FIGURE 2.8 Seven stages of the Norman model.

(El Sayed, Awadalla, Ali, & Mostafa, 2012), classify images (Lotfabadi, 2011), recognize the basic human gesture, face detection, and trucking humans in a natural cultured environment (Rahat, Nazari, Bafandehkar, & Ghidary, 2012).

Human–robot vocal interaction

The integration of speech recognition in robots enables them to start a speech conversation with a human (Feil-Seifer & Mataric, 2005). Communication in the form of speech is the usual way of interaction among people. The robot available in the museum is an example of this interaction.

Human–robot social interaction

Humans usually express social behavior through emotional facial expressions in the natural environment. Many investigations are ongoing concerning such facial actions like Avatars in the virtual humanoid look, frequently applied by Conversational Animated Agents, and interact with humans or other agents through a multimodal communicative behavior (Poggi, Pelachaud, de Rosis, Carofiglio, & De Carolis, 2005), for example, Greta (Pelachaud & Bilvi, 2003), REA (Cassell et al., 1999), Grace or Cherry (Lisetti, Brown, Alvarez, & Marpaung, 2004). Cherry is attached to a robot and orients people in the University of Central Florida.

Zoomorphic and anthropomorphic robots are two types of presently available physical robots designed for social interaction. Aibo (Sony), iCat (Philips) are toy zoomorphic robots having look like cats or dogs. On the other hand, anthropomorphic robots are better bestowed in the arena of social engagement and tend to share their thoughts.

From human–computer interaction to human–robot social interaction

It has been observed, HRI and HCI can supplement one another in the simulation of robot emotions (Lisetti et al., 2004; Thomaz & Breazeal, 2007). An example of this is illustrated in robots, which shows emotions like happiness, angriness and satisfaction (Chastagnol, Clavel, Courgeon, & Devillers, 2014) (Fig. 2.9).

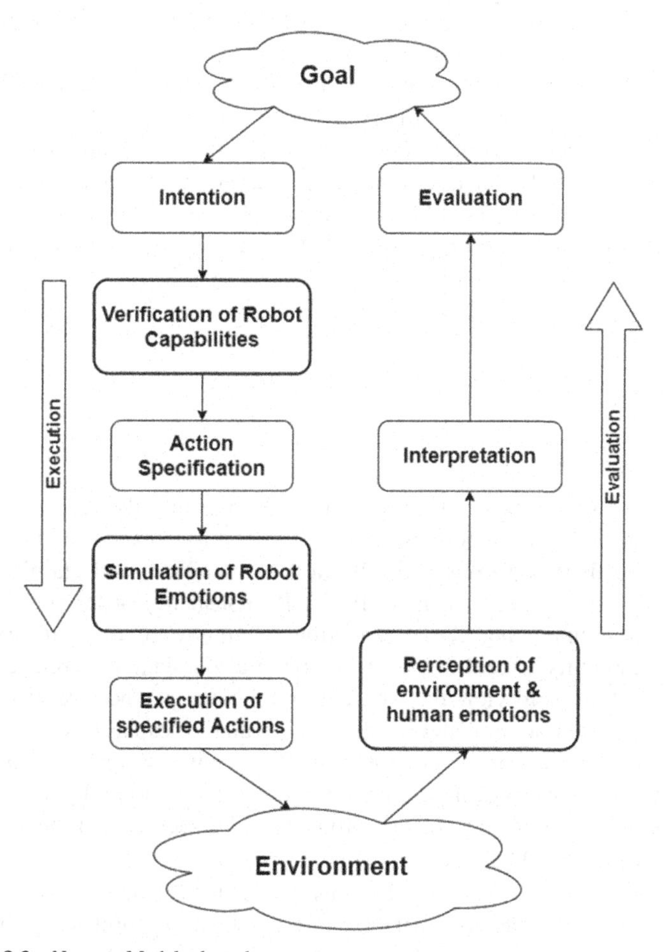

FIGURE 2.9 Norman Model adapted.

Cognitive robotics

Now we enter in an era where robots are felt in every day activity and even society is enjoying robot services in a number of ways. Cognitive robotics requires an adequate cognitive architecture for technical cognition and intelligence support in the humanoid robotic system. Initially, in the initial definition of the cognitive robotic system, the perception feature is included from learning, reasoning, knowledge representation, motor control, communication features, problem-solving and goal orientation. However, there is still a dispute about the addition of self-consciousness, motivation, and emotion feature included in the definition (Burghart et al., 2005). The AI gave the first framework structure for the development of the cognitive architecture for intelligent robot systems. The toy robotics are made on this architectural plan. To overcome the problem of explicit representation of goals and coordination of the component, a more advanced three-layered architecture was proposed (Dillmann, Becher, & Steinhaus, 2004).

The next advance architecture for cognitive robotics was subsumption architecture. In this plan, each functional behavior is represented by a different but interconnected component. Such architecture is supposed to have a human operator model for task-related, skill-based, rule-based, and planning-based components (Brooks, 1986; Kawamura, Rogers, & Ao, 2002).

The architectural plan proposed by the German Human Project has components for following feature:

- Perceive
- Learn
- Anticipate
- Act
- Adapt

Cognitive robots attain their objective by observing the real world, and being attentive in the important activities, intending what to do, predicting the result of their decisions and other agents' decision, and learning from the subsequent interaction (Firby, 1992). By constantly observing, analyzing and communicating the information, cognitive robotics resolve the intrinsic ambiguity of real world. A key feature of cognitive robots are predictive capabilities to view the world from a human perspective to forecast human's intended actions and needs. The predictive capabilities are working in direct interaction mode, for example, a robot assisting a doctor in operation theater or work in interaction mode well explained by an example of a robot arranging particular products at a designated space in a manufacturing industry.

The robots with cognitive features are not just like other physical machines but are machines having a sense of physical morphology, dynamics, kinematics properties of their immediate environment (Breazeal, 2001).

Cognitive robotics architecture

The perception control, task execution, recognition and interpretation of complex context, planning of intrinsic task, learning of behaviors, and understanding are basic characteristics of the cognitive robotics system (Burghart et al., 2005). In order to fulfill these demands, a three-layer architecture comprises parallel behavior-based components as shown in Fig. 2.10 (Kortenkamp, Bonasso, & Murphy, 1998). The sensors and actuators are the main components of the hardware part of the robot. The perceptual and task-oriented component are embedded in three layers. Fast reactive component in the lowest layer, recognition, and task coordination component in mid-level and components related to perceptual results and recognition of situations and contexed are embedded in the top-level-the dialog component, learning component, and task manager also present in the top layer. A global knowledge database has detailed information about the positioning of the component on a different layer as shown in Fig. 2.10 (Fig. 2.11).

Components of cognitive robotics system

Perceptual component

The perceptual component assists in the fast interpretation of sensor data without accessing the system knowledge database. The perception component also interprets the user's actions and reflect them as communicative or non-communicative behavior. The speech understanding, movement interpretation, gesture interpretation, and intention prediction are the perceptual component's outcomes (Soltau, Metze, Fugen, & Waibel, 2001; Stiefelhagen et al., 2004).

Dialog manager

The dialog manager performs the two-way communication between humans and robots. It helps in the interpretation of user context, answers a user question, and pass missing information to the task manager. Dialog manager also reports the emotion, system, and situational constraints of the user (Holzapfel, Fuegen, Denecke, & Waibel, 2002).

Task-oriented components

The task-oriented component operates in the real-time level using task knowledge. The task planer's function comes into the picture when the data from perception component and dialog manager are interpreted out successfully (Kobayashi, Nakatani, Takahashi, & Ushio, 2002; Ly, Regenstein, Asfour, & Dillmann, 2004). The task flow generator makes a sequence of actions from the knowledge base and generates a specified plan for demonstration or optimization of actions. The task planer also makes plan B in the case of emergencies and intercept actions during run-time. The final

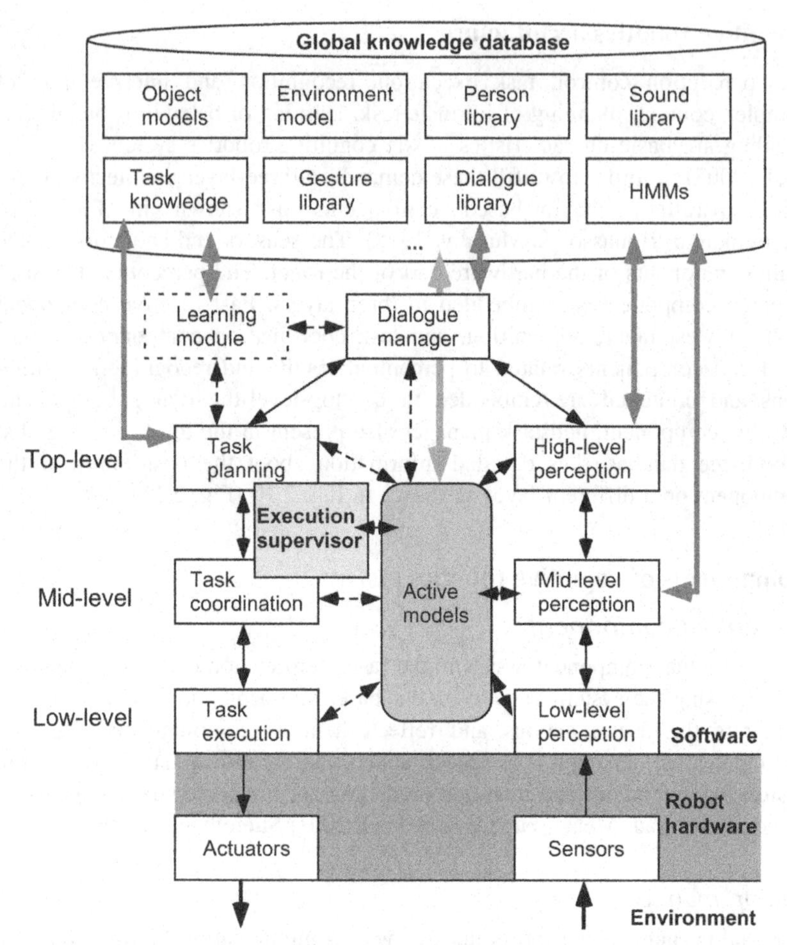

FIGURE 2.10 Cognitive architecture of the Karlsruhe Humanoid Robot.

scheduling of the task and resource management of robots is executed by the execution supervisor located on the task planer's output. Humanoid robots' chief resources are hardware subsystems with associated software modules, and perceptual software and resource management is a challenging task due to its complex organization. This can be illustrated with the example of a vision-based robot system.

Learning component

The ability to learn is a prerequisite characteristic for an intelligent cognitive system. The supervised and unsupervised are the two modes of learning in the robotic system. This helps them acquire new behaviors, memorize newly learned facts, tasks, and flow of actions in the knowledge base. The learning

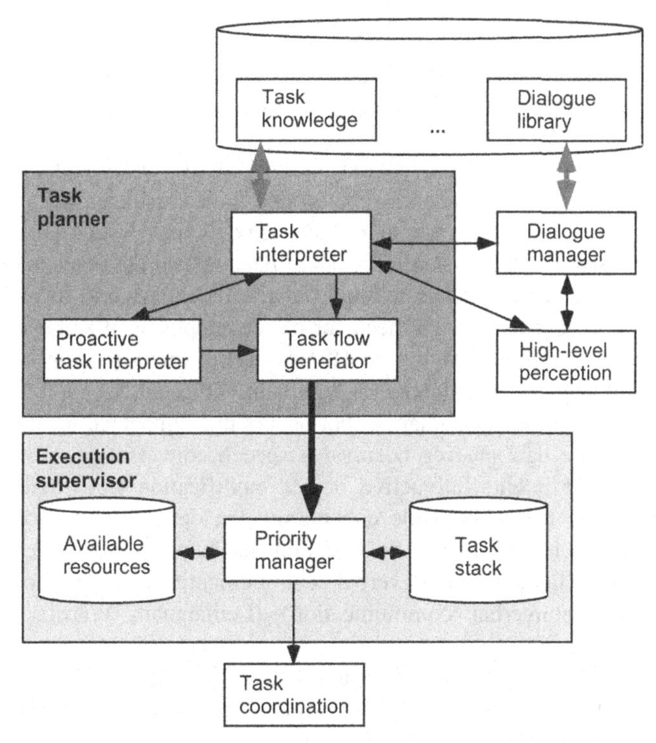

FIGURE 2.11 Task planner and execution supervisor of cognitive architecture of the Karlsruhe Humanoid Robot.

modes are triggered by the task planner as soon as the task interpreter has correctly recognized the command coming from the dialog manager. At present, the robotic system learns tasks and flows of actions in an off-line manner by programming by demonstration or by teleoperation (Robler & Hanebeck, 2004; Zollner, Asfour, & Dillmann, 2004).

Models in cognitive robotic system

The cognitive robotic system models are obligatory for recognition and interpretation of communication aspects like speech, gesture, human behavior, and dynamics of environment. All the models are stored in the knowledge base. The object ontologies and geometries, Hidden Markow Models (HMMs), kinematic models are subdatabases stored in the knowledge base. The active models are required for the fast processing of relevant data. An active model consists of the interfaces, the internal knowledge representation, inputs, and outputs for information extraction.

Cognition for human–robot interaction

Robots with cognitive abilities are necessary to support human–robot interaction (HRI) for fulfilling of necessities, capabilities, task, and demand of their human counterparts.

Cognitive-human–robot interaction is a very active and bourgeoning area of research focusing on human(s), robot(s) and their articulated activities as a cognitive system for developing algorithms, models and design guidelines to enable the design of such system. Thanks to phenomenal advances undertaken in human neuroscience, HRI is moving out from fiction world to real world context. Key to such HRI is the need for the development of suitable models capable to execute joint activity for HRI, a deeper understanding of human prospects and cognitive responses to robot actions (Fig. 2.12).

The interaction among human–robot occurs in a connected local environment furthermore data sharing occurs via speech conversation, motion gestures, and social intent. Interactive object modification is assumed to be accomplished by the robot while considering the determination, viewpoint, and beliefs of a human being. So that humans may interact with robot in explicit mode (human–robot verbal communication) or implicit mode (human–robot nonverbal communication) (Lemaignan, Warnier, Sisbot, Clodic, & Alami, 2017).

There are plans to extend the capabilities of this framework by developing a unified human–robot system that work in pro-active (by preparing and

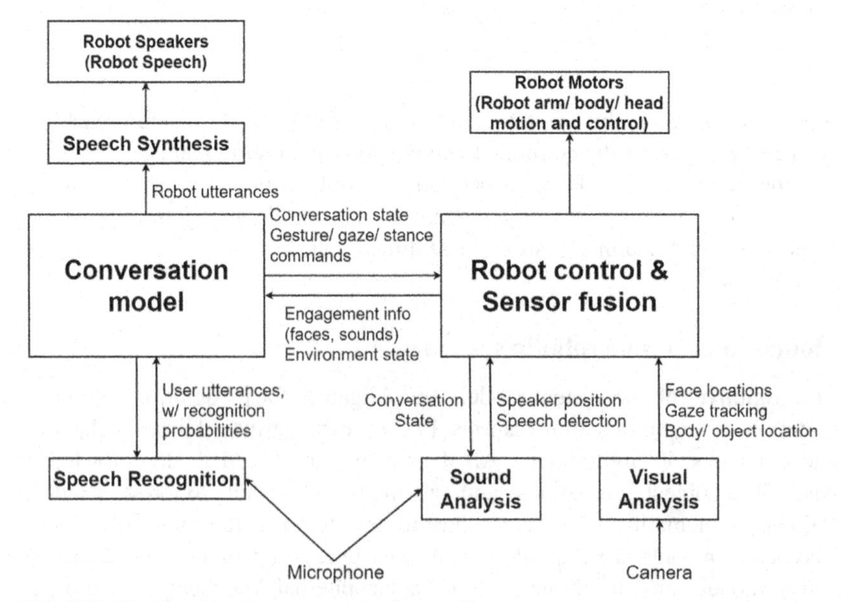

FIGURE 2.12 Cognition for human–robot interaction.

suggesting outcome schedule) and reactive aspect. The HRI should be safe, efficient, comprehensible with maintain adequate proxemics.

The robot must be skilled to undertake collaborative task, in pro-active (by preparing and suggesting outcome schedule) and reactive aspect. The robot should be allowed to function and operate safely and logically, following ethics of the society. The communication and joint action execution between the human and the robot indicates a need to take action in order to implement cognitive skills in robots (Knoblich, Butterfill, & Sebanz, 2011; Sebanz, Bekkering, & Knoblich, 2006). The actions are making a collaborative objective which has been earlier ascertained and consented, establishment of a realistic ecosystem in which the exteroceptive sensing skills of a robot is supplemented by conclusion extracted from earlier observations and creating a state of reality that include a deduced general knowledge, understanding and behavior for both the robots as well as its human partner involved.

Specifically, the deliberative architecture of a robot is designed to share space and tasks with humans, and to act and communicate in a praxis that reinforces the human's own actions and decisions.

Implementation of cognitive computing in human–robot interaction

Several cognitive components have been already incorporated and embedded into humanoid robot system. A humanoid robot prototype based on ARMAR consists of five subsystems from the kinematics control point of view. The right arm, left arm, head, torso and a movable platform are subsystems of a humanoid that provide 23 degrees of freedom to robots. The separate modules have also existed for the robot control system. Each subsystem of the robot has its own hardware, software and control module (Lemaignan et al., 2017).

The orchestration of various autonomous system components in a unified robotic framework design poses a number of technological challenges but also a sustainable design challenge in implementation of CC in HRI. Existing studies have relied that it is easier to achieve human-level interaction if the robot itself relies internally on human-level semantics (Fong et al., 2005; Trafton et al., 2013). The proposed technique relies on having substantial information illustration and modification. The interaction of software components occurs via first-order logic statements arranged in ontologies and whose semantics are similar as those of modified by users. A comprehensive overview of architecture is presented here in which an active knowledge base (Oro) act as a semantic blackboard for assembling most of the modules. The geometric reasoning module (Spark) generate a considerably high frequency symbolic claims illustrating the state of the robot environment and its evolution over time. The knowledge base also stores logical statements for future reference required by the language processing module (dialogs), the symbolic task planer (HATP) and execution controller (Shary or pyRobots). the language processing module outputs and the robot controller activities are stored back as symbolic statements. An example of this is illustrated by a cup lying on table might be picked by the Spark and represented in symbolic terms as <CUP 1type Cup, CUP1 is on TABLE>. Later the robot was subjected to process an additional sentence like "give me another cup?." To resolve this issue, the dialog module would then query the knowledge base and find (?obj type cup,? obj different from CUP 1), and inscribe assertion like <HUMAN desires GIVE_ACTION45, GIVE_ACTION45 acts on CUP2> to knowledge base. This would in turn trigger the execution controller Shary to prepare act. It would first call the HATP planer which uses Oro to initialize the planning domain (CUP 2 is at? location), and a full symbolic plan transferred to the execution controller. Finally, the controller must carry out the plan and track its achievement both for itself and for human. The above-mentioned architectural design philosophy is capable to nurture the robot's decision-making components with models of human conduct and expectations for the purpose of building an efficient artificial cognition for a robot that can assist and communicate with humans effortlessly. The main features of the architecture are Beliefs, Desires, Intentions (BDI) and known as BDI architecture (Weiss, 1999).

The knowledge model architecture of RDF, knowledge manipulation is performed by a central server (the Oro server). The users of the Oro server are in charge for making any change in the knowledge as server is pro-active in this regard. The Oro server depends on Description Logics (OWL) to illustrate and modify information. The advantages of description logic are good understanding of its trade off. The OpenCyc (Lenat, Guha, Pittman, Pratt, & Shepherd, 1990), DBPedia, RoboEarth (Waibel et al., 2011) and WordNet are open access online knowledge base. Beside above-mentioned architecture, the Prolog/ OWL combination relied upon by KnowRob (Tenorth & Beetz, 2009), answer Set Programming has been used in robotics to be practical, the first-order logic and OWL ontologies have basic, efficient and adequate symbolic structure for practical diligences. The Open Robots common sense ontology has been designed for fulfilment of our experimental needs.

Symbol grounding or anchoring is the role of building and sustaining a bidirectional connection among sub-symbolic (sensor data and actuation) representations and symbolic representation that can be modified and cleared out (Coradeschi & Saffiotti, 2003; Harnad, 1990). Taking the context of HRI the cognitive ability is of specific importance. The connection among the knowledge model to the perception and actuation capabilities of the robot is acquired through symbol grounding. The geometric reasoning and dialog processing modules are components of symbol grounding. These components assist robots in building and pushing new symbolic contents about the world to the knowledge base.

Application of cognitive computing

Cognitive computing and robot technology

In the past few decades, computer technology advancement gave birth to robots. After that, more than five decades of continuous advancement in this area resulted in the emergence of modern robots to relieve human tasks. How a country or society employed robots for performing various activities become a significant emblem for measuring the extent of innovation in science, technology, and the cutting-edge production processing capacity of a society or nation. The concept of a machine that performs human-like activity is scribbled in the mythological literature, and it had been a substantial aspiration of people to achieve this. In the present timeline, the relationship between robots and human is like a supervisor and slave. The advent of the CC system stimulates change relationships like symbiotic associations. The robots and humans complement each other to solve complex problems, and society enjoys their harmony of coexistence. The outcome of this is marked as a new feather in the cap new generation robots.

Emotional communication system

The advent of internet technology and their widespread application in society changed the world. The cooperative association between the physical and cyber world turns out to be a new arena of lifestyle. The example of Smart Home application may explain this. The Smart Home 1.0 version is equipped with technology in which networking of machine to machine (M2M) assist humans to control their home appliances from a far distance. Though in Smart Home 2.0 version additionally equipped with cognitive technology that integrates human emotions in performing task in visual distance environment.

Medical cognitive system

Human society is always under threat of chronic disease pandemics; however, the pharma industry's significant advancement overrules a greater part of the threat, but challenges are still there. A medical expert armed with a cognitive system can work more judiciously for diagnosis and making decisions with various data.

The medical field's success also lies in the correct interpretation of data because a slight deviation in data interpretation can lead to the end of life. The cognitive system components like AI, ML, and NLP play an influential role in interpreting data, learning from data, and choosing the best decision. Thus, cognitive system applications in the medical field become a life saver.

Cognitive computing with other technologies

The CC system can also be merged with other technologies and tools. This can combine and extend the capacity of existing information systems and integrate a task or domain-specific functions and interfaces as required.

Cognitive computing and big data analysis

The data is available in both structured and unstructured forms and increasing exponentially. In the CC system dictionary, this data is denoted by 5 V's that is Volume, Velocity, Variety, Value, and Veracity. Initially, big data analysis was a challenge for machine, but CC now tackles this issue efficiently. The big data analysis plays just like human brain-like thought process for CC. The human creature is constantly accumulating information form the environment and memorizing it as experience for the rest of life. This aspect is analogous to deep learning in the cognitive system. The competence of human being for big data analysis segregate them into different intellectual level. In this regard, the CC applications are smart city, smart home, health monitoring, and psychological support.

Additionally, in upcoming days the association of CC and big data will be reflected in many areas. CC is influenced by the knowledge exploration dimension of human beings, aids them in recognizing and distinguish an image very shortly, for example' Google Photos" application able to differentiate the image of cow form buffalo by learning a lot of embodiments. Still, there is space for improvement in the learning process and big data analysis in the cognitive system.

Cognitive computing and cloud computing

Virtualization in computing, storage, and bandwidth is the real meaning of cloud computing. This reduces the cost of software services as well as help in popularization of CC application. Cloud computing also add some new feature like flexibility, dynamic aspect, virtuality and storage capacity to the CC system. The ML is adopted to retrieve data from cloud storage space and extract useful information from this data for decision-making. The NLP component assists in the processing of information.

The cognitive services of IBM and Google focus on understanding brain-like cognition and judgment by deploying cloud services model to provide accurate assistant in decision-making.

Cognitive computing and reinforcement learning

Reinforcement learning is explained as learning from the environment and apply in conduct. Broadly, the ML methods are categorized into supervised learning and unsupervised learning. Reinforced learning is like the learning process of humans.

Cognitive computing and image understanding

In cognitive system problem-solving feature is based on the trained model. This model extracts useful information from original data for model training and deliver logical thinking ability to model. Further, the deep learning aspect delivers the visual thinking ability of the human brain to model. As the new heights in efficiency are attained by the computer application, the scientists consider that various real-world problems that are very simple to human brain understanding and analysis are difficult to rationalize machines for generating prediction models. This issue is somewhat resolved by image feature extraction that is image classification and image retrieval. The application of the face recognition tool best described this application.

Conclusion

CC is a nascent interdisciplinary domain. The confluence of the cognitive science, data science, neuroscience, and cloud computing are flown in the form of CC. Although, CC does not introduce a revolutionary novelty into the AI and Big Data research outcomes it enables digital solutions to meet human centric requirements. The machines become able to think, act, and behave like a human in order to achieve optimal synergy from human-machine interaction. Very soon any digital device will be evaluated by its cognitive abilities. CC would be a crucial step toward digital humanism. Traits like speech understanding, face detection, recommendations, risk assessment, medical diagnosis, sentimental analysis, psychometrics to identify psychological profiles are several examples of "human problems" that are tractable through CC. Adaptive, Interactive, Iterative and stateful, and Contextual are mandatory attributes of the CC system. Learning, Reasoning, and Understanding are ornaments of CC system. The ML, Speech Processing, Data Mining, Deep Learning, Computer Vision, Reasoning, Emotional Intelligence, NLP, and Human–Computer Interaction are some top technologies used by CC systems and make CC applications extremely captivating.

CC enhances human cognition, such that human beings can utilize past information, experience, figures, and facts to make self-assured and precise decisions. The goal of CC is to relieve humans in their everyday tasks and decision-making deprived of actually replacing them.

In a narrative form, it can be thought that a complete CC System performs as a superhuman who is aware, intelligent, understanding, helpful, emphatic, and accurate. The cognitive world welcomes humans to a forward thinking future.

References

Anderson, J. R. (2005). Human symbol manipulation within an integrated cognitive architecture. *Cognitive Science*, *29*(3), 313–341.

Anderson, J. R., & Lebiere, C. (2003). The Newell test for a theory of cognition. *Behavioral and Brain Sciences*, *26*(5), 587–601.

Bannat, A., Bautze, T., Beetz, M., Blume, J., Diepold, K., Ertelt, C., et al. (2010). Artificial cognition in production systems. *IEEE Transactions on Automation Science and Engineering.*, *8* (1), 148–174.

Beaudouin-Lafon, M. (2000). Instrumental interaction: An interaction model for designing post-WIMP user interfaces. In Proceedings of the SIGCHI conference on *human factors in computing systems* (pp. 446–453), Available from https://doi.org/10.1145/332040.332473.

Bellas, F., Duro, R. J., Faiña, A., & Souto, D. (2010). Multilevel darwinist brain (MDB): Artificial evolution in a cognitive architecture for real robots. *IEEE Transactions on Autonomous Mental Development.*, *2*(4), 340–354.

Breazeal, C. (2001). Socially intelligent robots: Research, development, and applications. *2001 IEEE International Conference on Systems, Man and Cybernetics. e-Systems and e-Man for Cybernetics in Cyberspace (Cat. No. 01CH37236)*, *4*, 2121–2126.

Brooks, R. (1986). A robust layered control system for a mobile robot. *IEEE Journal on Robotics and Automation*, *2*(1), 14–23.

Burghart, C., Burghart, C., Mikut, R., Stiefelhagen, R., Asfour, T., Holzapfel, H., Steinhaus, P., et al. (2005). A cognitive architecture for a humanoid robot: A first approach. *5th IEEE-RAS International Conference on Humanoid Robots, 2005*, 357–362.

Carroll, J. M. (2003). *HCI models, theories, and frameworks: Toward a multidisciplinary science*. Elsevier.

Cassell, J., Billinghurst, M., Campbell, L., Chang, K., Vilhjálmsson, H., Yan, H., et al. (1999). *Embodiment in conversational interfaces: Rea. Proceedings of the SIGCHI conference on human factors in computing systems* (pp. 520–527). New York, NY: Association for Computing Machinery.

Chandiok, A., & Chaturvedi, D. K. (2018). CIT: Integrated cognitive computing and cognitive agent technologies based cognitive architecture for human-like functionality in artificial systems. *Biologically Inspired Cognitive Architectures*, *26*, 55–79.

Chastagnol, C., Clavel, C., Courgeon, M., & Devillers, L. (2014). *Designing an emotion detection system for a socially intelligent human-robot interaction. Natural interaction with robots, knowbots and smartphones* (pp. 199–211)). Springer.

Chen, M., Herrera, F., & Hwang, K. (2018). Cognitive computing: Architecture, technologies and intelligent applications. *IEEE Access*, *6*, 19774–19783.

Coradeschi, S., & Saffiotti, A. (2003). An introduction to the anchoring problem. *Robotics and Autonomous Systems*, *43*(2–3), 85–96.

Dillmann, R., Becher, R., & Steinhaus, P. (2004). ARMAR II—A learning and cooperative multimodal humanoid robot system. *International Journal of Humanoid Robotics*, *1*(01), 143–155.

Duch, W., Oentaryo, R. J., & Pasquier, M. (2008). Cognitive architectures: Where do we go from here? In P. Wang, B. Goertzel, & S. Franklin (Eds.), *Frontiers in artificial intelligence and applications* (vol. 171). IOS Press.

El Sayed, M. S., Awadalla, M. H., Ali, H. E. I., & Mostafa, R. F. A. (2012). Interactive learning for humanoid robot. *International Journal of Computer Science Issues*, *9*(4), 331.

Feil-Seifer, D. & Mataric, M. J. (2005). Defining socially assistive robotics. In 9th *international conference on rehabilitation robotics*, 2005. ICORR 2005, (pp. 465–468). doi: 10.1109/ICORR.2005.1501143.

Feldman, S. E. (2012). The answer machine. *Synthesis Lectures on Information Concepts Retrieval and Services*, *4*(3), 1–137.

Ferrucci, D., et al. (2010). Building Watson: An overview of the DeepQA project. *AI Magazine.*, *31*(3), 59–79.

Firby, R. J. (1992). *Building symbolic primitives with continuous control routines. Artificial Intelligence Planning Systems* (pp. 62–69).

Fong, T., et al. (2005). The peer-to-peer human-robot interaction project. *Space, 2005*, 6750.

Franklin, S., Madl, T., D'mello, S., & Snaider, J. (2013). LIDA: A systems-level architecture for cognition, emotion, and learning. *IEEE Transactions on Autonomous Mental Development, 6*(1), 19–41.

Franklin, S., Strain, S., McCall, R., & Baars, B. (2013). Conceptual commitments of the LIDA model of cognition. *Journal of Artificial General Intelligence, 4*(2), 1–22.

Georgeon, O. L., Marshall, J. B., & Manzotti, R. (2013). ECA: An enactivist cognitive architecture based on sensorimotor modeling. *Biologically Inspired Cognitive Architectures, 6*, 46–57.

Goertzel, B., Ke, S., Lian, R., O'Neill, J., Sadeghi, K., Wang, D., et al. (2013). The cogprime architecture for embodied artificial general intelligence. In 2013 IEEE symposium on computational intelligence for human-like intelligence (CIHLI), (pp. 60–67). doi: 10.1109/CIHLI.2013.6613266.

Gray, W. D. (2007). *Integrated models of cognitive systems* (vol. 1). Oxford University Press.

Greenemeier, L. (2013). *Will IBM's Watson Usher in a new era of cognitive computing? Scientific American* (pp. 1860–1861).

Gudwin, R., et al. (2017). The multipurpose enhanced cognitive architecture (MECA). *Biologically Inspired Cognitive Architectures, 22*, 20–34.

Harnad, S. (1990). The symbol grounding problem. *Physica D (Nonlinear Phenomena), 42* (1–3), 335–346.

High, R. (2012). *The era of cognitive systems: An inside look at IBM Watson and how it works. IBM Corporation Redbooks* (pp. 1–16).

Holzapfel, H., Fuegen, C., Denecke, M., & Waibel, A. (2002). In Integrating emotional cues into a framework for dialogue management. In Proceedings fourth IEEE *international conference on multimodal interfaces*, (pp. 141–146). Available from https://doi.org/10.1109/ICMI.2002.1166983.

Hurwitz, J., Kaufman, M., Bowles, A., Nugent, A., Kobielus, J. G., & Kowolenko, M. D. (2015). *Cognitive computing and big data analytics*. Wiley Online Library.

Kawamura, K., Rogers, T. E., & Ao, X. (2002). *Development of a cognitive model of humans in a multi-agent framework for human-robot interaction. Proceedings of th*e first international joint conferenc*e on autonomous* agen*ts and* multiagent systems: Par*t 3* (pp. 1379–1386).

Kelly, J. E., III, & Hamm, S. (2013). *Smart machines: IBM's Watson and the era of cognitive computing*. Columbia University Press.

Kieras, D. E., & Meyer, D. E. (1997). An overview of the EPIC architecture for cognition and performance with application to human-computer interaction. *Human–Computer Interaction, 12*(4), 391–438.

Kirk, J., Mininger, A., & Laird, J. (2016). Learning task goals interactively with visual demonstrations. *Biologically Inspired Cognitive Architectures, 18*, 1–8.

Knoblich, G., Butterfill, S., & Sebanz, N. (2011). Psychological research on joint action: Theory and data. *Psychology of Learning and Motivation, 54*, 59–101.

Kobayashi, K., Nakatani, A., Takahashi, H., & Ushio, T. (2002). *Motion planning for humanoid robots using timed petri net and modular state net, IEEE* international conference on systems, man and cyber*netics* (vol. 6, p. 6-pp).

Kortenkamp, D., Bonasso, R. P., & Murphy, R. (1998). *Artificial intelligence and mobile robots: Case studies of successful robot systems*. MIT Press.

Laird, J. E. (2008). Extending the Soar cognitive architecture. *Frontiers in Artificial Intelligence and Applications, 171*, 224.

Laird, J.E., Kinkade, K.R., Mohan, S., Xu, J.Z. (2012). Cognitive robotics using the Soar cognitive architecture. In 8th international conference on cognitive robotics, (cognitive robotics workshop, twenty-sixth conference on artificial intelligence (AAAI-12)), Toronto, CA.

Lemaignan, S., Warnier, M., Sisbot, E. A., Clodic, A., & Alami, R. (2017). Artificial cognition for social human–robot interaction: An implementation. *Artificial Intelligence, 247*, 45–69.

Lenat, D. B., Guha, R. V., Pittman, K., Pratt, D., & Shepherd, M. (1990). CYC: Toward programs with common sense. *Communications of the ACM, 33*(8), 30–49.

Lisetti, C. L., Brown, S. M., Alvarez, K., & Marpaung, A. H. (2004). A social informatics approach to human-robot interaction with a service social robot. *IEEE Transactions on Systems, Man, and Cybernetics – Part C: Applications and Reviews, 34*(2), 195–209.

Lotfabadi, M. S. (2011). The presentation of a new method for image distinction with robot by using rough fuzzy sets and rough fuzzy neural network classifier. *International Journal of Computer Science Issues, 8*(4), 419.

Ly, D. N., Regenstein, K., Asfour, T., & Dillmann, R. (2004). *A modular and distributed embedded control architecture for humanoid robots, 2004 IEEE/RSJ* international conference on intelligent robots and syste*ms (IROS) (IEEE Cat. No. 04CH37566)* (vol. 3, pp. 2775–2780).

Mounier, G. (2014). Cognitive computing: Why now and why it matters. *Information Today, Inc.*

Newell, A. (1994). *Unified theories of cognition*. Harvard University Press.

Noor, A. K. (2015). Potential of cognitive computing and cognitive systems. *Open Engineering, 5*(1), 75–88. Available from https://doi.org/10.1515/eng-2015-0008.

Norman, D. A., & Draper, S. W. (1986). *User centered system design; new perspectives on human-computer interaction*. L. Erlbaum Associates Inc.

Ortony, A., Clore, G. L., & Collins, A. (1990). *The cognitive structure of emotions*. Cambridge University Press.

Pelachaud, C., & Bilvi, M. (2003). *Computational model of believable conversational agents. Communication in multiagent systems* (pp. 300–317). Springer.

Persiani, M., Franchi, A. M., & Gini, G. (2018). A working memory model improves cognitive control in agents and robots. *Cognitive Systems Research, 51*, 1–13.

Poggi, I., Pelachaud, C., de Rosis, F., Carofiglio, V., & De Carolis, B. (2005). Greta. a believable embodied conversational agent. In O. Stock, & M. Zancanaro (Eds.), *Multimodal intelligent information presentation. Text, speech and language technology* (vol. 27, pp. 3–25). Dordrecht: Springer. Available from https://doi.org/10.1007/1-4020-3051-7_1.

Pynadath, D. V., Rosenbloom, P. S., & Marsella, S. C. (2014). Reinforcement learning for adaptive theory of mind in the sigma cognitive architecture. In B. Goertzel, L. Orseau, & J. Snaider (Eds.), *Artificial general intelligence. AGI 2014. Lecture notes in computer science* (vol 8598, pp. 143–154). Cham: Springer. Available from https://doi.org/10.1007/978-3-319-09274-4_14.

Rahat, M., Nazari, M., Bafandehkar, A., & Ghidary, S. S. (2012). Improving 2D boosted classifiers using depth LDA classifier for robust face detection. *International Journal of Computer Science Issues, 9*(3), 35.

Reynolds, H., & Feldman, S. (2014). Cognitive computing: Beyond the hype. *KM World, 27*.

Robler, P., & Hanebeck, U. D. (2004). Telepresence techniques for exception handling in household robots. In 2004 IEEE *international conference on systems, man and cybernetics* (IEEE Cat. No. 04CH37583), (vol. 1, pp. 53–58).

Samsonovich, A. V. (2013). Emotional biologically inspired cognitive architecture. *Biologically Inspired Cognitive Architectures, 6*, 109–125.

Sato, K., & Lim, Y. (2000). Physical interaction and multi-aspect representation for information intensive environments. In Proceedings 9th IEEE *international workshop on robot and human interactive communication*. IEEE RO-MAN 2000 (Cat. No. 00TH8499), (pp. 436–443).

Sebanz, N., Bekkering, H., & Knoblich, G. (2006). Joint action: Bodies and minds moving together. *Trends in Cognitive Sciences, 10*(2), 70–76.

Soltau, H., Metze, F., Fugen, C., & Waibel, A. (2001). A one-pass decoder based on polymorphic linguistic context assignment. In IEEE *workshop on automatic speech recognition and understanding*. ASRU'01. 2001, (pp. 214–217). doi: 10.1109/ASRU.2001.1034625.

Starzyk, J. A., & Graham, J. (2015). MLECOG: Motivated learning embodied cognitive architecture. *IEEE Systems Journal, 11*(3), 1272–1283.

Stiefelhagen, R., Fugen, C., Gieselmann, R., Holzapfel, H., Nickel, K., & Waibel, A. (2004). Natural human-robot interaction using speech, head pose and gestures. In 2004 IEEE/RSJ *international conference on intelligent robots and systems* (IROS) (IEEE Cat. No. 04CH37566), (vol. 3, pp. 2422–2427).

Sun, R. (2009). Motivational representations within a computational cognitive architecture. *Cognitive Computation, 1*(1), 91–103.

Sun, R., Merrill, E., & Peterson, T. (2001). From implicit skills to explicit knowledge: A bottom-up model of skill learning. *Cognitive Science, 25*(2), 203–244.

Tenorth, M., & Beetz, M. (2009). KnowRob—knowledge processing for autonomous personal robots. In 2009 IEEE/RSJ international conference on intelligent robots and systems, (pp. 4261–4266).

Thomaz, A. L., & Breazeal, C. (2007). Robot learning via socially guided exploration. In *2007 IEEE 6th international conference on development and learning*. (pp. 82–87), doi: 10.1109/DEVLRN.2007.4354078.

Toumi, T., Zidani A. (2014). From human-computer interaction to human-robot social interaction. *arXiv Prepr. arXiv1412.1251*.

Trafton, J. G., Hiatt, L. M., Harrison, A. M., Tamborello, F. P., Khemlani, S. S., & Schultz, A. C. (2013). ACT-R/E: An embodied cognitive architecture for human-robot interaction. *Journal of Human-Robot Interaction, 2*(1), 30–55.

Trappl, R., Petta, P., & Payr, S. (Eds.), (2002). *Emotions in humans and artifacts*. MIT Press.

Vernon, D., Metta, G., & Sandini, G. (2007). A survey of artificial cognitive systems: Implications for the autonomous development of mental capabilities in computational agents. *IEEE Transactions on Evolutionary Computation, 11*(2), 151–180.

Waibel, M., Beetz, M., Civera, J., D'Andrea, R., Elfring, J., Gálvez-López, D., et al. (2011). RoboEarth, robotics automation magazine. *IEEE Robotics & Automation Magazine, 18*(2), 69–82.

Wang, Y., Zhang, D., & Kinsner, W. (2010). *Advances in cognitive informatics and cognitive computing* (vol. 323). Springer.

Weiss, G. (1999). *Multiagent systems: A modern approach to distributed artificial intelligence*. MIT Press.

Zollner, R., Asfour, T., & Dillmann, R. (2004). Programming by demonstration: Dual-arm manipulation tasks for humanoid robots. In 2004 IEEE/RSJ *international conference on intelligent robots and systems* (IROS) (IEEE Cat. No. 04CH37566), (vol. 1, pp. 479–484).

Chapter 3

Cognitive computing in human activity recognition with a focus on healthcare

S. Ravi Shankar[1], Gopi Battineni[2] and Mamta Mittal[3]

[1]Department of CSE, Vidya Academy of Science and Technology, Thrissur, India, [2]Telemedicine and Tele Pharmacy Centre, School of Medical Products and Sciences, University of Camerino, Camerino, Italy, [3]Department of CSE, G. B. Pant Government Engineering College, New Delhi, India

Introduction

Artificial intelligence (AI) has been remote since the time beginning of computers, and consistently it appears to draw nearer and closer to that objective with novel cognitive computing models (Ligeza, 1995). Originating from the combination of cognitive science and dependent on the fundamental reason for recreating the concept, the behaviors of human thinking and cognitive computing will undoubtedly have extensive effects on our private lives, yet additionally businesses like medical care, protection, and many other. The benefits of cognitive computing are well and a stage beyond the traditional AI frameworks (Demirkan et al., 2017).

Cognitive computing simply presents as a mixture of AI, machine learning (ML), neural networks, text mining, sentiment analysis, and natural language processing (NLP) to tackle everyday issues as same as a human (Battineni et al., 2020). IBM characterizes cognitive computing as a comprehensive system that learns at scale, reason with reason, and associates with people in a characteristic structure. The objective behind cognitive computing not only to mimic human behavior instead to assemble machines that can analyze a big amount of data but also learns similarly (Boden, 2006). We can exploit their qualities to assist us with tackling troublesome issues and settle on the best choices dependent on more information than any one individual would want to hold in their mind. The cognitive computing frameworks can adjust and figure out data in an unstructured format such as

Cognitive Computing for Human-Robot Interaction. DOI: https://doi.org/10.1016/B978-0-323-85769-7.00006-9

speech and images (Bardram et al., 2006). This enables us to recognize and understand patterns and gives them meaning.

Cognitive computing is a kind of research that investigates the different varieties of data and generate rich data insight interpretations (Miller and Brown, 2018). It involves many techniques and tools which include predictive analytics, big data, ML, NLP, and the Internet of things (Singh R, 2019). The main features behind cognitive computing are skills learning, hypothesis, development analysis, and knowledge improvement with no reprogramming (Noor, 2015). These methods are mainly categorized as (1) event observation, (2) interpretation, (3) evolution, and (4) conclusion or decision making (Gupta et al., 2018). In the medical industry, a lot of different approaches or methods are involved to support patient treatment and health. Because medical profile or electronic health records consists of patient history and biometric data. At the same time, the patient life cycle can involve daily habits, obesity, health behavior, social behavior, hobbies, etc. By involving all these players to make access to various data sources, the government initially controls and manages requirements monitoring (Patel & Kannampallil, 2015). Fig. 3.1 presents the conceptual model in the utilization of cognitive computing and data stored in the cloud for healthcare can be presented as structured, unstructured, and semi-structured.

To help clinical expert in better treatment of infections, and improve persistent results, medical care has achieved an advance of cognitive computing. The cognitive computing systems had processed the tremendous measures of information right away to answer an explicit inquiry and make altered keen suggestions (Behera et al., 2019; Croskerry, 2013). These systems in medical care connect the working of humans and machines where the human brain and

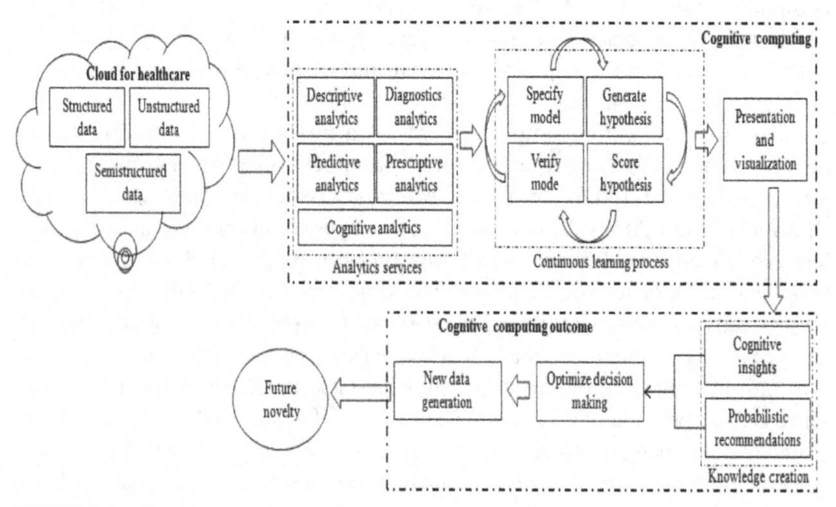

FIGURE 3.1 Cognitive computing modeling in healthcare (Behera et al., 2019).

computers are covered to improve human dynamics. Therefore, the analysis of cognitive computing technologies associated with medical care will provide a comprehensive view of recent advancements that happened in this area.

However, cognitive computing in medicine still in yet infancy stage since there are some works have been published but none of them are given complete insight on this emerging domain. It is hard to understand the actual potentiality of cognitive computing in medicine for providing quality care. Therefore, in this chapter authors have reviewed the basic concepts of ML-based human activity recognition (HAR), applications of healthcare robots, and future perspectives of cognitive computing in healthcare.

Human activity recognition

In this section, the authors had reviewed the different ML algorithms that are associated with HAR. HAR has been becoming a focal point for a decade because of its significance in contemplating numerous zones, including medical care, intelligent gaming, sports, and observing frameworks for general purposes (Aggarwal & Ryoo, 2011). The association between HAR and ML approaches can be observed in Fig. 3.2. HAR intends to perceive human exercises in controlled and uncontrolled conditions. Despite bunch applications, HAR calculations face numerous difficulties, including (1) intricacy and an assortment of everyday exercises, (2) intersubject and intra-subject changeability for a similar action, (3) trade-off among execution and protection, (4) computational proficiency in portable and embedded devices, and (5) complexities in data annotation.

The training and testing data for HAR algorithms are normally acquired from two primary sources such as (1) surrounding sensors (for example a

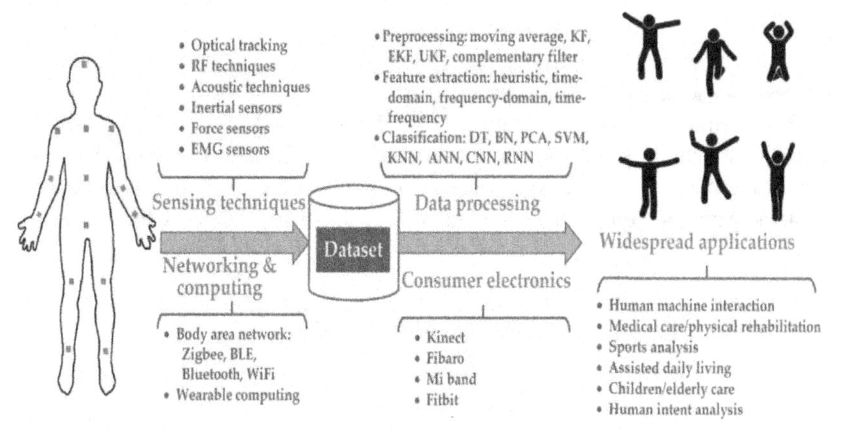

FIGURE 3.2 Human activity recognition with the inclusion of ML approaches (Meng et al., 2020).

surveillance camera) and (2) inserted sensors (for example independent sensors or accelerometers on a smartwatch). Encompassing sensors can be natural sensors or video cameras situated at explicit points in the environment. Inserted sensors are incorporated into individual gadgets, for example, cell phones and smartwatches or are coordinated into garments or other specific clinical hardware (Brezmes et al., 2009). Cameras have been generally utilized in the HAR applications, anyway, gathering video information presents numerous issues concerning protection and computational necessities. While videocams produce rich relevant data, protection issues restrictions have driven numerous specialists to work with other embedded and ambient sensors, including deep learning images as an elective for privacy protection.

The traditional ML algorithms were extensively used for HAR over the past decade. The sensor data from wearable devices and mobile devices were used to identify the activity recognition. These sensor data were pre-processed by removing noise, then normalized also segmented as per the requirement. Then any classical parametric or non-parametric ML algorithms were used for classification. Of late, deep learning-based architectures were extensively used for HAR. A literature review of various deep learning algorithms for HAR revealed the usage of deep neural network (DNN), convolution neural network (CNN), recurrent neural network (RNN) capsule network, generative adversarial network (GAN), etc (Kim et al., 2010; Aggarwal A, 2021). Also, much research was conducted on combining all these algorithms. The features were classified into spatial and temporal. In many such works, CNN was used for spatial feature extraction and RNN for temporal feature extraction.

A deep learning model (Xu et al., 2019) combining inception CNN (Alom et al., 2020) and RNN was used for HAR. The spatial features were extracted using inception CNN, which was then fed to a gated recurrent unit (GRU) for the extraction of temporal features is presented in Fig. 3.3. Inception CNN is a pretrained model, widely applied for computer vision problems. Fined tuned Inception neural networks using batch normalization and factorization methods were also used. GRU was used for better performance in case of execution time.

The extracted temporal features may be then fed to a fully connected dense layer network and perform the activity classification. HAR using this hybrid method got a better F-measure compared to the DNN and CNN based models. Another approach presented in Fig. 3.4 was to extract the temporal

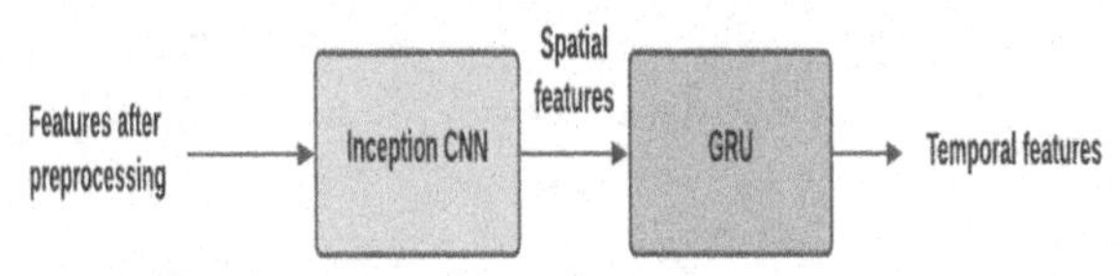

FIGURE 3.3 Inception convolution neural network-gated recurrent unit based feature extraction for human activity recognition.

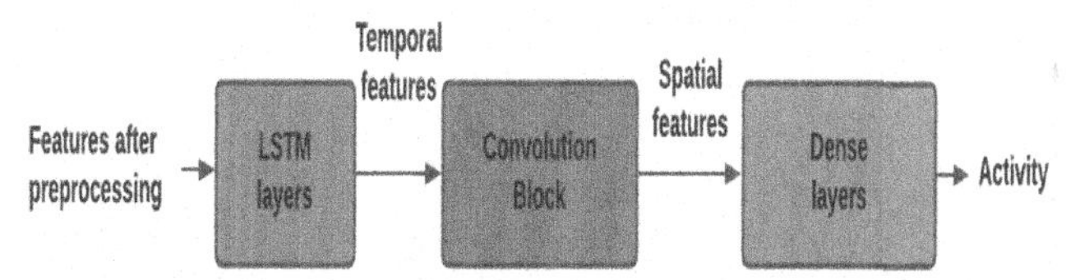

FIGURE 3.4 Long short-term memory-convolution neural network hybrid model for human activity recognition.

features first using long short-term memory (LSTM). Then the output of the LSTM was fed to CNN. The LSTM network is robust as it is not prone to the vanishing gradient descent problem. The pre-processed input features were passed through two LSTM layers having 32 neurons. The output was given to a convolution block consisting of two convolution layers and one pooling layer (Xia et al., 2020). The output of the CNN was fed to a global average pooling layer. A batch normalization layer was added to reduce the training time and its output was fed to a final fully connected SoftMax layer for classification of activity. This hybrid model seemed to have worked well compared to the traditional DNN and CNN models in terms of F-measure.

Apart from the hybrid models, several types of research were carried out using various ML models. The sensory data from smartphone accelerometers was processed and fed to various ML models for HAR (Wan et al., 2020). An approach using the bidirectional LSTM (BLSTM) model was also used in addition to DNN, CNN, and other traditional methods like support vector machine. BLSTM-based approach used one LSTMs to capture past and future information. Capsule networks were also used for HAR from wearable device sensors (Pham et al., 2020). Capsule networks were used to identify activities from time-series data by capturing temporal features. The capsule network model used a dynamic routing approach to extract spatial features.

Applications of healthcare robot services

Healthcare robotics has been prominent for many years. A lot of research has been done to find the application of robotics into medical applications since 1974. But the healthcare sector took a lot of time to adapt despite the hype, media stories, and theoretical research studies. However, the recent advancements in the area of robotics have led to a significant impact in the medical field. Also, the use of collaborative robots was found useful in operating theaters too.

Why robot take-up has been delayed in medical services?

The utilization of surgical robots in the healthcare field has been very limited although many popular ones were available. One such famous example is the da Vinci Surgical System.

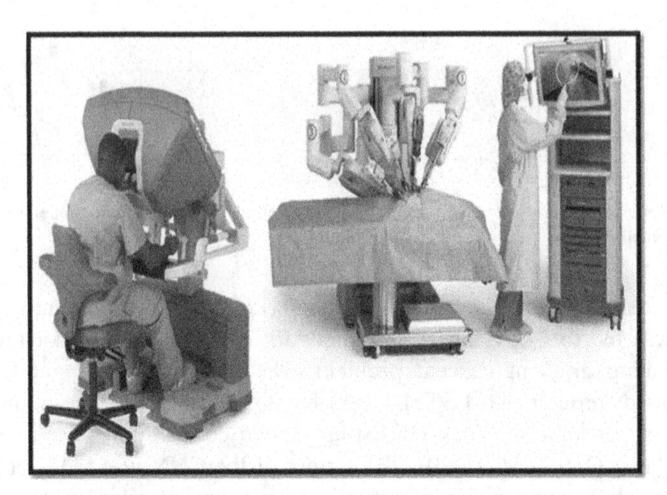

FIGURE 3.5 Key components of da Vinci surgical systems (O'Sullivan & O'Reilly, 2012).

The robot-helped laparoscopic medical procedure was well known during the 1980s (Tse & Ngan, 2015). The primary automated medical procedure was directed in 1985 utilizing a Puma 650 robot to improve accuracy in neurosurgical biopsies. Even though robotic surgery was popular in the 1980s, only 131 hospitals had the da Vinci surgical system in 2010.

This moderate take-up is halfway because of the requirement for medical services robots to be affirmed by administering bodies. The da Vinci was just Food and Drug Administration (FDA)-endorsed in the year 2000 and by 2014 it was as yet the main FDA-affirmed surgical robot for both head and neck operations (Hagen & Curet, 2014). The main components of the da Vinci surgical system are presented in Fig. 3.5.

The governing bodies were reluctant to approve the healthcare robots and it has partially led to its slow uptake. By 2014, the main FDA-endorsed surgical robot for head and neck medical procedure was the da Vinci, which was affirmed by 2000.

But the usage of surgical robots has become increasingly popular recently. In 2017, around nine surgical and clinical robots were endorsed by FDA and practically 45% of the United States medical centers had obtained the da Vinci robot.

Types of robots used in the healthcare industry

Surgical robots were the most discussed and widely used in the healthcare sector. However, besides to surgical robots, numerous different zones where robotic technology discovers its application incorporate rehabilitation, telepresence, transportation of medical devices, prescription dispensing, and disinfection (Burgner-Kahrs et al., 2015). The discussion here is limited to collaborative robots and applications.

Modern healthcare cobots were designed to perform specific tasks whereas general purpose robots were extensively used in other areas like manufacturing industries.

The tasks which require high precision in the medical field are a challenge for collaborative robots. But a few other tasks can be done more effectively by a cobot because it can reduce the contamination risks. The tasks like medical device packaging require a completely sterile environment, where cobots are extremely useful. Most importantly, in five ways the cobots were used in medicine and healthcare.

Medical device packaging

Cobots are extremely useful for packaging applications. Cobots were extensively used for a variety of industries like manufacturing, consumer goods, and logistics (Bauer, 2009).

Packaging medical devices is a difficult task as sterilization is a vital aspect. The high risk of contamination of such devices prevails if humans perform such tasks which would affect the integrity of the products.

Lab automation

Lab automation is another application in which cobots could play a crucial role (Vilela et al., 2016). In 2014, more than 277 million test requests came in the United Kingdom alone. In 2014, in excess of 277 million test demands came in the United Kingdom alone. All these lab tests required a series of similar procedures, where the cobots were discovered to be incredibly valuable.

There were 3000 tests conducted on blood samples per day in a laboratory at Copenhagen University. Lab automation was experimented with by adding two UR robots and productivity was increased beyond expectations. The results of 90% of the tests were delivered with a response time of 1 hour despite an increase in workload by 20%.

Neurosurgery

In addition to the medical device packaging and lab automation, cobots were used for neurosurgery also. A cobot-based product was developed for neurosurgery by a company named synaptive medical (Klinger et al., 2018). They have used an UR robot for the same.

The surgery was not solely performed by the cobots. All things considered, the cobots were utilized to play out a couple of undertakings like to situate a digital microscope, which would help neurosurgeons get the best perspective on the medical procedure and improve their accuracy.

On the off chance that the specialist needs to move the microscope during the operation, they can do so either by moving the robot or by guiding it to move by means of the software. The neurosurgeons have various options to move the digital microscope. They could physically move the robot or give instructions to the robot to move with the help of the software interface.

Cutting bone

Cobots were also used to perform the actual surgery. Bone removal is a method wherein some portion of the bone is removed to eliminate a favorable development. This was finished with the assistance of a cold ablation robot-guided laser osteotome (CARLO) framework (Augello et al., 2018). The CARLO framework utilized a cobot or a KUKA lightweight arm to play out the bone removal measure. This system was the world's first medical robot, which can cut bone without contact. It used cold laser technology for performing surgery. The CARLO system was used for performing bone ablation due to its high precision.

Therapeutic massage cobots

Cobots were used for therapeutic purposes also. A massage robot was designed by a Californian based start-up, which could take the role of a physiotherapist. The UR robots were used to give full-body massages to the patients using the round end effectors. The whole setup resembles the full-sized version of "back massage" gadgets available in the market.

The application of cobots in the therapeutic massage is not a critical one compared to the other three. It does not address an existing healthcare problem. Hence, those start-ups existing in these areas have not received much funding over the years.

Place of robots in healthcare

In the healthcare sector, robots are going to perform tasks like surgery assistance, disinfecting rooms, dispensing medication, etc. And these are going to be deployed soon in hospitals, pharmacies, or doctor's offices. Collaborative robots are going to make a significant impact in every field of medicine. An overview of robotics in healthcare is presented in this section to provide an awareness to everyone working in this field.

Metallic allies for the benefit of the vulnerable

The use of machines can replace some of the monotonous tasks like cleaning the hospital floor, delivering medicines up to the top floor of the building, etc. Even though there are several concerns over replacing human jobs with machines, many advantages exist as such monotonous and repetitive jobs could be easily done by machines (Weaver et al., 2008). And humans could control these machines and thereby increase their work productivity.

The monotonous jobs and other administrative tasks shall be carried out by the healthcare robots, thereby saving the time and effort of healthcare professionals. Meanwhile, they could concentrate on caring for the sick and the vulnerable by spending their time fully.

Thus, despite using healthcare robots, the service of healthcare professionals stays relevant. A discussion of some of the most useful robots is presented in the following section.

Metalheads for surgical precision

Surgery is one of the key areas where manpower and resources lay a crucial role. The surgery may be waitlisted considering the availability of manpower and resources. Hence, surgical robots can be the prodigies of surgery. The surgical robotics sales are expected to be almost double to $6.4 billion in 2020 according to the market analysis (Newswire, 2017).

The da Vinci surgical system was the most popular surgical robot, which came into existence about 15 years ago (Hagen and Curet, 2014). It consisted of a magnified 3D high definition vision system along with tiny wristed instruments that may be bent and rotate to a greater degree than the human hand. The surgeon is in control of the robotic system and can carry out precise operations, which was thought impossible earlier.

Many industries were also interested in developing surgical robots. In 2015, Google in collaboration with pharma giants like Johnson & Johnson (J & J), had also started working on the development of surgical robots. Google's co-founder Sergey Brin used the surgical robot to suture synthetic tissue, in early 2018. The companies named Auris and Orthodoxy were acquired by J & J. Auris has been in the area of developing robotic technologies and they mainly developed robotic technologies related to lung cancer. Orthodoxy was popular in developing software-enabled surgical technologies.

Blood-drawing and disinfector robots help put "care" back into healthcare

The patients, while they stay at hospitals, interact most of their time with nurses than other healthcare professionals. The nurses are involved with blood sample collection, monitor the patient's condition, check vital signs, take care of a patient's hygiene, etc. They may even work 12-hour shifts and most of these tasks could be physically and mentally daunting (Jeon et al., 2017).

The monotonous task will be carried out by robots instead of nurses. Thus, the human staff can handle issues which require decision making,

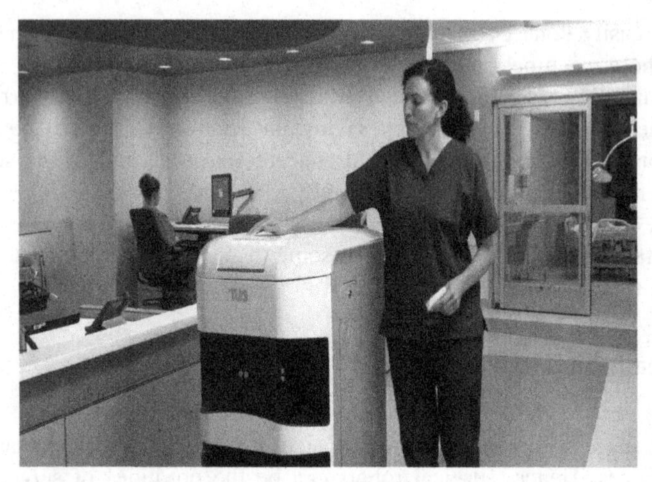

FIGURE 3.6 TUG autonomous robots for medical support.

creativity, etc. Shortly blood-drawing robots may be deployed so that it can also perform laboratory tests without the intervention of nurses.

The TUG autonomous mobile delivery robots are also widely used in the health sector, which was found very much useful for nurses. The labor force is greatly reduced due to the intervention of robots which could take up the human task. The TUG finishes the task requested and goes to the charging dock for further charging. The TUG has been very common in the United States. Tasks like carrying loads, carrying hazardous or sensitive materials are nowadays taken up by robots. Fig. 3.6 presents the TUG autonomous mobile robots in the hospital works.

Robotic assistance for a better life

Apart from using robots in laboratories and blood-drawing, remote-controlled robots may be used in the health sector for patient caretaker interaction. This gives us a solution for tedious home visits. The caretakers can check for further appointments and determine the conditions. This kind of medical robots thus draws attraction in the healthcare domain.

The first robotic concierge named Sam was developed by a company called Luvozo. It was first used in a living community in Washington, DC. with a lot of senior residents. Those robots were best in cutting-edge technology and they were human-sized and smiling ones. It was able to provide frequent check-ins and nonmedical care for patients. These robots were developed to spend time with elderly people. The costs of taking care of elderly people were reduced to a great extent and hence increased the satisfaction index of patients.

Telemedical network for increasing accessibility

The medical professionals may not be able to reach on time in case of emergency cases like accidents (Mirmoeini et al., 2019). It may not be the case in developed countries. But in other parts of the world, millions were deprived of emergency services as of 2019.

Some of the companies offering services to serve emergency cases are With InTouch Health, Doctor on Demand, Health Tap, American Well, Tel, Avizia, Babylon, etc. They provide services like high-quality emergency consultations for cardiovascular problems, dermatological problems, heart attacks, etc. Most of these services may be a mobile-based application which can be accessed comfortably by both patients and doctors.

The power of exoskeletons

The exoskeletons were familiar to people via movies and video games (de Looze et al., 2016). Those exoskeletons were developed by few companies, which give a sense of invincibility to people as it helps humans to move around and lift heavy objects. As an example, a paralyzed person named Matt Ficarra was helped by a gait-training exoskeleton suit on his wedding day. So, the public including healthcare professionals, soldiers, etc would soon be using exoskeletons daily to extend their muscle power, stamina, and weight lifting skills.

The exoskeleton suit has been already helping healthcare professionals by aiding them in surgery by reducing the time required. There were also reports from the BBC about a French citizen, who was able to move using an exoskeleton suit despite all the limbs being paralyzed. FDA has also given recognition to exoskeletons for their rehabilitation utility. In 2019, a pioneer company in this field named ReWalk Robotics has come up with exoskeleton suits for commercial purposes. They have made it available to rehabilitation centers across the United States.

Robots in the supply chain

Robots are also used for doing potentially hazardous tasks, thereby helping humans in this regard. Robots were used for moving heavy boxes or testing kits in addition to performing the monotonous and repetitive tasks. A company named Boston Dynamics came up with robots that were designed for testing chemical protection clothing for the United States army. These robots are also useful during emergencies which are so dangerous to be handled by humans.

Robotics also finds application in pharmaceutical distribution chains in addition to it being placed in potentially dangerous situations for humans (Fitzgerald & Quasney, 2017). Many robotic medical dispenser systems and medication management solutions exist. For example, the PharmASSIST TOBOTx may assist in providing such solutions. These robots were also equipped with data mining features so that they could give valuable insights on traffic and efficiency. Thus, the use of medical robots would assist pharmacists to concentrate on important tasks like providing awareness and advice on preventive measures to the people and thus ensure proper care in the healthcare sector.

Robots disinfecting hospital rooms

One of the leading causes of death in the United States is hospital acquired infections (HAI) (Dellinger, 2016). One in every 25 patients gets HAI, according to Centre for disease control and prevention (CDC) statistics used by Xenex. Among them, 1 in 9 dies.

The HAI also accounts for a financial loss of approximately 30 billion dollars a year. The robot manufactured by a Texas-based company called Xenex produces a unique robot for disinfection. It uses ultraviolet light to quickly and efficiently disinfect any space in the healthcare facility. It reduces the number of HAIs by causing cellular level damage to the microorganisms. In addition to disinfection, the workload of the hospital staff also gets decreased to an extent.

Nanorobots swimming in blood

While we have not arrived at the time of nanotechnology, patterns point towards innovation getting increasingly huge. With the rise of absorbable and advanced pills, we draw nearer to nanorobots bit by bit. On that front: analysts from the Max Planck Institute have been trying different things with uncommonly miniature measured – more modest than a millimeter – robots that swim through our natural liquids and could be utilized to convey drugs or other clinical alleviation in an exceptionally focused on manner. These scallop-like micro-bots are intended to swim through non-Newtonian liquids, like our circulation system, around the lymphatic system, or over the tricky goo on the outside of our eyeballs.

Social companion robots to cheer up and for company

Social companion robots have emerged up due to the advancement of robotics and AI (Dautenhahn, 2007). They are human or animal-like shaped and can carry out various tasks along with interaction. The tasks which could be performed by these robots include helping the humans in the kitchen, housekeeping, house guarding, etc. They could also teach children and become their companions. These robots may also take care of elderly ones by reminding them of their medication as well as by removing their loneliness.

Examples of social companion robots are Jibo, Pepper, Paro, Zora, and Buddy. These companion robots are equipped with electronic sensors, cameras, microphones, etc. Thus, it truly provides a company for their owners in addition to setting reminders for medication.

The various types of robots mentioned will cause a drastic disruption in the healthcare sector (Battineni et al., 2020). They exist in different shapes and sizes. They aid healthcare professionals and are efficient and cost-effective. It would change the life of medical professionals by cutting down their workload and thereby bring more laurels to them. Medical robots are going to play a crucial role shortly. So, the world should adapt quickly to reap the benefits out of it.

Future scope and limitations of cognitive computing in healthcare

Future perspectives

The fate of medical care is centered around giving people a total image of the numerous elements that influence their well-being. Looking for similar comforts they access in different enterprises, the present patients need customizing, straightforward, coordinated, and excellent consideration. To give the experience engaged purchasers request, medical care associations need better approaches to take advantage of and interpret medical data continuously (Arefin et al., 2017). This continuous examination permits specialists, scientists, guarantors, case managers, and different partners to settle on the most educated choices, while additionally giving patients more noteworthy command over their consideration. Nonetheless, this involved coordination requires critical time and can burden the assets of even the most flexible associations.

Like some other industry, medical care is being disturbed and changed by a dramatic development in information. More medical services information is delivered today than before. By 2020, the impression of clinical information will twofold at regular intervals: an expected 80% of this information will be unstructured. While clinical experts approach medical data, the volume is excessively huge for them to consume, break down, and apply in manners that are significant for patients. Also, the development and movement of clinical exploration, clinical preliminaries, and treatment alternatives are undecided. It is assessed that it would take a specialist 150 hours every week to peruse each bit of substance distributed in their field of interest.

These developing pools of medical care related data, including electronic well-being records, clinical exploration, pathology reports, lab results, radiology pictures, voice chronicles, and exogenous information, are hard to share since they are divided. Moreover, these data sources don't promptly fuse basic data about a person's nonclinical conditions, which may have a solid bearing on well-being. Subsequently, patients and their medical services suppliers are compelled to settle on choices dependent on a restricted pool of proof.

Scientific progressions in the medical services industry, the expanding weight of constant illnesses, and a prod in the appropriation of third processing stages are driving the medical services psychological registering market. It esteemed at $1,722,000 out of 2017 and it is relied upon to progress at a 34% during the estimated time frame (2018–24) (Newswire, 2017). Cognitive computing alludes to the formation of human deduction in a modernized model utilizing NLP, data mining, and pattern recognition.

The increasing associations among driving players are moving in the cognitive computing market of medical care. For research identified with constant sicknesses, psychological processing innovation is broadly being utilized. This has been made conceivable by the expanding coordinated efforts among assembling organizations and exploration establishments. For example, NVIDIA Corporation and Nuance Communications Inc. declared in November 2017 that they were intending to prepare medical care information researchers and radiologists with AI. This association was pointed toward improving patient consideration by featuring key clinical discoveries.

The regions of the geology portion of the medical services intellectual processing market are Latin America, Europe, the Middle East, and Africa, North America, and the Asia-Pacific. The biggest giver in the market in 2017 was North America. By 2024, it is anticipated to represent 44.3% of the piece of the overall industry. Rising medical care consumption, expanding coordinated efforts among medical services firms, data innovation firms, and exploration focuses, and broad innovative work exercises by drug organizations are making North America the worldwide leading market.

Limitations

Apart from the benefits of cognitive computing, it is also important to highlight the limitations of cognitive computing models. The cognitive frameworks failed at breaking down the danger which is not present in the formats of unstructured data. This incorporates financial elements, culture, worlds of politics, and individuals. For instance, a model that predicts an area to do surgery of brain damage. At that time, the specialist has come to an action to understand the precise area that was damaged. Therefore, cognitive computing has to consider such issues and it is obvious for doctors to estimate the danger and ultimate decision making.

At first, cognitive frameworks need preparing information to comprehend the cycle and improve. The demanding cycle of preparing cognitive systems is undoubtedly the purpose behind its moderate selection. In IBM Watson, the way towards preparing Watson for use by the safety net provider incorporates auditing the content on each clinical approach with IBM engineers.

The nursing staff continues taking care of cases until the framework comprehends a specific ailment. Also, the complicated and costly process of utilizing cognitive frameworks makes it even worse.

The scope of present cognitive technology is restricted to decision and commitment. Cognitive computing frameworks are best as partners which are more like insight enlargement rather than manmade brainpower. It supplements human reasoning and examination yet relies upon people to take basic choices. Sharp associates and chatbots are genuine models. Instead of big business wide reception, such ventures are a viable route for organizations to begin utilizing psychological frameworks.

Conclusions

This chapter presented the cognitive computing applications and recent developments in the healthcare industry. Cognitive computing is unquestionably the next stage in computing that began via robotization. It sets a benchmark for computer frameworks to arrive at the degree of the human brain. Yet, it has a few restrictions which make AI hard to apply in circumstances with an elevated level of vulnerability, fast change, or innovative requests. The unpredictability of the issue develops with the number of information sources. It is trying to integrate, aggregate, and analyze such unstructured data. A complex cognitive solution had to do numerous advances that exist together to give profound insights into this healthcare domain.

References

Aggarwal A., et al. (2021). Generative adversarial network: An overview of theory and applications. *International Journal of Information Management Data Insights*, *1*(1), 100004. Available from https://doi.org/10.1016/j.jjimei.2020.100004.

Aggarwal, J. K., & Ryoo, M. S. (2011). Human activity analysis: A review. *ACM Computing Surveys*, *43*(3), 16. Available from https://doi.org/10.1145/1922649.1922653.

Alom, M. Z., Hasan, M., Yakopcic, C., Taha, T. M., & Asari, V. K. (2020). Improved inception-residual convolutional neural network for object recognition. *Neural Computing and Applications*, *32*, 279–293. Available from https://doi.org/10.1007/s00521-018-3627-6.

M.S. Arefin, T.H. Surovi, N.N. Snigdha, M.F. Mridha, M.A. Adnan, Smart health care system for underdeveloped countries. *2017 IEEE International conference on telecommunications and photonics (ICTP)* (pp. 28-32), 2017. doi: 10.1109/ICTP.2017.8285926.

Augello, M., Baetscher, C., Segesser, M., Zeilhofer, H. F., Cattin, P., & Juergens, P. (2018). Performing partial mandibular resection, fibula free flap reconstruction and midfacial osteotomies with a cold ablation and robot-guided Er:YAG laser osteotome (CARLO®)—A study on applicability and effectiveness in human cadavers. *Journal of Cranio-Maxillo-Facial Surgery: Official Publication of the European Association for Cranio-Maxillo-Facial Surgery.*, *46*(10), 1850–1855. Available from https://doi.org/10.1016/j.jcms.2018.08.001.

Bardram, J. E., Mihailidis, A., & Wan, D. (Eds.), (2006). *Pervasive computing in healthcare*. CRC Press, ISBN 9780367389888.

Battineni, G., Chintalapudi, N., & Amenta, F. (2020). AI Chatbot design during an epidemic like the novel coronavirus. *Healthcare*, *8*(2), 154. Available from https://doi.org/10.3390/healthcare8020154.

Battineni, G., Sagaro, G. G., Chinatalapudi, N., & Amenta, F. (2020). Applications of machine learning predictive models in the chronic disease diagnosis. *Journal of Personalized Medicine*, *10*(2), 21. Available from https://doi.org/10.3390/jpm10020021.

Bauer, E., (2009). Medical device packaging, in *Pharmaceutical packaging handbook.*

Behera, R. K., Bala, P. K., & Dhir, A. (2019). The emerging role of cognitive computing in healthcare: A systematic literature review. *International Journal of Medical Informatics*, *129*, 154–166. Available from https://doi.org/10.1016/j.ijmedinf.2019.04.024.

Boden, M. (2006). Mind as machine: A history of cognitive science. Oxford University Press.

Brezmes, T., Gorricho, J. L., & Cotrina, J. (2009). Activity recognition from accelerometer data on a mobile phone. In S. Omatu, et al. (Eds.), *Distributed computing, artificial intelligence, bioinformatics, soft computing, and ambient assisted living. IWANN 2009. Lecture notes in computer science* (vol. 5518). Berlin, Heidelberg: Springer. Available from https://doi.org/10.1007/978-3-642-02481-8_120.

Burgner-Kahrs, J., Rucker, D. C., & Choset, H. (2015). Continuum robots for medical applications: A survey. *IEEE Transactions on Robotics*, *31*(6), 1261–1280. Available from https://doi.org/10.1109/TRO.2015.2489500.

Croskerry, P. (2013). From mindless to mindful practice—Cognitive bias and clinical decision making. *The New England Journal of Medicine*, *368*(26), 2445–2448. Available from https://doi.org/10.1056/nejmp1303712.

Dautenhahn, K. (2007). Socially intelligent robots: Dimensions of human-robot interaction. *Philosophical Transactions of the Royal Society B*, *362*(1480), 679–704. Available from https://doi.org/10.1098/rstb.2006.2004.

Dellinger, E. P. (2016). Prevention of hospital-acquired infections. *Surgical Infections (Larchmt).*, *17*(4), 422–426. Available from https://doi.org/10.1089/sur.2016.048.

de Looze, M. P., Bosch, T., Krause, F., Stadler, K. S., & O'Sullivan, L. W. (2016). Exoskeletons for industrial application and their potential effects on physical work load. *Ergonomics*, *59* (5), 671–681. Available from https://doi.org/10.1080/00140139.2015.1081988.

Demirkan, H., Earley, S., & Harmon, R. R. (2017). Cognitive computing. *IT Professional*, *19*(4), 16–20. Available from https://doi.org/10.1109/MITP.2017.3051332.

Fitzgerald, J., Quasney, E. (2017). Using autonomous robots to drive supply chain innovation, Deloitte.

Gupta, S., Kar, A. K., Baabdullah, A., & Al-Khowaiter, W. A. A. (2018). Big data with cognitive computing: A review for the future. *International Journal of Information Management*, *42*, 78–89. Available from https://doi.org/10.1016/j.ijinfomgt.2018.06.005.

M.E. Hagen, M.J. Curet, The da Vinci surgical® systems. In G. Watanabe (Ed.), *Robotic surgery* (pp. 9-19), 2014.

S. Jeon, J. Lee, J. Kim, (2017). Multi-robot task allocation for real-time hospital logistics. In 2017 IEEE International Conference on Systems, Man, and Cybernetics (SMC) (pp. 2465-2470). doi: 10.1109/SMC.2017.8122993.

Kim, E., Helal, S., & Cook, D. (2010). Human activity recognition and pattern discovery. *IEEE Pervasive Computing*, *9*(1), 48–53. Available from https://doi.org/10.1109/MPRV.2010.7.

Klinger, D. R., Reinard, K. A., Ajayi, O. O., & Delashaw, J. B. (2018). Microsurgical clipping of an anterior communicating artery aneurysm using a novel robotic visualization tool in lieu of the binocular operating microscope: operative video. *Operative Neurosurgery*, *14*(1), 26–28. Available from https://doi.org/10.1093/ons/opx081.

Ligeza, A. (1995). Artificial intelligence: a modern approach. *Neurocomputing*, *9*(2), 215–218. Available from https://doi.org/10.1016/0925-2312(95)90020-9.

Meng, Z., et al. (2020). Recent progress in sensing and computing techniques for human activity recognition and motion analysis. *Electronics (Switzerland)*, *9*(9), 1357. Available from https://doi.org/10.3390/electronics9091357.

Miller, D. D., & Brown, E. W. (2018). Artificial intelligence in medical practice: the question to the answer? *American Journal of Medicine*, *131*(2), 129–133. Available from https://doi.org/10.1016/j.amjmed.2017.100.035.

Mirmoeini, S. M., Shooshtari, S. S. M., Battineni, G., Amenta, F., & Tayebati, S. K. (2019). Policies and challenges on the distribution of specialists and subspecialists in rural areas of Iran. *Medicina (Kaunas)*, *55*(12), 783. Available from https://doi.org/10.3390/medicina55120783.

P.R. Newswire. (2017). Abdominal surgical robots: Market shares, market strategies, and market forecasts, 2016 to 2022, REPORTBUYER-Surgical.

Noor, A. K. (2015). Potential of cognitive computing and cognitive systems. *Open Engineering*, *5*(1), 75–88. Available from https://doi.org/10.1515/eng-2015-0008.

O'Sullivan, O. E., & O'Reilly, B. A. (2012). Robot-assisted surgery:-impact on gynaecological and pelvic floor reconstructive surgery. *International Urogynecology Journal*, *23*(9), 1163–1173. Available from https://doi.org/10.1007/s00192-012-1790-3.

Patel, V. L., & Kannampallil, T. G. (2015). Cognitive informatics in biomedicine and healthcare. *Journal of Biomedical Informatics*, *53*, 3–14. Available from https://doi.org/10.1016/j.jbi.2014.120.007.

Pham, C., et al. (2020). SensCapsNet: Deep neural network for non-obtrusive sensing based human activity recognition. *IEEE Access*, *8*, 86934–86946. Available from https://doi.org/10.1109/ACCESS.2020.2991731.

Singh R., et al. (2019). IoT based intelligent robot for various disasters monitoring and prevention with visual data manipulating. *Int. J. Tomogr*, *32*(1), 90–99.

Tse, K. Y., & Ngan, H. Y. (2015). The role of laparoscopy in staging of different gynaecological cancers. *Best Practice & Research. Clinical Obstetrics & Gynaecology*, *29*(6), 884–895. Available from https://doi.org/10.1016/j.bpobgyn.2015.010.007.

Vilela, D., Romeo, A., & Sánchez, S. (2016). Flexible sensors for biomedical technology. *Lab on a Chip*, *16*(3), 402–408. Available from https://doi.org/10.1039/c5lc90136g.

Wan, S., Qi, L., Xu, X., Tong, C., & Gu, Z. (2020). Deep learning models for real-time human activity recognition with smartphones. *Mobile Networks and Applications*, *25*(2), 743–755. Available from https://doi.org/10.1007/s11036-019-01445-x.

Weaver, A. J., Koenig, H. G., & Flannelly, L. T. (2008). Nurses and healthcare chaplains: Natural allies. *Journal of Health Care Chaplaincy*, *14*(2), 91–98. Available from https://doi.org/10.1080/08854720802129042.

Xia, K., Huang, J., & Wang, H. (2020). LSTM-CNN architecture for human activity recognition. *IEEE Access*, *8*, 56855–56866. Available from https://doi.org/10.1109/ACCESS.2020.2982225.

Xu, C., Chai, D., He, J., Zhang, X., & Duan, S. (2019). InnoHAR: A deep neural network for complex human activity recognition. *IEEE Access*, *7*, 9893–9902. Available from https://doi.org/10.1109/ACCESS0.2018.2890675.

Chapter 4

Deep learning-based cognitive state prediction analysis using brain wave signal

D. Devi[1], S. Sophia[1] and S.R. Boselin Prabhu[2]
[1]Sri Krishna College of Engineering and Technology, Coimbatore, India, [2]Surya Engineering College, Mettukadai, Erode, India

Introduction

In the era of modern technologies, students are exposed to cutting-edge online learning platforms. Various factors indulge in this mode of teaching/learning and activities to monitor their behavioral state which is quite difficult (Dutta et al., 2019). Determining a student's state and behavior engaged in a daily routine is complex and it is sorted by the use of brain computer interface technology which determines an individual's capabilities that recognizes their cognitive state based on their tasks which they carryout (Huang et al., 2016). The cognitive assessment is limited to one's working memory capacity. It is the memory in which the information is retained and manipulated by the brain by considering various aspects. These aspects range from performing a certain task to crucial decision-making manipulation that is being processed by the brain (Knyazev et al., 2017). The consideration of individuals' working memory plays a viable role since the increase in such memory would result in a state of confusion or it makes the learning ability less important. This may lead the subjects to be affected through behavioral and mental state processing. With the help of state-of-the-art tools like electroencephalogram (EEG) device are being implemented to analyze these signals in real-time. The purpose is to build an effective model that serves the goal for understanding the user perception, personal interceptions, and provide a predictive analysis subsequently that is designed tediously (Appriou et al., 2018; Lin & Kao, 2018). The conceptual model illustrates and bridges the existing gap by identifying and addressing cognitive issues that provides a feasible performance.

The EEG signals have several frequency ranges (Fig. 4.1) ranging from lowest to the highest frequency levels. The delta (0.5–4 Hz) waves are

Cognitive Computing for Human-Robot Interaction. DOI: https://doi.org/10.1016/B978-0-323-85769-7.00017-3

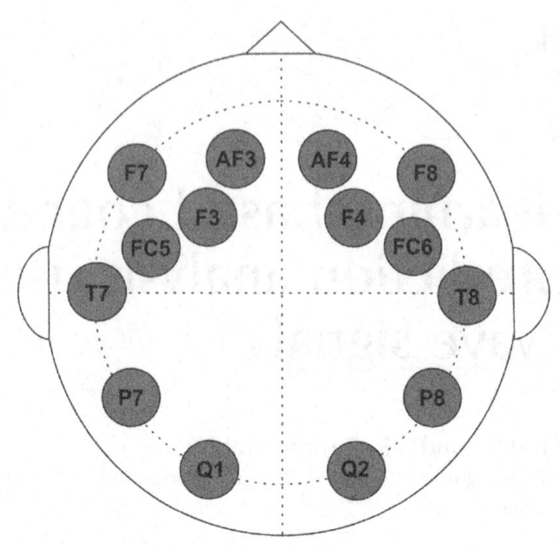

FIGURE 4.1 Electroencephalogram frequency bands.

determined during deep meditation and theta (4–8 Hz) waves occur mostly in sleep which paves further intuitions and senses are the lowest frequency ranges. The coherent range states alpha and beta have peak levels in EEG signal ranging around 8–13 Hz and 13–30 Hz, respectively. The alpha range is a viable parameter that capsulates the cognitive pattern (Xue et al., 2016). Each attributes play a special role in cognitive activities. EEG signal reception is a combination of tangible information that includes several artifacts. Unresolved components are filtered using least mean squares (LMS) through a pretreatment. discrete wavelet transform (DWT) decomposition is used to nonstatic signals to obtain the spectral and statistical factors such as entropy, energy, mean value, etc., (Ilyas et al., 2016; Noshadi et al., 2016; Xue et al., 2016; Zhiweiand & Minfen, 2017). Clustering is done via a fuzzy fractal dimension (FFD) measure, provided that extracted size parameters are being indulged. Implementing advanced techniques of deep learning, classification can be performed to enhance neural networks like convolutional neural network (CNN) that are able to derive various constraints as the concentrated level (i.e., higher and lower). Therefore, the purpose of the chapter is toward improving the learning system that is present in the recent days through online by analyzing the signals obtained from the brain via EEG with various learning tasks.

Materials and methods

Illustration of dataset

Detailed information on participants, empirical challenges, methodology, and validation is determined here. The experiment involved about 20 individuals

(20–21 years old), provided they are healthy. The subjects do not have any knowledge of various tasks involved and it is only that during the experiment, the subjects were provided with corresponding information. The approval was authorized by UTP management.

Illustration of EEG recordings

The EEG device is used to record EEG signals via MUSE sensing headband which comprises of various waves not limited to delta, beta, gamma, and alpha. To obtain a signal range between 0 and 100 Hz, a bandpass is being used where the frequency of sampling is about 250 Hz. The main intention of employing fast Fourier transform is towards extracting features for pretreatment of unprocessed EEG signals. The main tasks that were employed for this experiment are as follows.

Eyes closed task

It is one of the initial tasks that is being implemented at the earlier stage of the experiment. For relaxation the individuals were allowed to have their eyes closed during this experimentation. The samples were recorded for 5 minutes.

Reading technical article

It is the succeeding task that is being implemented after the first analysis. The individuals were informed to read articles which were provided throughout this experiment. The EEG samples were taken for 10 minutes which was carried out during this experiment. It is consecutively made repetitive for the second and third cycles.

Question/answering task (Recall)

The individuals will be assigned a task for 10 minutes duration where each 30 seconds were allocated for each respectively. The questions were obtained via modern technical studies, reports and reviews having 30 multiple choice questions in total with four distinct choices to treat the individual in a decision-making cognitive task. The EEG samplings were recorded for 10 minutes.

Research methodology

Actual EEG data might include numerous artefacts that can misrepresent information about the signal. Preprocessing is performed using adaptive filtering approach to extract this feature from raw EEG signals. In extracting two spectral features based on entropy for measurement, FFD measure and DWT were employed for extracting the features. So as to categorize the level of concentration (i.e., higher and lower), ECNN classifier was employed.

Layers of ECNN

ECNN has many layers like input, hidden, convolutional, pooling/subsampling, fully connected, and output layers, and these layers are explained in subsection below and pictorially represented in Fig. 4.2.

Input layer

The work flow of CNN begins with the input layer of ECNN. It does not consider input from the previous model layer. It contains several neurons which are equal to next features in the dataset.

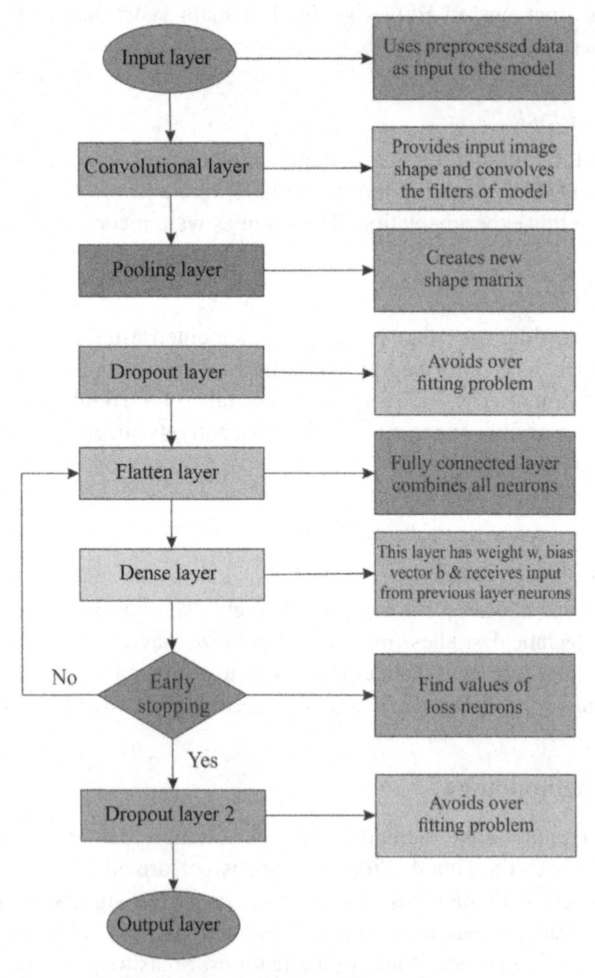

FIGURE 4.2 Layers of enhanced convolutional neural network.

Hidden layer

It is the second layer model of this network. It has three layers (i.e., convolutional, fully connected and subsampling layers). This passes the input values to the hidden layer and typically possesses different number of neurons. The output of this layer is the matrix multiplication of weight and previous layer's output.

Convolutional layer (Cx)

This layer performs the computations of network model. It consists of several filters that were successfully employed subsequently for carrying out filtering mechanism. Thereby, the computation shall be performed and the corresponding output is passed to the next layer to the network.

Pooling layer/subsampling layer (Sx)

Pooling layer combines the output of the neurons. This greatly reduces the parameters and computation of the network model. Pooling layer is further classified into three layers, the sum pooling, average pooling and max pooling.

Fully connected layer

This takes the outputs of both convolution and pooling process, and subsequently uses them to provide labels of images by classification.

Output layer

This layer provides the output of particular classification. Hidden layer values are considered as input of this layer.

Proposed methodology

Adaptive filter

It is linear filter system that consists of restrained transfer function with various parameters and means corresponding to optimization techniques. Adaptive linear filters are dynamically linear by flexible structures and the constraints to modify their properties (i.e., processing of input signal to generate output is without noise and undesirable components). Adaptive linear filter is the maximum prevalent resolution for reducing the signal losses triggered by probable/impulsive artifacts. It is widely used in large number of real-time applications. The basic concept of this filter is to filter the input signal in the way that it should match with the preferred output signal.

The preferred signals get decreased over output filtered signals for producing noise error signals which produces filter coefficients that shall reduce the error signal. This minimization design techniques are referred as Finite and

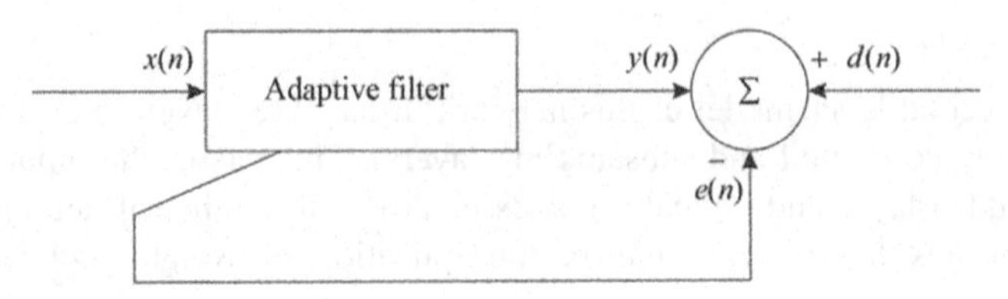

FIGURE 4.3 A typical adaptive mean filter.

Infinite impulse response filters. They possess two units (i.e., digital filter and adaptive algorithm for updating of the parameters that needs to be optimized). The most commonly used algorithmic version of adaptive filters is the LMS strategy. The application of this filter provides smart solution to the problem which typically occurs whenever there is a result that operates from unknown environment. Fig. 4.3 shows the articulation of a typical adaptive mean filter. The basic components of adaptive filters is specified as: x(n) corresponds to the input signals, y(n) corresponds to the filtered signals, d(n) corresponds to the desired signal, and e(n) corresponds to the error output signals.

Adaptive LMS algorithm

To attain finest response from filter, it necessitates the employment of efficient algorithms. LMS approach was used in this experiment as a preprocessing technique as it has low computation complexity and maximum convergence rate. LMS algorithm generally relates the filter coefficients with the generated error signal. LMS algorithm uses straight forward approach for artifact removal which was first introduced by Windrow. This estimates the time varying signal by using gradient origin.

LMS algorithm can also be used by the researchers in hardware implementation due to its simplicity in structure. The LMS algorithm towards artifacts is specified by the three equations (i.e., weight evaluation, filtering output, and error estimation).

In LMS algorithm, the weight evaluation is given by,

$$w_i(n+1) = w_i(n) + n' \times e(n) \times x(n-i) \tag{4.1}$$

Correspondingly, the filtering output of LMS is given by,

$$y(n) = \sum_{i=0}^{M-1} w_i(n) \times x(n-i) \tag{4.2}$$

The error estimation is generally computed as,

$$e(n) = d(n) - y(n) \tag{4.3}$$

Here, x(n) is vector input and w(n) represents adaptive filter weight tap vector. The approximation of equation employs these weight vectors. LMS algorithm uses iterative functions. It is clear that, each iteration gives information about x(n), w(n), and e(n), thereby iteration initiates with w(0). The parameter μ is the power spectral density step size of input x(n) and length of the filter M−1. It controls stability of the filter and speed of convergence in LMS algorithm.

Adaptive filtering algorithm

LMS is a preprocessing technique, since this does have the low complexity in computation and convergence rates will be more. LMS is based on the concept of the linear filtering. This calibrates the relationship between the incoming and outgoing signals. This is done in an iterative manner. By using LMS algorithm, this filter itself self adjusts the weighted filter coefficients and compares them with desired signal. This is related with the generated error signals. Even this is based on the contemporary instance of time.

Algorithm steps

1. Attributes: p = order of the filter, μ = iteration size
2. Implementation: $\hat{W}$ (0) = zeros(p)
3. Calculation: for n = 0,1,2. . ..., n = Sequence length

$$e(n) = d(n) - \widehat{W}^{h}(0) \times x(n) \tag{4.4}$$

$$\widehat{h}(n+1) = \widehat{h}(n) + \mu e \times (n)x(n) \tag{4.5}$$

Fuzzy fractal dimension measures

Signal complexity and structural complexity can be described by FFD. Using the correlation method or even by using the box counting method, they can be represented. Even the data of the smaller sets are used by this fractal dimensions. Using fuzzy sets, these dimensions can be computed. To make all the calibration and to find the efficiency of the regression models, these algorithm frameworks can be used. This approach is represented by the frequently used fuzzy sets.

Accordingly, using this research, FFD measures extract the dimensions and even this will be predicted in the form of classifiers from dimensions. This is given as,

$$\mathrm{D} = \mathrm{lim}_{M\to\infty} \lim_{\varepsilon\to 0} \frac{\log\left(g_{\frac{\varepsilon}{M^2}}\right)}{\log\varepsilon} \tag{4.6}$$

$$D0 = \lim_{\varepsilon\to 0} \frac{\log\left(g_{\varepsilon/M^2}\right)}{\log\varepsilon} \tag{4.7}$$

Here, D corresponds to the exponent and N signifies the series length.

Steps

- Attributes: l max, input series
- Launch first input series, and also second input l max.
- Creation of sub series
- Sub series length calibration.
- Measuring dimension and forming the group.

Discrete wavelet transform

The most stable and relevant method for transformation of signals in various applications is the DWT feature extraction. It breaks down the non-stationary signals into frequency signals. The actual wavelet functions, dilations and scaling factors are segregated using DWT. The degree of decimation relies on the sampling frequency of the signals given as input. Here, 4-level decimation was used. Cascaded DWT shall be expressed as,

$$\phi(xi) = \sum_{l=-\infty}^{\infty} a_l \phi(S_x - l) \tag{4.8}$$

Feature extraction parameters

Characteristics extraction technique is done in this research to find the mean, sample and estimated entropy attributes. The chaotic characteristics consist of both biggest and smallest values. Mean is defined as the measure location of a value from a dataset that identifies the central position and the value of mean is,

$$m(n) = \frac{1}{n}\sum_{n=1}^{N} x_i(n) \tag{4.9}$$

The SampEn is manipulated by,

$$\text{Sample Entropy } (T, W) = \text{Lim}_{N\to\infty}\left[\ln\left(\Phi^T(W)\right) - \left(\Phi^{T+1}(W)\right)\right] \tag{4.10}$$

ApEn approximate entropy is achieved by,

$$\text{Approximate entropy } (T, W) = \text{Lim}_{N\to\infty}\left(\left(\Phi^T(W)\right) - \left(\Phi^{T+1}(W)\right)\right) \tag{4.11}$$

Moreover,

$$\text{Approximate entropy } (T, W, N) = \left(\Phi^T(W)\right) - \left(\Phi^{T+1}(W)\right) \tag{4.12}$$

For each respective short noisy data, the approximate entropy is applied. It is put forwarded to quantify the recurrence of a time-series signal. The count of template matching is present in addition to it, it provides high discerning values.

Enhanced convolutional neural network

For accelerating the training, a novel disparity of CNN was introduced which is referred as ECNN. It is an integrated convolution neural network, and combines the knowledge of actual CNN by the benefit of RBF similar to neurons in every three layers. The major property of ECNN is weight sharing and subsampling. This improves accuracy and durability against the given inputs. In this experiment, the levels of concentration are classified on the basis of preprocessed input values using ECNN.

The initial one should be subsampling and the last is completely associated. In addition to it, CNN offers excellent binary classification of the corresponding signal input.

Experimental results and discussions

Inputs were taken as the four frequency band EEG signals from dataset. The alpha band (8–13 Hz) was used in this research. The EEG signals consist of 150 input signals, among which 80% of the signals were used as training sets and 20% were taken as testing sets. These signals are simulated and data were recorded using Matlab R2013-version.

Spectral entropy

It is the computation of spectral power distribution along with forecastability of time-series signal. This entropy is based on Shannon and information entropy in the information data. The spectral entropy of the signal is given by,

$$\mathrm{SE}(F) = -\frac{1}{\log N_u \sum_u \left(P_u(F)\log_e P_u(F)\right)} \tag{4.13}$$

$$\mathrm{SSH}(F) = -\sum_u \left(P_h(F)\log_e P_h(F)\right) \tag{4.14}$$

Where, $P_u(F)$ represents the power spectral density function, $P_h(F)$ represents the estimation of the Shannon entropy (SSH (F)), and N_u signifies the entire frequencies.

Specificity

The specificity is generally referred as the true negative rate (TNR). It states that the ratio of true negative values that is detected correctly. Its computation is given by,

$$\text{Specificity} = \mathrm{TN}/(\mathrm{TN} + \mathrm{FP}) \tag{4.15}$$

Sensitivity

This is represented as a true positive rate (TPR). It is the measure of the proportion of actual positives rightly identified as such. The sensitivity corresponding to true positive rate is mathematically expressed as,

$$\text{Sensitivity/recall} = \text{TP}/(\text{TP} + \text{FN}) \tag{4.16}$$

Precision

Precision takes only the positive parameters values into account. It is ratio of appropriately predicted positives values to whole predicted positive parameters. It is represented mathematically as,

$$\text{Precision} = \text{TP}/(\text{TP} + \text{FP}) \tag{4.17}$$

F1-score

This corresponds to the mean of precision and recall. Thereby, this considers both false negative and positive values for the prediction. It is typically expressed as,

$$\text{F1 Score} = 2\frac{(\text{Recall} \times \text{precision})}{(\text{Recall} + \text{precision})} \tag{4.18}$$

Accuracy

Accuracy specifies the overall correctness detected by the classifier model. The parameters are computed by the real classifier parameters identification (i.e., TP + TN) which is then separated by sum of all classified parameters (i.e., TP + TN + FP + FN). TPR and TNR are the two main measures considered for the model accuracy prediction. Accuracy is given by,

$$\text{Accuracy} = \frac{\text{TP} + \text{TN}}{\text{TP} + \text{FP} + \text{FN} + \text{TN}} \tag{4.19}$$

The advantages in measuring the performance metrics is that, it measures the statistical quality of the machine learning model. Subsequently, TN, TP, FN and FP are described as given below.

True positive value

True positive value indicates the positive class outcomes predicted by the model correctly. This perfectly identifies the anomalous data in the classification (i.e., predicts the condition when there is a condition).

True negative value

True negative value indicates the negative class outcomes predicted by the model correctly. This perfectly and successively identifies the data that are not being anomalous in the classification process (i.e., true negative value does not predicts the condition when there is no condition).

False positive value

False positive value indicates the positive class outcomes predicted by the model incorrectly. This imperfectly identifies the anomalous data as such in the classification (i.e., predicts the condition when there is no condition).

False negative

False negative value indicates the negative class outcomes predicted by the model incorrectly. This imperfectly identifies the data that are not being anomalous in the classification (i.e., does not predicts the condition when there is condition). The advantages of confusion matrix are that, it is often used to determine the performance of a classifier with a set of testing data for which the true value parameters are known. The visualization of performance of an algorithm is described by confusion matrix.

Fig. 4.4 describes the input alpha band EEG signal. The procedure of preprocessing the input EEG signals by adaptive LMS filter is shown in

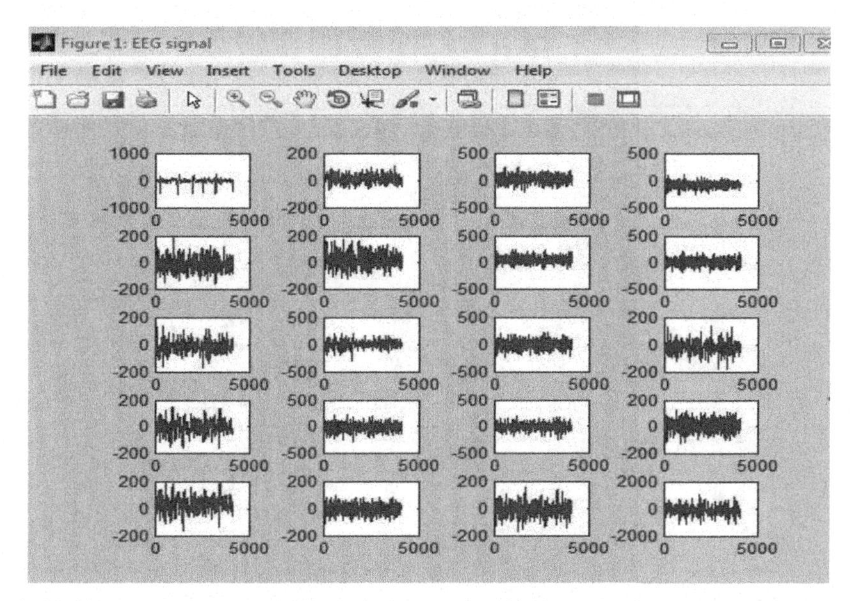

FIGURE 4.4 Characteristics of alpha wave electroencephalogram signal.

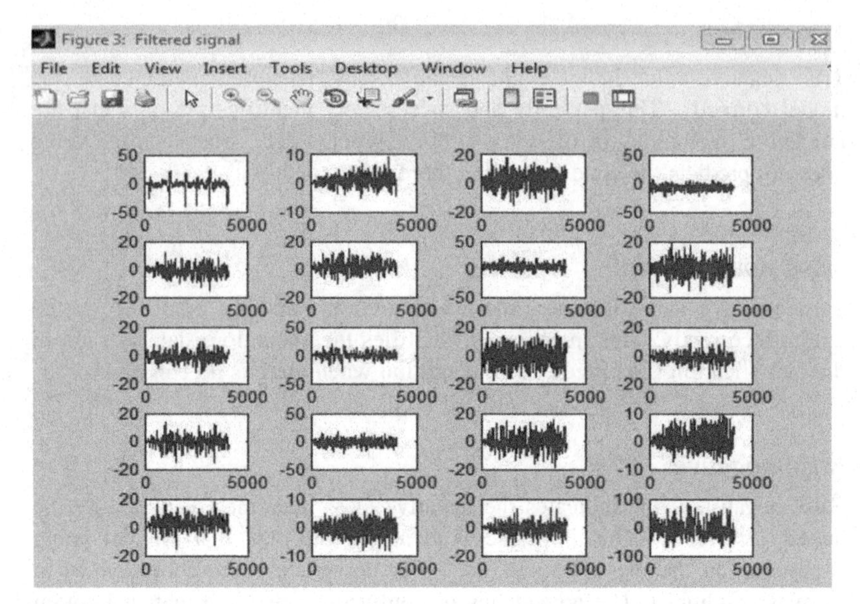

FIGURE 4.5 Preprocessed signal using least mean squares filter.

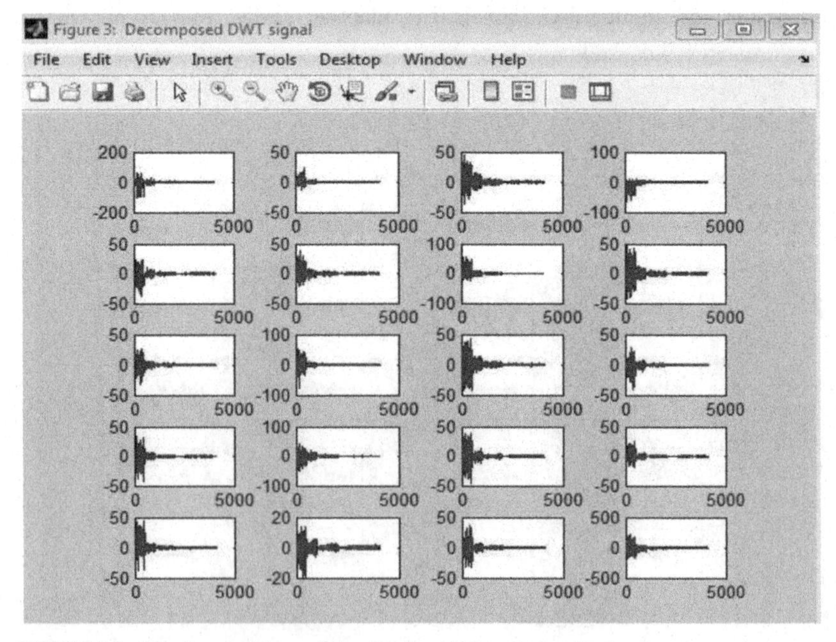

FIGURE 4.6 The decomposition of the signals with four decimation levels.

TABLE 4.1 Feature extraction parameters values.

Input samples	SampEn	ApEn	Average value (m)
25	−1.734	1.1789	1.7
50	103.7	1.1376	1.3
75	−131.9	1.0783	1.3
100	166	1.0955	2
125	123	1.3257	1.5
150	109	1.1350	1.4

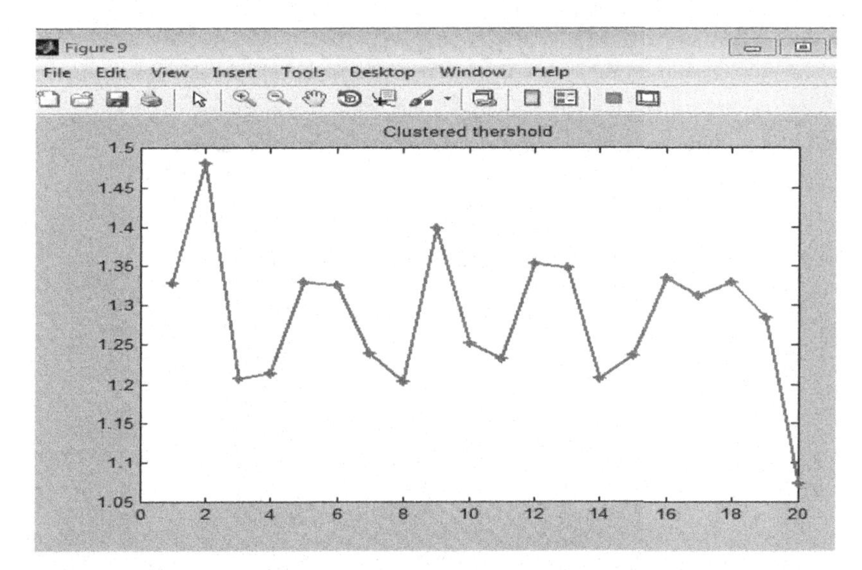

FIGURE 4.7 Clustering the extracted parameters using fuzzy fractal dimension measure.

Fig. 4.5. The decomposition of the signals with four decimation levels illustrated by cascaded DWT (Fig. 4.6).

Table 4.1 represents the feature extraction parameter values. The sample entropy corresponding to the samples 20, 40, 60, 80, 100, 120, 140 are −1.743, 103.7, −141.9, 176, 133, −142, 119 respectively. On the other hand, the approximate entropy corresponding to the samples 20, 40, 60, 80, 100, 120, 140 are 1.1898, 1.1476, 1.0873, 1.0755, 1.5247, 1.0887, 1.1450 respectively. Also, from the table it could be inferred that the mean values correspond to the samples 20, 40, 60, 80, 100, 120, 140 are 1.8, 1.4, 1.4, 2.0, 1.5, 1.5, 1.4 respectively. Fig. 4.7 represents clustering the extracted parameters using FFD measure in this experiment.

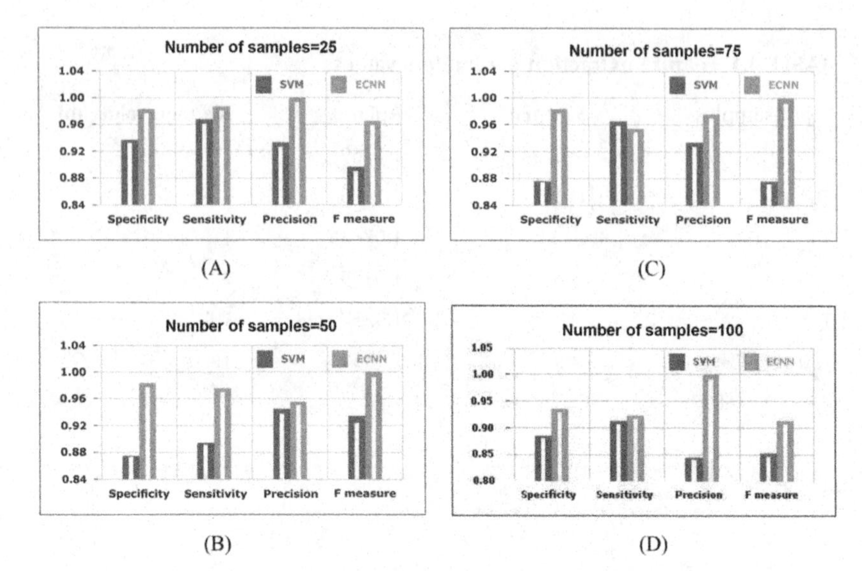

FIGURE 4.8 Performance evaluation of classifier (25, 50, 75, and 100 samples).

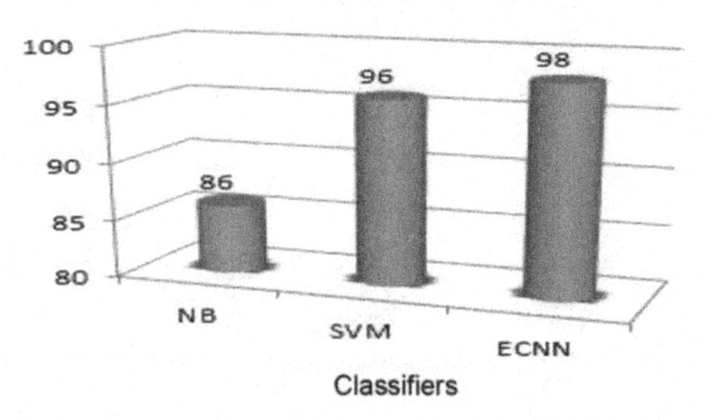

FIGURE 4.9 Performance comparison of accuracy with existing classifier.

Fig. 4.8 represents the evaluation of performance of classifier with 25, 50, 75 and 100 samples respectively. The specificity of ECNN is comparatively higher when 25, 50 and 75 samples were taken into consideration. On the other hand, the sensitivity of ECNN is typically higher when 25–50 samples were taken into consideration. Alternatively, the precision of ECNN is relatively better when 25–100 samples were considered. As a result, the F-measure of ECNN is relatively higher when 50–75 samples were considered. Almost in all the cases, the performance of ECNN was higher when compared with Support Vector Machine (SVM).

Fig. 4.9 enumerates the performance comparison of accuracy of ECNN with the existing classifiers. It was inferred that, the accuracy of SVM classifier is 96% and the accuracy of ENN classifier is 98%. Hence, ECNN classifier offers better accuracy when compared to its peers.

Concluding remarks and future recommendations

EEG-based assessment toward cognitive states over various learning assignments has been covered in this chapter. This research design has been demonstrated by implementing high precision strategies. In this, discrete wavelet transform is being employed for extracting the frequency/time domain parameter like mean, samples, approximate entropies. FFD measurement has been employed for providing the extracted features over dimensions for clustering. Enhanced deep learning algorithm, the ECNN was utilized as classification technique for classifying the extracted parameters onto a different concentration level. ECNN classification method is presumed as highly reliable when compared to other current classifiers, and thereby offers a feedback model for controlling the cognitive states. The future study shall be concentrated towards analysis and evaluation of cognitive state stress levels during learning process by employing numerous state-of-the-art methods.

References

Appriou, A., Cichocki, A., & Lotte, F. (2018). *Towards robust neuroadaptive HCI: Exploring modern machine learning methods to estimate mental workload from EEG signals. Extended abstracts of the 2018 CHI conference on human factors in computing systems* (p.p. LBW615) ACM.

Dutta, S., Hazra, S., & Nandy, A. (2019). Human cognitive state classification through ambulatory EEG signal analysis. Available from https://doi.org/10.1007/978-3-030-20915-5_16In L. Rutkowski, R. Scherer, M. Korytkowski, W. Pedrycz, R. Tadeusiewicz, & J. Zurada (Eds.), *Artificial intelligence and soft computing. ICAISC 2019. Lecture notes in computer science* (vol. 11509). Cham: Springer.

Huang, D., et al. (2016). Combining partial directed coherence and graph theory to analyse effective brain networks of different mental tasks. *Frontiers in Human Neuroscience*, *10*, 235.

Ilyas, M.Z., Saad, P., Ahmad, M.I., & Ghani, A.R.I., Classification of EEG signals for brain-computer interface applications: Performance comparison. In *2016 International conference on robotics, automation and sciences (ICORAS)* (pp. 1–4), Nov 2016.

Knyazev, G. G., Savostyanov, A. N., Volf, N. V., Liou, M., & Bocharov, A. V. (2017). EEG correlates of spontaneous self-referential thoughts: A cross-cultural study. *International Journal of Psychophysiology: Official Journal of the International Organization of Psychophysiology*, *86*(2), 173–181.

Lin, F. R., & Kao, C. M. (2018). Mental effort detection using EEG data in e-learning contexts. *Computers & Educaton*, *122*, 63–79.

Noshadi, S., Abootalebi, V., Sadeghi, M. T., & Shahvazian, M. S. (2016). Selection of an efficient feature space for EEG-based mental task discrimination. *Biocybernetics and Biomedical Engineering*, *34*(3), 159–168.

Xue, J.-Z., Zhang, H., Zheng, C.-X., & Yan, X.-G. (2016). Wavelet packet transform for feature extraction of EEG during mental tasks. *Proceedings of the 2003 international conference on machine learning and cybernetics*, 360–363).

Zhiwei, L & Minfen, S. (2007), Classification of mental task EEG signals using wavelet packet entropy and SVM. In *International conference on electronic measurement and instruments, 2007*. ICEMI'07. (pp. 3–906–3–909).

Chapter 5

Electroencephalogram-based cognitive performance evaluation for mental arithmetic task

Debatri Chatterjee[1], Rahul Gavas[1], Roopkatha Samanta[2] and Sanjoy Kumar Saha[2]

[1]TCS Research and Innovation, Tata Consultancy Services, Kolkata, India, [2]Department of Computer Science and Engineering, Jadavpur University, Kolkata, India

Introduction

Assessment of educational performance is usually done through classical approaches like examination or task results. Such assessments are usually expressed numerically based on comparison of student results against established performance criteria. With recent advances in Internet technology, e-learning is becoming one of the preferred alternate teaching-learning tools. The most important aspect of such e-learning tools are a measurement of student engagement and assessment of their learning outcomes. Several approaches are available for assessing student performance, each having their own advantages and disadvantages. However, successful student assessment in real time is still an issue.

In recent times, the study of human cognitive abilities and performance is gaining attraction and importance. Cognitive psychologists' study various cognitive functions like attention, memory, learning, thinking, perception and so on to understand how we acquire and process information. Effective functioning of these cognitive abilities results in better cognitive performance in any task. These cognitive capabilities are brain-based skills which are required for completing a task successfully. There are various socioeconomic, psychological, and environmental factors that directly and indirectly affect the cognitive performance of an individual (Danili & Reid, 2006; Mushtaq & Khan, 2012). Mainly there are three factors that affect the task performance of an individual—(1) task difficulty, (2) repeated practice, and (3) mental workload. Hence, in order to improve the cognitive performance of an individual, it is extremely important to study the underlying brain

Cognitive Computing for Human-Robot Interaction. DOI: https://doi.org/10.1016/B978-0-323-85769-7.00014-8

dynamics. Periodic data capture and subsequent analysis can help in understanding the cognitive capabilities of an individual. On the other hand, progression of performance can also be studied.

There are various methods that are widely used for assessing cognitive capabilities and performance. Psychologists mostly rely on standard psychometric test batteries or self-report questionnaires for this purpose. Self-report questionnaires provide a qualitative measure of the capabilities and often there are very less correlation between various scales measuring a cognitive ability (Stankov, Sabina, & Jackson, 2014). In the case of standard psychometric tests, the evaluation or assessment is done based on task performance-based measures like accuracy, task completion time, error rate, etc. These are also derived measures of cognitive performance and they frequently have lower reliability (Stankov, Sabina, & Jackson, 2014). Moreover, these tests normally use stimuli which relate to social customs, norms, and vocabulary of a particular culture. Hence, selecting a suitable test uniformly acceptable across different cultures might be challenging (Hughes & Bryan, 2003).

Our mental state is closely associated with various physiological parameters. The cognitive response to an external stimuli or task can be measured by using physiological parameters like heart rate, skin conductance, brain activation, pupil dilation, etc. (Aniruddha et al., 2015; Gavas, Chatterjee & Sinha, 2017; Kilseop & Rohae, 2005; Sinha et al., 2016). Human brain controls all the cognitive activities. Thus, by analyzing the brain activation dynamics, it is possible to have the detailed insight of the cognitive abilities of an individual. Moreover, it also facilitates assessing student performance in real time.

In the present work, we have tried to analyze the variations in brain activations while performing a cognitive task (mental arithmetic task). In order to do so, we have used a publicly available dataset (Zyma et al., 2019). The brain signals have been recorded using a commercially available 19 channel electroencephalography (EEG) device. The dataset contains brain activation recordings from the two groups of participants (i.e., good and bad performers) for a mental serial subtraction task. We have proposed a novel approach for analyzing the brain activations based on the statistical properties of the recorded EEG signal. The main objectives of this study are:

1. To study the brain activation signals recorded using an EEG device while performing a mental cognitive task.
2. To propose a novel statistical feature-based approach for assessing cognitive performance of an individual which in turn can be used for differentiating the brain activations for good and bad performers.

The major contribution of the proposed work lies in formulating the descriptor at the signal level. The salient aspects are as follows.

1. A set of conventional time and frequency domain features like various EEG band powers, Hjorth parameters are considered after validating their capability through hypothesis test.

2. A novel technique is proposed to summarize the window level features that exploit the feature distribution across the windows. Summarized feature values represent the statistical properties of the brain signal and are taken as the signal level descriptor.

We have validated our approach for analyzing cognitive performance only. This proposed approach mostly relies on the statistical property of brain activations rather than the domain specific EEG features in absolute terms. Thus, it can also be used for assessing other cognitive abilities like attention, perception, etc. Moreover, similar approach can be adopted for analyzing other physiological signals like galvanic skin conductance, heart rate, breathing signals, etc.

The chapter is organized as follows. Previous works section details existing state of the art works. It is followed by the proposed approach in Proposed methodology section. Results of our findings have been discussed in Results and discussions section, followed by conclusion and future scope in last section.

Previous works

Study of brain activations associated with various cognitive processes are gaining lot of interest both for clinical as well as nonclinical applications. Cognitive neuroscience is the field which analyzes brain functions underlying various cognitive processes. In order to do so, researchers use various neuroimaging techniques like functional magnetic resonance imaging (Poldrack, 2008), positron emission tomography (Frith et al., 1992), magnetoencephalography (Bialystok et al., 2005), etc. These techniques are mostly expensive and require expert intervention. EEG is another technique which is used for recording electrical activities of brain. This technique is widely available, inexpensive and has excellent temporal resolution. Moreover, recent invention of low resolution commercial grade EEG devices has resulted in increased usage of this technology for non-clinical applications and research.

Researchers have used EEG signals for studying the relationships between various cognitive phenomena and associated activity of brain (Sarter, Berntson, & Cacioppo, 1996). Cognition basically results from dynamic interaction of various brain areas operating as networks (Bressler & Menon, 2010). Prior works suggest that different EEG subbands reflects different activities. For example, alpha band reflects attentional demands whereas beta band reflects emotional state and cognitive processes (Ray & Harry, 1985). Researchers have investigated the relationship between ability in solving problems and brain activations recorded with 28-lead EEG device (Jaušovec, 2000). They have used alpha power and asymmetry in alpha band as the features. In some works, authors have reported increased beta band activity for increased attention span and alertness (Engel & Fries, 2010; Kamiński et al., 2012). It has also been reported that hippocampal theta activity is associated with cognitive performance (Kahana, Seelig, & Madsen, 2001). In a nutshell, all EEG sub-bands together control the human cognition and cognitive abilities. Apart from these band powers, other approaches like detrended fluctuation analysis (Márton et al., 2014), wavelet transform (Murata, 2005), principal component analysis (Chaouachi, ImèneJraidi, & Frasson, 2011), etc. have also been explored for studying the cognitive performances and abilities. An increased prefrontal brain activity for a decision-making task has also been observed (Fleming et al., 2012). Another (EEG) based study (Boldt & Yeung, 2015) shows a well characterized neural correlates of error awareness which is indicative of decision confidence. Most of these studies have been performed using high-resolution EEG devices and hence different brain regions were tapped for assessment. However, researches are being conducted using low resolution EEG devices also. In Aniruddha et al. (2015) and Sinha et al. (2016), authors used single lead EEG devices for recording brain activations for assessment of cognitive flow state and performance evaluation. Wong, Chan, and Mak (2014) analyzed activities recorded using single channel EEG device from frontal brain region for motor acquisition task. In Papakostas et al. (2017), researchers used another low-cost EEG device from Muse for predicting the sequential learning task performance, before the user completes the task.

Few researchers have come up with some novel metrics which they have used for quantification of a cognitive performance and mental workload. Berka, Levendowski, and Cvetinovic (2004) have defined a workload index for measuring mental workload for various tasks like arithmetic calculation, memory task, etc. Later they have applied same index for analyzing student performance in a problem-solving task (Stevens R., Galloway, & Berka, 2007; Stevens R.H., Galloway, & Berka, 2007). Multiple metrics like, EEG-distraction, EEG-workload and EEG-engagement have been defined for analyzing student performance (Stevens, Galloway, & Berka, 2006). Feature like, EEG band-power based engagement index for estimating mental workload and task engagement for mathematical problem solving are also taken up (Galán & Beal, 2012). In Stikic et al. (2011), authors used the metric proposed by Berka et al. (2004) in association with reaction time and accuracy and derived a measure for predicting present and future task performance for various cognitive tasks. Recently, researchers are also using neural network-based approaches for assessment and classification of mental states of an individual during execution of a cognitive task (Baldwin & Penaranda, 2012). Thus, there are multiple approaches and several EEG features which can be used for assessing cognitive abilities and performance.

These findings motivated us to study brain activations for assessing cognitive performance associated with a task. In the present study, we have used a dataset for studying the brain activations during arithmetic task. Instead of directly using traditional EEG domain-based feature, we have summarized the distribution of those feature values computed over several time-windows. This summarization helps to capture the overall trend of the feature. The proposed descriptor is a generic one and can be applied to other physiological sensor data (like heart rate, galvanic skin conductance, respiratory signal, etc.) also.

Proposed methodology

In the present work we have tried to investigate the brain activations of participants during performance of a mental arithmetic task. In order to do so, we have used a publicly available dataset from physionet (Zyma et al., 2019).

Dataset description

The dataset (Zyma et al., 2019) contains EEG recording of subjects before and during the performance of mental arithmetic task. Brain activations were recorded using 19 channel Neurocom EEG device. The electrodes were placed on recording sites according to international 10/20 scheme. The sampling rate of the device is 500 Hz. Participants were asked to perform a serial subtraction of two numbers. At the beginning of each trial a 4-digit minuend and a 2-digit subtrahend were communicated orally. Participants performed serial subtraction using these two numbers. Hence, this task is accompanied by intensive cognitive activity and is also associated with mental workload as stated in Zyma et al. (2019). Prior to the task, participants were asked to relax with closed eyes. This duration (3 minutes) is taken as the rest period. Each trial of arithmetic task lasted for a duration of 4 minutes. The dataset contains 3 minutes of rest data and first 1 minute of task data for 36 participants. The participants were also grouped into two categories viz. "Group B" (bad performers) and "Group G" (good performers), based on their performance score (number of arithmetic operations executed per minute). Out of 36 participants, 10 participants belonged to Group B, who faced difficulty in performing the task. Remaining 26 participants belonged to Group G, who performed the task without difficulty. The EEG data collected were preprocessed for removing power line noise and eye blinks. The dataset contains event files and task levels for each participant.

Proposed approach—EEG signal processing and feature extraction

In the present study, we have used the EEG signal provided in the dataset for investigating the correlation between brain activation and cognitive performance. The overall signal processing and proposed feature extraction methodology is depicted in Fig. 5.1. The process was repeated for the signal corresponding to every channel excepting two reference channels.

Firstly, the EEG signal corresponding to task interval is subdivided into windows of 2 second duration each. The task data provided in the dataset is

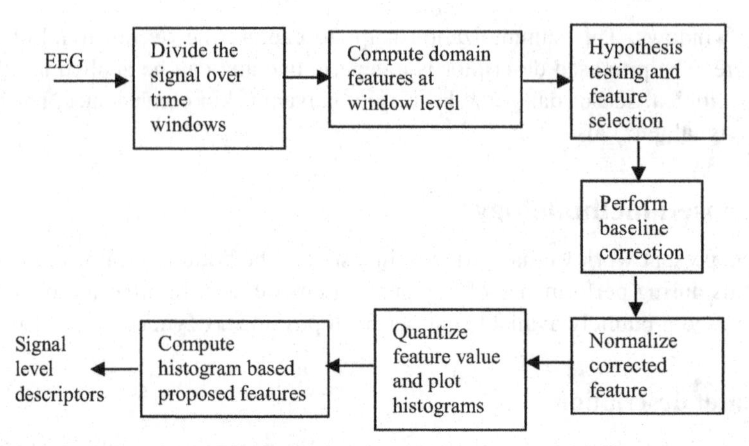

FIGURE 5.1 Proposed feature extraction approach.

of duration 1 minute. Thus, 30 such windows of 2 second durations are obtained for a subject corresponding to the task. On each of these windows a set of domain specific EEG features are calculated.

Literatures suggest that there are number of EEG domain specific features which are in use for assessment of various cognitive states. Initially, we have identified 14 most widely used features for our present study. These features include both time-domain features like mean, variance, skewness, kurtosis, Hjorth parameters, zero crossing rate of EEG signals and some frequency domain features like mean frequency, various signal band powers, engagement index.

In the given dataset, participants have been divided into two categories "group B" and "group G," based on the number of arithmetic operations they performed and on the correctness of the answers. Group G performed the task with relative ease while group B faced some difficulties while executing the task. This indicates the conclusion that the mental workload was less for group G than for group B. Thus, we hypothesized that the brain activation patterns would be different for these two groups. These changes in patterns are supposed to get reflected in EEG features also. In order to test this hypothesis and to check the discriminating nature of the identified features, we have used Wald-Wolfowitz test (Friedman & Lawrence, 1979). Wald-Wolfowitz test is a nonparametric test that checks if cumulative distributions of two samples are significantly different with respect to the features. For 8 out of 14 features, the p-values are found to be less than 0.05 which indicates that these features were significantly different for good and bad performers.

Details of the results of Wald-Wolfowitz test are presented in the Results and discussions section. The features having p-value less than 0.05 were selected for further analysis. These selected features are various EEG band powers, Hjorth parameters and engagement index. For ready reference, these features are elaborated as follows.

1. EEG band powers—EEG signal is composed of various frequency components. The frequencies which are mainly present during awake state are alpha (7−12 Hz), beta (12−30 Hz), theta (4−7 Hz) and gamma (30−100 Hz). We have calculated these frequency components on each window. Hence, in total, four band power features $\left(E_\alpha, E_\beta, E_\theta, E_\gamma\right)$ were calculated on each window.
2. Hjorth parameters—these are a set of parameters which indicate the statistical property of the signal. These parameters are activity (H_a), mobility (H_m) and complexity (H_c). Statistical features like variance, proportion of standard deviation and change in frequencies of the EEG signal are calculated by these parameters (Hjorth & Elema-Schönander, 1970).
3. Engagement index—EEG signal is widely used for measuring task specific engagement level. In frequency domain, spectral power of various frequency bands gives an indication of engagement level associated with a task (McMahan et al., 2015). The engagement index is defined as

$$EI = \frac{E_\beta}{(E_\alpha + E_\theta)} \tag{5.1}$$

Thus, in total we have used eight EEG domain features for each channel as given in Eq. (5.2).

$$F = \left\{E_\alpha,\ E_\beta,\ E_\theta,\ E_\gamma,\ H_a,\ H_c,\ H_m,\ EI\right\} \tag{5.2}$$

The EEG task data of duration 60 seconds was subdivided into 30 windows of duration 2 seconds. We calculated these eight domain specific EEG features on each of these windows. Thus, for each feature (f) we got 30 feature values ($f_1, f_2, f_3, \ldots, f_{30}$) corresponding to 30 windows.

Next, we performed the baseline correction for each subject to remove subject specific offsets as shown in Fig. 5.1. The baseline data provided in the dataset is of duration 3 minutes. We have selected 60 seconds of baseline data from the middle of this 3 minutes data.

We calculated all the features mentioned above on the baseline data and performed offset correction for each feature (f) using Eq. (5.3).

$$f_{\text{corrected}} = \frac{f_{\text{task}} - f_{\text{baseline}}}{f_{\text{baseline}}} \tag{5.3}$$

where, f_{task} is the feature value in a window of task interval and f_{baseline} is the feature value for 60 seconds baseline interval. Consequently, for each feature, we get 30 such baseline corrected feature values corresponding to 30 task windows.

In order to remove subject specific differences, we have normalized these features using subject specific minimum and maximum feature values.

Derivation of proposed signal level descriptor

Researchers have used the EEG domain specific features for classification of various mental states of an individual. Using these raw domain features for the assessment of mental states is only one aspect of the analysis. In order to analyze the organization and trend of these features for two classes of performers, we are proposing to use a descriptive statistical approach. The proposed approach would help us to find the difference between brain activation patterns for good and bad performers.

As features are computed over several small time-windows of the signal, finding out the signal level descriptor is challenging. The common practice is to consider the average and standard deviation of the feature values over the windows. In this context we have proposed a simple technique consisting of two steps, like quantization of the real valued features followed by histogram formation and computation of statistical parameters from the histogram. Parameters thus obtained are taken as signal level descriptors.

First, we quantized the baseline corrected features. In order to do so, we calculated the mean (μ) and standard deviation (σ) of these 30 baseline corrected feature values.

Next, based on these mean and standard deviation values we defined 10 bins such as

$$\text{bin}_1 = \text{feature value} < (\mu - 2\sigma)$$

$$\text{bin}_2 = (\mu - 2\sigma) \leq \text{feature value} < (\mu - 1.5\sigma)$$

$$\text{bin}_3 = (\mu - 1.5\sigma) \leq \text{feature value} < (\mu - \sigma)$$

.

.

.

$$\text{bin}_9 = (\mu + 1.5\sigma) \leq \text{feature value} < (\mu + 2\sigma)$$

$$\text{bin}_{10} = \text{feature value} > (\mu + 2\sigma)$$

Thus, the feature values are mapped to a fixed set of bins. It also helps to smooth out the real valued features. Once the quantization is over, a histogram of quantized feature value is formed where x-axis denotes the bin number and y-axis stands for count of windows. Thus, the histogram reflects the distribution of feature value over the bins. Finally, we have computed four statistical parameters, viz first moment, second moment, third moment and entropy in order to capture the nature of the distribution. It is done as follows.

First moment (m_1) is computed as

$$m_1 = \sum_{i=1}^{N} \frac{i}{N} \times y_i \tag{5.5}$$

where, i is the bin number, y_i is the normalized count for the ith bin and N is the total number of bins (in our case, it is 10). So, m_1 is same as the average quantized value (m') of the feature. To keep the feature value within [0,1], the bin numbers are also normalized by dividing i by N.

Second order moment (m_2) around mean is computed as,

$$m_2 = \sum_{i=1}^{N} \left(\frac{i}{N} - m'\right)^2 \tag{5.6}$$

Third moment (m_3), is computed as,

$$m_3 = \sum_{i=1}^{N} \left(\frac{i}{N} - m'\right)^3 \tag{5.7}$$

Thus, the first moment (m_1) gives the mean or central tendency of the quantized feature value, whereas second (m_2) and third moment (m_3) reflect the variance and skewness of the quantized feature value respectively.

Finally, these parameters are augmented with normalized entropy (E_n). It is computed as

$$E_n = -\frac{\sum_{i=1}^{N} y_i log y_i}{log N} \tag{5.8}$$

Entropy captures the randomness of the distribution well. With increase in randomness in the distribution, value of E_n increases with a maximum value of log$N$.

Thus, a signal level descriptor (F_H) consisting of four parameters is computed from a domain feature

$$F_H = \{m_1, m_2, m_3, E_n\} \tag{5.9}$$

For every participant, eight domain features are computed on each window of the signal corresponding to a particular channel. Each such feature is summarized at the signal level by a four-dimensional parameter, F_H. Thus, the signal of a channel is represented by 32-dimensional parameter vector. EEG signal of a participant consists of 19 channels (excluding the two reference channels). Thus, the brain activation pattern of a participant is summarized by 608-dimensional parameter vector.

Classification of good and bad performers

Finally, we have tried to classify good and bad performers based on the proposed signal level descriptors. The dataset provides EEG signals for 10 bad performers and 26 good performers. In order to remove this class imbalance, first we performed Synthetic minority oversampling (SMOTE) technique. SMOTE (Chawla et al., 2002) is a statistical approach that is used for generating synthetic samples for the under-sampled class. This is done based on nearest neighbors judged by Euclidean distance between data points in feature space. Next, we have tried several classifiers for classification of good and bad performers. The details of the classification accuracy for each of these classifiers are presented in the Results and discussions section.

Results and discussions

This section details the results obtained for the present study. We hypothesized that the brain activation patterns would be different for good and bad performers and these differences should be reflected in various EEG domain features.

Hypothesis testing and feature selection

In order to test the hypothesis, we identified 14 most widely used EEG time and frequency domain features. There were eight time-domain features and remaining six were frequency-domain features. We computed these features at window level for both good and bad performers. Finally, we performed Wald-Wolfowitz test to check the significance of these features for discriminating the brain activation patterns for good and bad performers. The *p*-values obtained for these features are shown in Table 5.1.

It is evident from the results, except Hjorth parameters and variance, other widely used time-domain features were not significant enough and hence should be rejected.

On the other hand, most of the frequency domain features we identified were significant enough with $P > .05$. Out of six frequency domain features, only mean frequency was found to be non-discriminating feature with $P = .08$.

Based on these findings, we selected eight (three time-domain and five frequency-domain features) features named alpha, beta, theta, gamma band powers, engagement index, activity, mobility, and complexity.

Classification of good and bad performers

Next, we used these features for further analysis. We performed baseline correction and subject specific normalization as explained in Proposed methodology section. For each feature, we calculated mean and standard deviation and quantized the feature values as explained in Proposed methodology section. Two such normalized histograms obtained for beta band power and Hjorth activity are shown in Fig. 5.2. It is visible that the distribution of windows in bins for good and bad performers are significantly different. Similar plots were obtained for other features also.

Finally, we calculated signal level descriptors for each distribution. Thus, we obtained 608-dimensional signal level descriptors for each participant.

In the given dataset, there were 10 bad performers and 26 good performers. To remove this class imbalance, we applied SMOTE. After performing

TABLE 5.1 *P*-values obtained for various EEG domain features using Wald-Wolfowitz test.

Domain	Features	*P*-value	Comments
Time-domain	Mean	0.08	Similar distribution
	Variance	0.002	Different distribution
	Skewness	0.5	Similar distribution
	Kurtosis	0.08	Similar distribution
	Zero crossing rate	0.2	Similar distribution
	Hjorth parameter, activity	0.0001	Different distribution
	Hjorth parameter, mobility	0.006	Different distribution
	Hjorth parameter, complexity	0.001	Different distribution
Frequency domain	Theta band power	0.0025	Different distribution
	Alpha band power	0.001	Different distribution
	Beta band power	0.0001	Different distribution
	Gamma band power	0.0001	Different distribution
	Engagement index	0.0004	Different distribution
	Mean frequency	0.08	Similar distribution

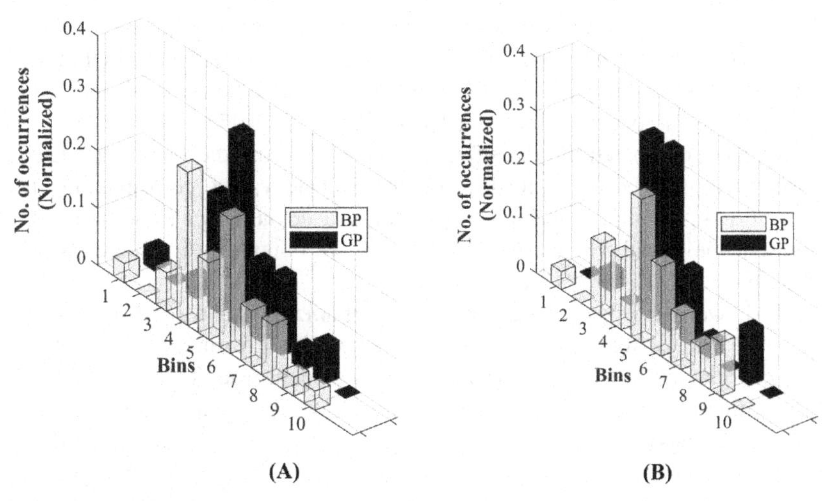

FIGURE 5.2 Normalized histogram obtained from (A) beta power and (B) Hjorth activity for good and bad performers.

TABLE 5.2 Classification accuracy for various classifiers using proposed approach.

Classifier	Mean accuracy	Mean F1 score	Standard deviation of F1 score
LR	0.77	0.57	0.34
SVM	0.83	0.62	0.35
GNB	0.85	0.63	0.43
RF	0.75	0.50	0.42
MLP	0.74	0.54	0.33
GB	0.70	0.46	0.38

SMOTE, a number of samples for both bad and good performers were made same. Finally, we tried to classify the good and bad performers based on these signal level descriptors using fivefold cross validation approach. The details of the results are shown in Table 5.2.

In the present study we have tried linear regression (LR) (Ng & Michael, 2002), support vector machine (SVM) (Keerthi et al., 2000), Gaussian naïve bayes (GNB) (Ng & Michael, 2002), random forest (RF) (Rodriguez-Galiano et al., 2012), multilayer perceptron (MLP) (Hampshire, John, & Pearlmutter, 1991), and gradient boosting (GB) (Mayr et al., 2014) classifiers. We found that maximum accuracy is obtained with Gaussian gradient boosting classifier. The mean accuracy across participants is 85%. However, it may be noted that for other classifiers the classification performance is satisfactory.

Next, we compared our proposed methodology with state of the art summarization techniques. In order to do so, we have considered commonly used summarization technique in the form of average and standard deviation of feature values across the windows (Papakostas et al., 2017).

We have compared the classification accuracy using the same eight EEG domain features. For each feature we obtained 30 values corresponding to 30 windows. We calculated mean and standard deviation of these values to represent the signal. Hence, for a participant we obtained 16 parameters per EEG channel and 16×19 features in total.

Next, we performed participant level normalization and tried to classify good and bad performers based on these normalized feature values. Details of the results are presented in Table 5.3. It is evident from Table 5.3 that for the commonly used summarization approach we got maximum accuracy with GNB classifier with a mean accuracy of 81% (shown in bold in Table 5.3) which is less compared to the mean classification accuracy of the proposed approach. Moreover, the classification performance for other classifiers is also better for the proposed way of summarization (Table 5.3).

Finally, we have tried to combine our signal level descriptors with EEG domain features. The details of classification accuracy obtained using various classifiers are shown in Table 5.4. We observed that this increases the classification accuracy further for all the classifiers. Maximum classification accuracy of 92% is obtained with GNB with a standard deviation of 0.06.

The performance of all six classifiers with proposed features, state of the art EEG domain features and fusion is shown in Fig. 5.3. Thus, except for MLP, in all other classifiers the proposed approach outperforms the conventional one. The proposed signal level descriptor can be used to discriminate cognitive performance successfully.

TABLE 5.3 Classification accuracy for various classifiers using state of the art approach.

Classifier	Mean accuracy	Mean F1 score	Standard deviation of F1 score
LR	0.73	0.46	0.29
SVM	0.73	0.46	0.27
GNB	**0.81**	**0.59**	**0.35**
RF	0.73	0.52	0.31
MLP	0.77	0.49	0.30
GB	0.69	0.46	0.22

TABLE 5.4 Classification accuracy after fusing proposed feature with EEG domain features.

Classifier	Mean accuracy	Mean F1 score	Standard deviation of F1 score
LR	0.83	0.63	0.33
SVM	0.83	0.62	0.27
GNB	0.92	0.70	0.36
RF	0.77	0.57	0.31
MLP	0.77	0.49	0.30
GB	0.76	0.54	0.3

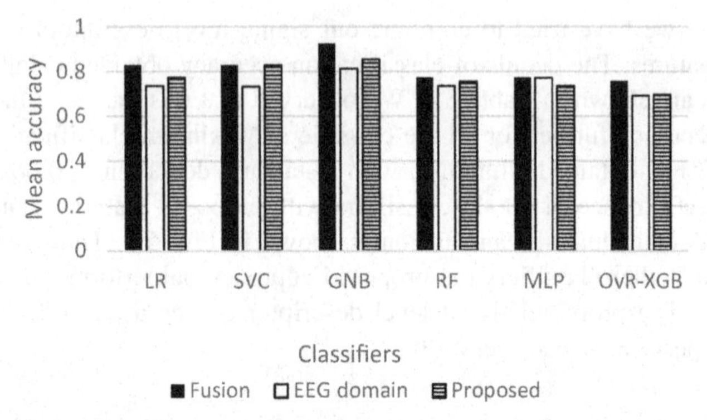

FIGURE 5.3 Classification performance for various approaches.

Conclusion and future scope

This study presents an approach for assessing cognitive performance-based on brain activations and to derive a novel feature for the same. So far, various EEG domain specific features have been used for assessing and quantifying various mental states. Here we have used EEG domain features and derived a signal level descriptor from those features. The new feature set has been obtained by a novel process of summarizing the set of window level features. These descriptors are simple yet effective for studying cognitive performance. We have applied our approach on publicly available dataset. Classification results suggest that our proposed feature can classify good and bad performers with a mean accuracy of 85% and it outperforms the commonly used approach of summarization of window level features. Fusion of our signal descriptors with traditional features results in further increase in accuracy to 92%. A proposed approach of summarization is generic in nature and hence can be used for studying other cognitive states like attention, stress, executive functions, etc. The analysis has been carried out on a small dataset as the work is a proof of concept of the approach. In the future, we would like to validate our findings on a larger population and for various tasks like decision-making, problem solving, and memory-related tasks.

References

Aniruddha, S., Rahul, G., Debatri, C., Rajat, D., & Arijit, S., (2015) Dynamic assessment of learners' mental state for an improved learning experience. In 2015 IEEE *frontiers in education conference (FIE) - dynamic assessment of learners' mental state for an improved learning experience*. 1–9.

Baldwin, C. L., & Penaranda, B. N. (2012). Adaptive training using an artificial neural network and EEG metrics for within-and cross-task workload classification. *Neuroimage*, *59*(1), 48–56.

Berka, C., Levendowski, D. J., Cvetinovic, M. M., Petrovic, M. M., Davis, G., Lumicao, M. N., Zivkovic, V. T., Popovic, M. V., Olmstead, R., et al. (2004). Real-time analysis of EEG indexes of alertness, cognition, and memory acquired with a wireless EEG headset. *International Journal of Human-Computer Interaction, 17*, 151–170.

Bialystok, E., Craik, F. I. M., Grady, C., Chau, W., Ishii, R., Gunji, A., & Pantev, C. (2005). Effect of bilingualism on cognitive control in the simon task: Evidence from MEG. *Neuroimage, 24*(1), 40–49.

Boldt, A., & Yeung, N. (2015). Shared neural markers of decision confidence and error detection. *Journal of Neuroscience, 35*(8), 3478–3484.

Bressler, S. L., & Menon, V. (2010). Large-scale brain networks in cognition: Emerging methods and principles. *Trends in Cognitive Sciences, 14*(6), 277–290.

Chaouachi, M., Jraidi, I., & Frasson, C. (2011). Modeling mental workload using EEG features for intelligent systems. In J. A. Konstan, R. Conejo, J. L. Marzo, & N. Oliver (Eds.), *User modeling, adaption and personalization. UMAP 2011. Lecture notes in computer science*. (vol. 6787). Berlin, Heidelberg: Springer. Available from https://doi.org/10.1007/978-3-642-22362-4_5.

Chawla, N. V., Bowyer, K. W., Hall, L. O., & Kegelmeyer, W. P. (2002). SMOTE: Synthetic minority over-sampling technique. *Journal of Artificial Intelligence Research, 16*, 321–357.

Danili, E., & Reid, N. (2006). Cognitive factors that can potentially affect pupils' test performance. *Chemistry Education Research and Practice, 7*(2), 64–83.

Engel, A. K., & Fries, P. (2010). Beta-band oscillations—signaling the status quo. *Current Opinion in Neurobiology, 20*(2), 156–165.

Fleming, S. M., Huijgen, J., & Dolan, R. J. (2012). Prefrontal contributions to metacognition in perceptual decision making. *Journal of Neuroscience, 32*, 6117e25.

Friedman, J. H., & Lawrence, C. R. (1979). Multivariate generalizations of the Wald-Wolfowitz and Smirnov two-sample tests. *The Annals of Statistics, 7*, 697–717.

Frith, C. D., Friston, K. J., Liddle, P. F., & Frackowiak, R. (1992). PET imaging and cognition in schizophrenia. *Journal of the Royal Society of Medicine, 85*(4), 222.

Galán, F. C., & Beal, C. R. (2012). *EEG estimates of engagement and cognitive workload predict math problem solving outcomes. International conference on user modeling, adaptation, and personalization*. Berlin, Heidelberg: Springer.

Gavas R., Chatterjee D., Sinha A. (2017). Estimation of cognitive load based on the pupil size dilation. *2017 IEEE international conference on systems, man, and cybernetics (SMC)*, 1499–1504. Available from https://doi:10.1109/SMC.2017.8122826.

Hampshire, I. I., John, B., & Pearlmutter, B. (1991). Equivalence proofs for multi-layer perceptron classifiers and the Bayesian discriminant function. *Connectionist Models*, 159–172. Available from https://doi.org/10.1016/B978-1-4832-1448-1.50023-8.

Hjorth, B., & Elema-Schönander, A. B. (1970). EEG analysis based on time domain properties. *Electroencephalography and Clinical Neurophysiology, 29*, 306–310. Available from https://doi:10.1016/0013-4694(70)90143-4.

Hughes, D., & Bryan, J. (2003). The assessment of cognitive performance in children: Considerations for detecting nutritional influences. *Nutrition Reviews, 61*(12), 413–422.

Jaušovec, N. (2000). Differences in cognitive processes between gifted, intelligent, creative, and average individuals while solving complex problems: An EEG study. *Intelligence, 28*(3), 213–237.

Kahana, M. J., Seelig, D., & Madsen, J. R. (2001). Theta returns. *Current Opinion in Neurobiology, 11*(6), 739–744.

Kamiński, J., Brzezicka, A., Gola, M., & Wrobel, A. (2012). Beta band oscillations engagement in human alertness process. *International Journal of Psychophysiology, 85*(1), 125–128.

Keerthi, S. S., Shevade, S. K., Bhattacharyya, C., & Murthy, K. R. K. (2000). A fast iterative nearest point algorithm for support vector machine classifier design. *IEEE Transactions on Neural Networks, 11*(1), 124–136.

Kilseop, R., & Rohae, M. (2005). Evaluation of mental workload with a combined measure based on physiological indices during a dual task of tracking and mental arithmetic. *International Journal of Industrial Ergonomics*, *35*, 991–1009.

Márton, L. F., Brassai, S. T., Bako, L., & Losonczi, L. (2014). Detrended fluctuation analysis of EEG signals. *Procedia Technology*, *12*(1), 125–132.

Mayr, A., Binder, H., Gefeller, O., & Schmid, M. (2014). The evolution of boosting algorithms-from machine learning to statistical modelling. *Methods of Information in Medicine*, *53*(06), 419–427. Available from https://doi:10.3414/ME13-01-012.

McMahan, T., Parberry, I., & Parsons, T. D. (2015). Evaluating player task engagement and arousal using electroencephalography. *Procedia Manufacturing*, *3*, 2303–2310.

Murata, A. (2005). An attempt to evaluate mental workload using wavelet transform of EEG. *Human Factors*, *47*(3), 498–508.

Mushtaq, I., & Khan, S. N. (2012). Factors affecting students academic performance. *Global Journal of Management and Business Research*, *12*(9).

Ng, A. Y., & Michael, I. J. (2002). *On discriminative versus generative classifiers: A comparison of logistic regression and naive bayes*. Advances in neural information processing systems (pp. 841–848). Cambridge: MIT Press.

Papakostas, M., K. Tsiakas, T. Giannakopoulos, F. Makedon. Towards predicting task performance from EEG signals. In *2017 IEEE international conference on big data (Big Data)*. 2017. 4423-4425. Available from https://doi:10.1109/BigData.2017.8258478.

Poldrack, R. A. (2008). The role of fMRI in cognitive neuroscience: Where do we stand? *Current Opinion in Neurobiology*, *18*(2), 223–227.

Ray, W. J., & Harry, W. C. (1985). EEG alpha activity reflects attentional demands, and beta activity reflects emotional and cognitive processes. *Science*, *228*(4700), 750–752.

Rodriguez-Galiano, V. F., Ghimire, B., Rogan, J., Chica-Olmo, M., & Rigol-Sanchez, J. P. (2012). An assessment of the effectiveness of a random forest classifier for land-cover classification. *ISPRS Journal of Photogrammetry and Remote Sensing*, *67*, 93–104.

Sarter, M., Berntson, G. G., & Cacioppo, J. T. (1996). Brain imaging and cognitive neuroscience: Toward strong inference in attributing function to structure. *The American Psychologist*, *51*, 13–21.

Sinha, A., P. Das, R. Gavas, D. Chatterjee, S.K. Saha (October, 2016) Physiological sensing-based stress analysis during assessment. *2016 IEEE frontiers in education conference (FIE)* Eire, PA, USA.

Stankov, L., Sabina, K., & Jackson, S. A. (2014). Measures of the trait of confidence. In G. J. Boyle, D. H. Saklofske, & G. Matthews (Eds.), *Measures of personality and social psychological constructs* (pp. 158–189). Academic Press, Chapter 7.

Stevens, R., Galloway, T., & Berka, C. (2007). EEG-related changes in cognitive workload, engagement and distraction as students acquire problem solving skills. In C. Conati, K. McCoy, & G. Paliouras (Eds.), *User modeling 2007. UM 2007. Lecture notes in computer science* (vol 4511). Berlin, Heidelberg: Springer. Available from https://doi.org/10.1007/978-3-540-73078-1_22.

Stevens, R., Galloway, T., & Berka, C. (2006). Integrating EEG models of cognitive load with machine learning models of scientific problem solving. *Augmented Cognition: Past, Present and Future*, *2*, 55–65.

Stevens, R.H., Galloway, T., & Berka, C., (2007). Integrating innovative neuro-educational technologies (I-Net) into K-12 science classrooms. In *Proceedings of the* 3rd international conference on foundations of augmented cognition. Beijing, China, 47–56.

Stikic, M., Johnson, Robin R., Levendowski, Daniel J., Popovic, Djordje P., Olmstead, Richard E., & Berka, Chris (2011). EEG-derived estimators of present and future cognitive performance. *Frontiers in Human Neuroscience*, *5*, 70.

Wong, S. W., Chan, R. H., & Mak, J. N. (2014). Spectral modulation of frontal EEG during motor skill acquisition: A mobile EEG study. *International Journal of Psychophysiology*, *91*(1), 16–21.

Zyma, I., Tukaev, S., Seleznov, I., Kiyono, K., Popov, A., Chernykh, M., & Shpenkov, O. (2019). Electroencephalograms during mental arithmetic task performance. *Data*, *4*(1), 14. Available from https://doi.org/10.3390/data4010014.

Chapter 6

Trust or no trust in chatbots: a dilemma of millennial

Shivani Agarwal
KIET School of Management, KIET Group of Institutions, Delhi-NCR, Ghaziabad, India

Introduction

Human beings are considered as social animals for ages. To fulfill the social needs, interactions between humans exist over long distances for thousands of years (Agarwal, Jindal, Garg, and Rastogi, 2017). The social world is now shifting toward the virtual world. So, the way of interaction has also evolved. In the social and virtual world, human beings interacted as human–human interaction, then human–machine interaction which is further refined as human- robot interaction and now industry invented new integration tool for interactions with humans which is considered as "Chatbot." As the name suggest, chatbot interacts with human beings and provide a better customer experience. Chatbot further helps the industry to provide better customer engagement by reducing customer service cost. It is a computer-based program that allows communicating with human beings and generally based on natural language via text or speech (Agarwal & Linh, 2021).

In the 21st century, millennials are increasingly being subjected to competitive environments. In the virtual world, human beings are facing trust issues among themselves. With the advent of technology, trust issues come across not only between human beings but also between machines and human beings. There is ample amount of chatbots which organizations now a day are using as the first interaction between their organizational platform and the customers.

Digitalization is all over the world and in all the sectors and industries also. To provide quick and better 24/7 services to customers, organizations are adopting chatbots. In the past few years, chatbots are considered as a talk of the town where both organizations and customers were on board with the revolution and evolution of technology.

Cognitive Computing for Human-Robot Interaction. DOI: https://doi.org/10.1016/B978-0-323-85769-7.00007-0

Chatbots are typically adopted by different industries for different usage and purposes. Following are the industries which are using chatbots in their organizations:

1. E-Commerce: The most booming sector until 2017 was E-Commerce. It is typical to handle it through phone calls, email, and social media as the number of people using Internet is increasing exponentially. So, to combat this war of handling queries for this sector, chatbot is considered as a tool that helps in customer retention and acquisition as well. Also, it provides satisfactory information to the users or customers also.
2. Medicine: In the current scenario, where pandemic situation is in front of all over the world. The most affected sector is health sector where doctors, nurses and attendants are working tirelessly to save the life of living beings. Chatbots plays a prominent role in providing the information related to admissions, transfer requests, consultation requests, discharge, etc.
3. Human resource (HR): HR is the most essential part for any industry as from hiring to firing of any employees, all the responsibilities are handled by HR. Organizations profits and loss is sheer dependent on the employees an organization have. In a research, it was found that 40%–50% of HR time has been consumed in talent management and handling the internal queries. Where, that time could be saved by chatbots and HR can utilize that time in strategic decision of organizations.
4. Travel: In the global world, human beings are exploring more and more. For the same, tourism sector is also booming. In that situation, chatbots permits to generate travelers leads round the clock by handling their queries day or night. There is no such compulsion of time with chatbots; there is no time lag also. It provides a detail information regarding tickets, hotels, nearby places, food, culture, etc.
5. Real estate: Customers have query like locality, nearby schools, grocery shops, mode of transportation, hospitals, etc. So, all these answers can be handled single handedly by chatbots. Real estate sectors are also using chatbots for a better customer experience.
6. Banking: To resolve the quest of customers like loan service, wealth manager, personal banker, ATM locator, chatbot can serve the purpose in a single click.

The present research discusses the dilemma that millennials are facing regarding the perception of trust, which is very much based on one's own cognitive judgment. In all the facets of the environment, interpersonal trust plays an important role viz; digital marketing, virtual management, sociology (interaction between machine and human being), psychology (counseling Apps), and economics, etc. Trust is the most vulnerable and salient feature in the organizations.

So, from the above discussion, in all the facets of environment, chatbots are implemented by the industries for their customers and earn more and more profits.

Millennials

Millennials are defined as the people born between 1981 and 1996 (ages 23–40 in 2020). People born from 1997 onward are a part of a new generation. To know the exact meaning of the term "millennial," the above chart helps in understanding what actually millennials are. When we label people on the basis of their year of birth and death it is known as generation name or generation cohorts. The human race started identifying the generation name from 1890 to 2025 now. It started from the lost generation till Gen alpha. But in this chapter, the focus is on millennials whose date of birth lie between 1980 and 1994. Millennials are considered as digital natives as they are generally noticeable by usage of and familiarity with the Internet, mobile devices, and social media. All the generation cohorts are mentioned in Table 6.1 which is below:

All the industries are captured by chatbots for providing better interface to the users/customers. But from the customer's viewpoint, question arises, being a digital native or millennial faces the dilemma regarding the perception of trust which is very much based on one's own judgment to interact with machine or chatbots.

Scope of the study

This chapter empirically investigates what dilemma millennials have regarding the perception of trust while interacting with chatbots and how perception of trust can be measured in terms of this research context.

The present study has adopted the concept of trust which has been propounded by McAllister (1995) and defined the term trust as "The extent to which a person is confident in and willing to act on the basis of, the words, actions and

TABLE 6.1 Generation cohorts.

Generation cohort	Born year start	Born year end
The lost generation	1890	1915
The greatest generation	1910	1924
The silent generation	1925	1945
Baby boomers generation	1946	1964
Generation (X)	1965	1979
Generation (Y) or millennial	1980	1994
Generation (Z)	1995	2012
Gen alpha	2013	2025

decisions, of another." The major aspects of trust that have been studied in the present research and which, in combination, constitute the trust were:

The two factors of trust are:

1. Cognitive-based trust with chatbot: It is defined as the level of trust under which situations and circumstances we create on other to develop a relationship (Lewis & Weigert, 1985; McAllister, 1995).
2. Affect-based trust with chatbot: Human beings maintain trust relationships and do emotional investments as well. Basically, the relationship between individuals through emotional investment is known as affect-based trust.

Literature review

Gille, Jobin, and Ienca (2020) found that there is still dearth in understanding exactly what comprises user trust in AI and what the necessities are for its realization.

Sharan and Romano (2020) study tested the impact of personality characteristics and locus of control on trust behavior, and the degree to which people trust the suggestion from an AI-based algorithm, more than humans, in a decision-making card game.

Braganza, Chen, Canhoto, and Sap (2020) found that psychological contracts had a significant, positive effect on job engagement and on trust. Yet, with AI adoption, the positive effect of psychological contracts fell significantly.

Marsh et al. (2020) explores that trust can be calculated and can be contrast and rated, and that trust is of worth when we consider entities from data, through artificial intelligences, to humans, with side trips along the way to animals.

Xu, Shieh, van Esch, and Ling (2020) shows that it depends on the difficulties of the task whether to prefer AI or human centered interventions are required.

Carter et al. (2020) assessed the level of trust for the apparent solutions of AI provided to the patients in healthcare.

Ameen, Tarhini, Reppel, and Anand (2020) findings point to the significant role of trust and perceived sacrifice as factors mediating the effects of perceived convenience, personalization, and AI-enabled service quality. The findings also reveal the significant effect of relationship commitment on AI-enabled customer experience

Methodology

Participants and procedure

In this research, the participants are the people whose birth date lies between 1980 and 1994. With the help of Google form, data has been collected from

TABLE 6.2 Demographical details.

Demographic $n = 350$		No. of respondents	Percentage (%)
Age (in years)	Young (26–33)	154	44
	Middle age (33–40)	196	56
Gender	Male	228	65.14
	Female	122	34.85
Education	Graduation	260	74.28
	Post-graduation	90	25.72
Work experience	Less than 10	137	39.14
	10–20	213	60.85

the millennial belongs to the NCR, India. Data has been collected and further used for analysis from 350 millennials. The reliability of the questionnaire was measured and reported as 0.712. A demographical profile of millennials is shown in Table 6.2.

Measures

For this research, the author adapted a questionnaire for the current study, which was originally developed by McAllister (1995), and consists of 11 questions on a 7-point Likert scale which further talks about the overall trust among individuals.

Objectives of the study

The main objective of the study is to analyze the perception of trust of the millennial regarding sharing information with chatbot.

The objectives were further divided into three subobjectives:

1. To analyze that the level of perception of trust as cognition based among millennials regarding sharing information with chatbot.
2. To analyze that the level of perception of trust as affect-based among millennials regarding sharing information with chatbot.

Result and discussion

The findings of the data were represented in the form of a pie chart to easily indulge for the large number of scholars and researchers. The first finding of the result was shown in Fig. 6.1. A majority of 37.7% millennials strongly agree that they can share their ideas, thoughts and queries with chatbots, 36.6% agree that they can share their ideas, thoughts and queries with chatbots and 7% is strongly disagree that they can share their ideas, thoughts and queries with chatbots. Out of 350 millennials, there are 291 people millennials who agree that they can share their ideas, thoughts and queries with chatbots. On the other hand, 35 millennials are disagreeing that they can share their ideas, thoughts and queries with chatbots and 24 millennials are neutral that they can share their ideas, thoughts, and queries with chatbots.

The second finding of the result was shown in Fig. 6.2. A majority of 20% millennials strongly agree that they can talk about complexity facing at job, 39% agree that they can talk about complexity facing at job with chatbots and 0.9% is strongly disagree that they can talk about complexity facing at job with chatbots. Out of 350 millennials, there are 273 millennials who agree that they can talk about complexity facing at job with chatbots. On the other hand, 24 millennials disagree that they can talk about complexity facing at a job with chatbots and 53 millennials are neutral that they can talk about complexity facing at a job with chatbots.

The third finding of the result was shown in Fig. 6.3. A sum of 9.4% millennials strongly agree that they could experience a loss if chatbot could no longer work with them, 35.4% agree that they could experience a loss if chatbot could no longer work with them, and 3.4% is strongly disagree that they could experience a loss if chatbot could no longer work with them. Out of 350 millennials, there are 188 millennials who agree that they could experience a loss if chatbot could no longer work with them. On the other hand, 52 millennials disagree that they could experience a loss if chatbot could no

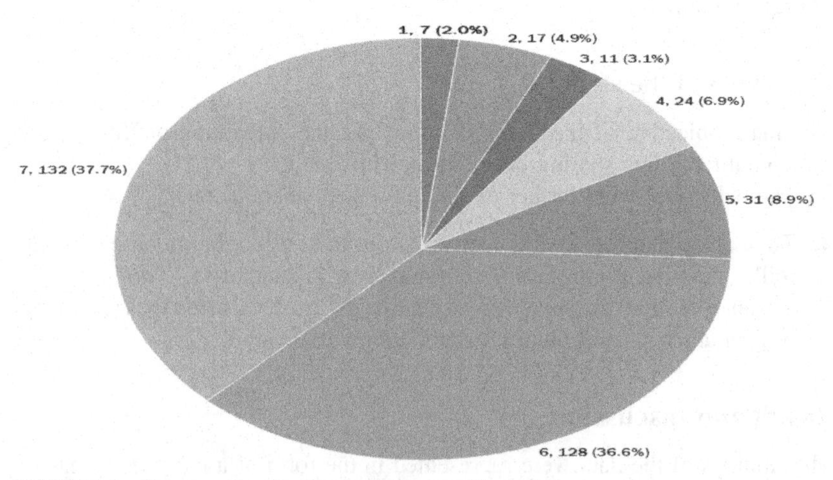

FIGURE 6.1 Share my ideas, feelings, and queries with chatbots.

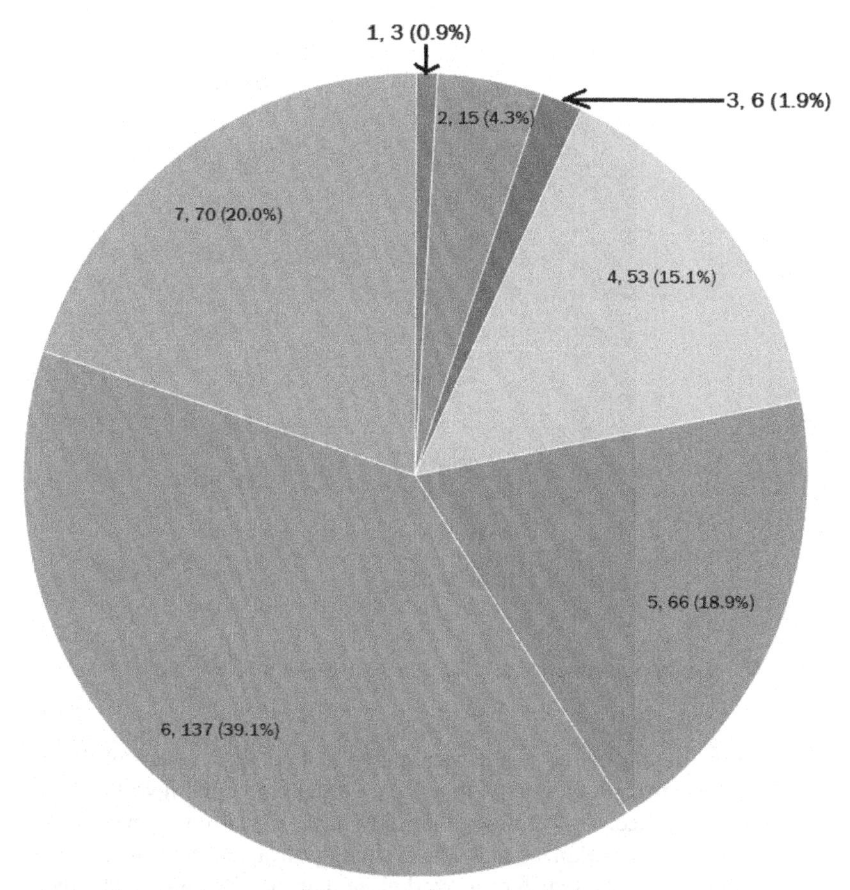

FIGURE 6.2 Talk about complexity facing at job with chatbot.

longer work with them and 110 millennials are neutral about if they could experience a loss if chatbot could no longer work with them.

The fourth finding of the result was shown in Fig. 6.4. A sum of 22% millennials strongly agree that they shared problems with this chatbot, and it would respond constructively and caringly, 58% agree that they shared problems with this chatbot, and it would respond constructively and caringly and 1.1% is strongly disagree that they shared problems with this chatbot, and it would respond constructively and caringly. Out of 350 millennials, there are 306 millennials that shared problems with this chatbot, and it would respond constructively and caringly. On the other hand, 23 millennials disagree that they shared problems with this chatbot, and it would respond caringly and constructively and 21 millennials are neutral about sharing problems with this chatbot, and it would respond constructively and caringly.

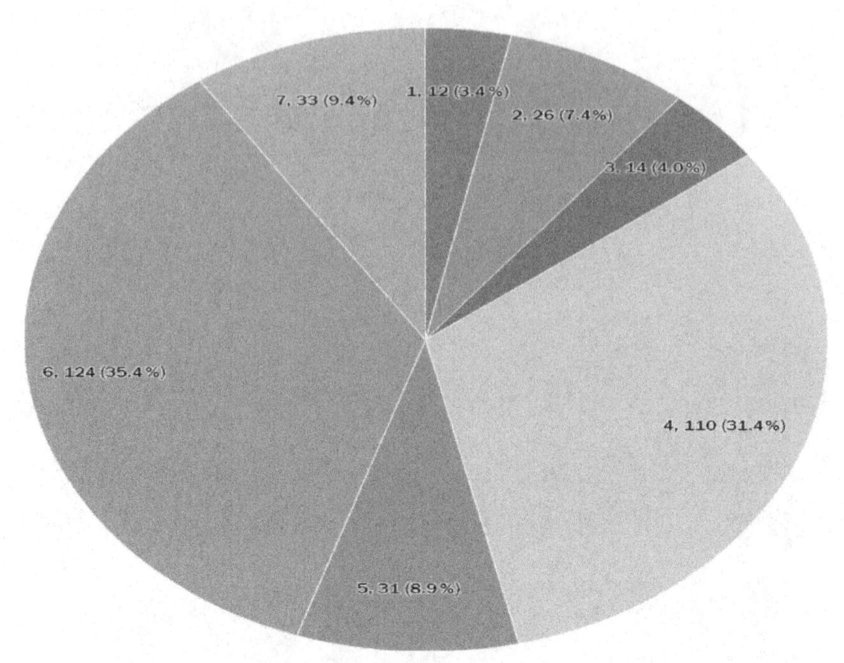

FIGURE 6.3 I could experience a loss if chatbot could no longer work with them.

The fifth finding of the result was shown in Fig. 6.5. A sum of 10.6% millennials are strongly agree that they have virtual emotional investments in chatbot at workplace, 48.9% agree that they have virtual emotional investments in chatbot at workplace and 2% is strongly disagree they have virtual emotional investments in chatbot at workplace. Out of 350 millennials, there are 237 millennials they have virtual emotional investments in chatbot at workplace. On the other hand, 35 millennials disagree they have virtual emotional investments in chatbot at workplace and 78 millennials are neutral about having virtual emotional investments in chatbot at workplace.

The sixth finding of the result was shown in Fig. 6.6. A sum of 22.6% millennials are strongly agree that chatbot approaches its job with professionalism and dedication, 61.1% agree that chatbot approaches its job with professionalism and dedication and 0.9% is strongly disagree that chatbot approaches its job with professionalism and dedication. Out of 350 millennials, there are 312 millennials that chatbot approaches its job with professionalism and dedication. On the other hand, 11 millennials are that chatbot approaches its job with professionalism and dedication and 27 millennials are neutral about how chatbot approaches its job with professionalism and dedication.

The seventh finding of the result was shown in Fig. 6.7. A sum of 21.4% millennials strongly agree that, they see no reason to doubt chatbot

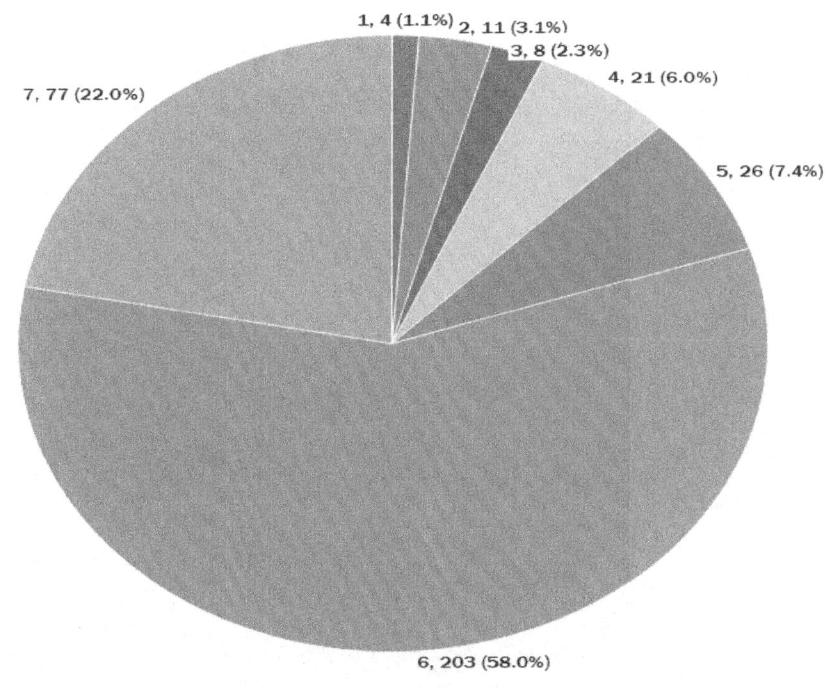

FIGURE 6.4 If I shared my problems with chatbot and discern it will act caringly and constructively.

competence and preparation for the job, 57.1% agree that they see no reason to doubt chatbot competence and preparation for the job and 1.4% is strongly disagree that they see no reason to doubt chatbot competence and preparation for the job. Out of 350 millennials, there are 302 millennials that see no reason to doubt chatbot competence and preparation for the job. On the other hand, 17 millennials see no reason to doubt chatbot competence and preparation for the job and 31 millennials are neutral that they see no reason to doubt chatbot competence and preparation for the job.

The eighth finding of the result was shown in Fig. 6.8. A sum of 30.6% millennials strongly agree that they believe that chatbot will not make my job complicated by sloppy job, 45.1% agree that they can rely on this chatbot will not make the job complicated by sloppy job, and 2% is strongly disagree that they can rely on this chatbot will not make the job complicated by sloppy job. Out of 350 millennials, there are 292 millennials that can rely on this chatbot will not make the job complicated by sloppy job. On the other hand, 29 millennials say they can rely on this chatbot will not make the job complicated by sloppy job, and 29 millennials are neutral that they can rely on this chatbot will not make the job complicated by sloppy job.

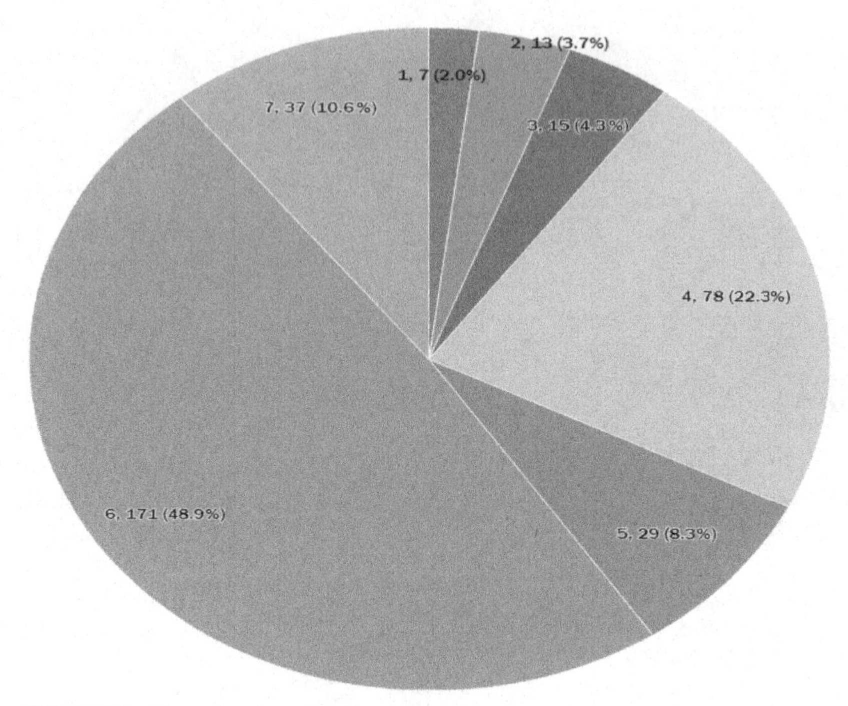

FIGURE 6.5 I have virtual emotional investments in Chatbot at workplace.

The ninth finding of the result was shown in Fig. 6.9. A sum of 22% millennials are strongly agree that they trust chatbot as a coworker, 48% agree that they trust chatbot as a coworker and 2% is strongly disagree that they trust chatbot as a coworker. Out of 350 millennials, there are 303 millennial that they trust chatbot as a coworker. On the other hand, 23 millennials say that they trust chatbot as a coworker and 24 millennials are neutral that they can trust chatbot as a coworker.

The tenth finding of the result was shown in Fig. 6.10. A sum of 21.7% millennials strongly agree that other peers of millennials who interact with chatbot individually consider it as trustworthy, 48% agree that other peers of millennial who interact with chatbot individual consider it as trustworthy and 1.1% is strongly disagree that other peers of millennials who interact with chatbot individuals consider it as trustworthy. Out of 350 millennials, there are 296 millennial peers of millennials who interact with chatbot individual consider it as trustworthy. On the other hand, 20 millennials that are peers of other millennials who interact with chatbot individual consider it as trustworthy and 34 millennials are neutral that other peers of millennials who interact with chatbot individuals consider it as trustworthy.

The last finding of the result was shown in Fig. 6.11. A sum of 19.1% millennials strongly agree that if they have more information regarding

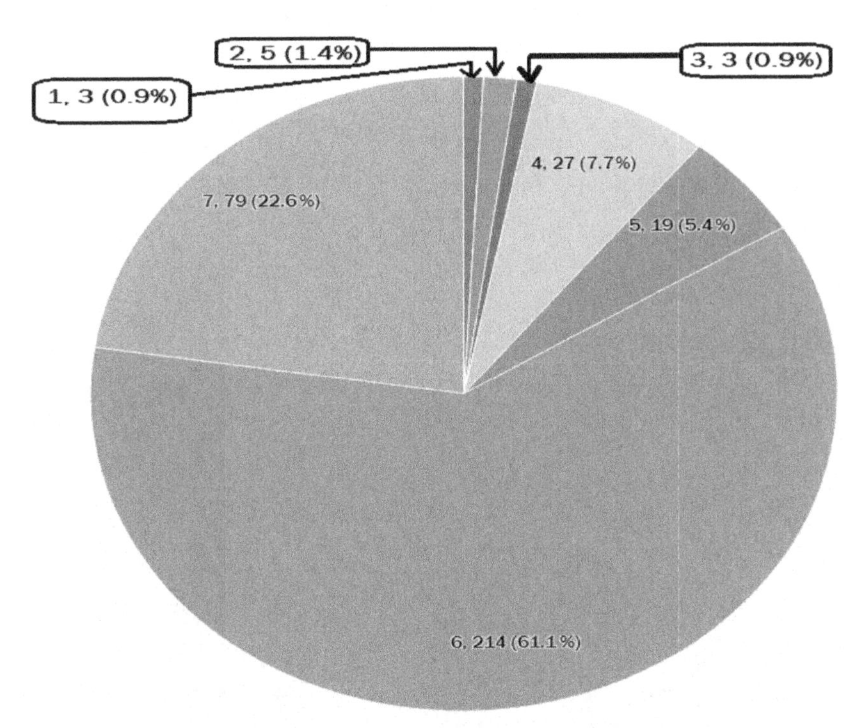

FIGURE 6.6 Chatbot approaches its job with professionalism and dedication.

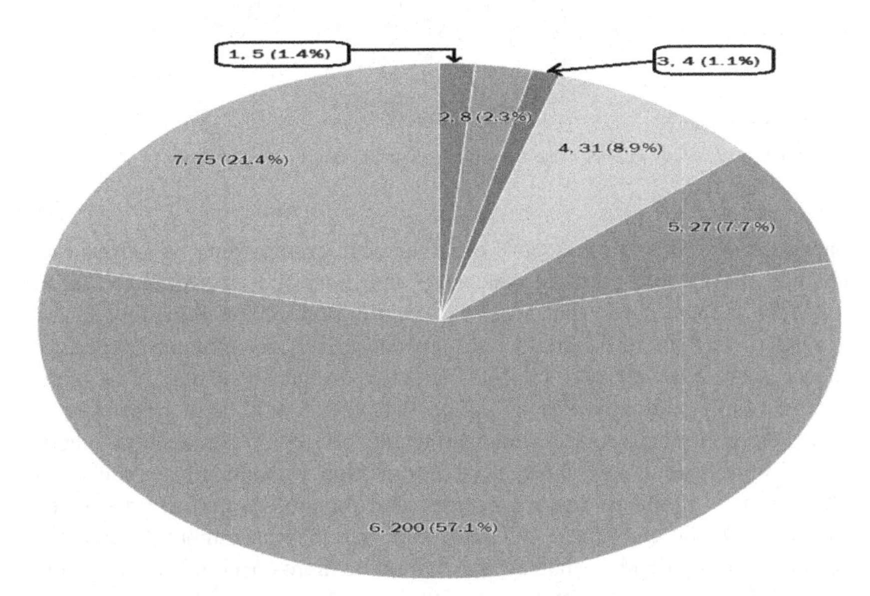

FIGURE 6.7 I see no reason to doubt chatbot competence and preparation for the job.

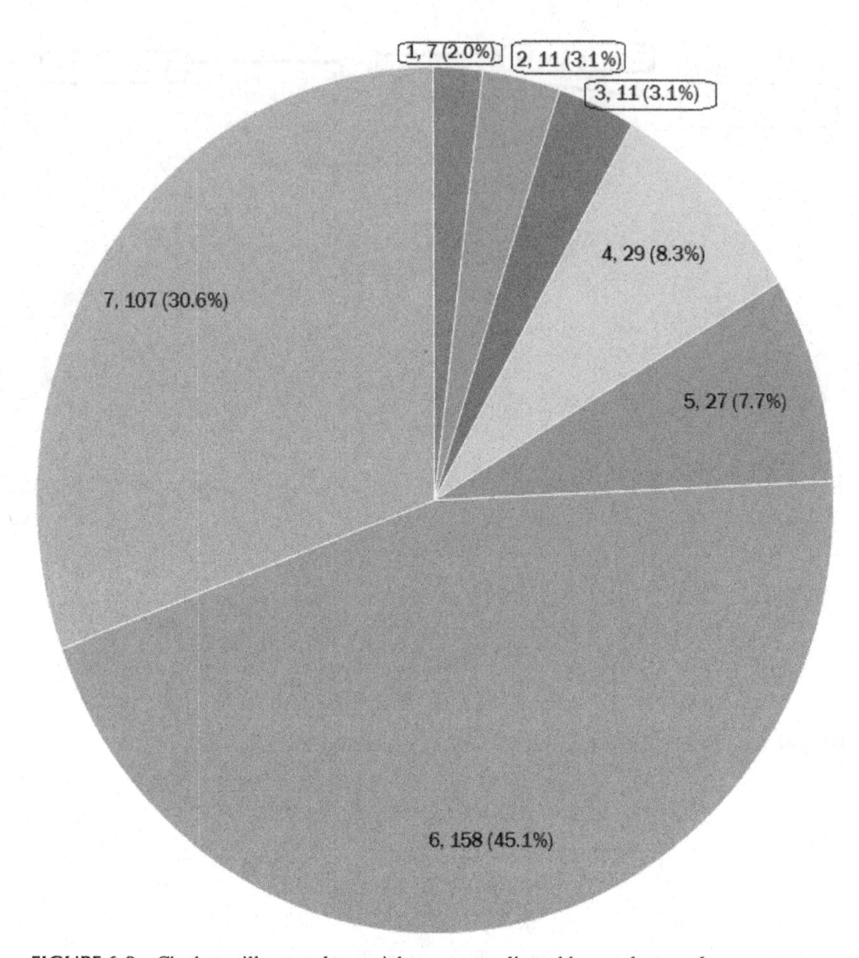

FIGURE 6.8 Chatbot will not make my job more complicated by careless work.

background of chatbot; they have more information regarding background of chatbot; they would be more concerned and monitor its performance more minutely, 13.4% agree that they have more information regarding background of chatbot; they would be more concerned and monitor its performance more minutely and 11.4% is strongly disagree that they have more information regarding background of chatbot; they would be more concerned and monitor its performance more minutely. Out of 350 millennials, there are 128 millennial they have more information regarding background of chatbot; they would be more concerned and monitor its performance more minutely. On the other hand, 177 millennials are they have more information regarding background of chatbot; they would be more concerned and monitor its performance more minutely and 45 millennials are neutral that they have

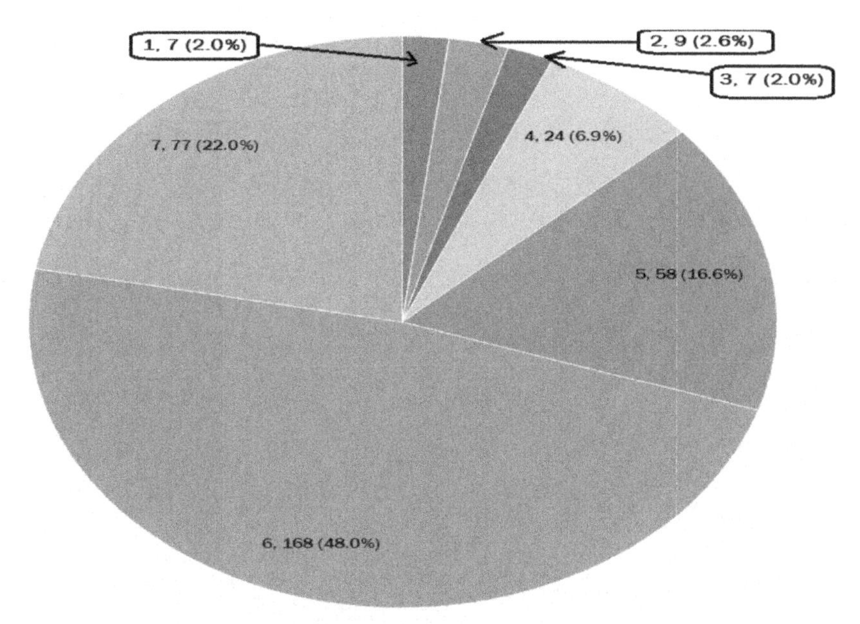

FIGURE 6.9 People trust and respect chatbot as a coworker.

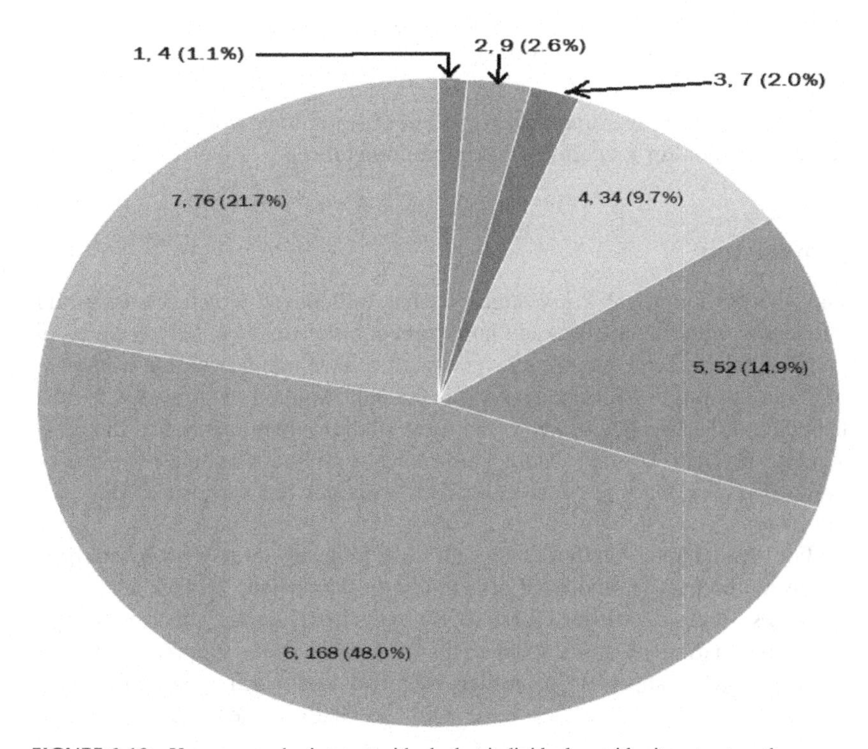

FIGURE 6.10 Your peers who interact with chatbot individual consider it as trustworthy.

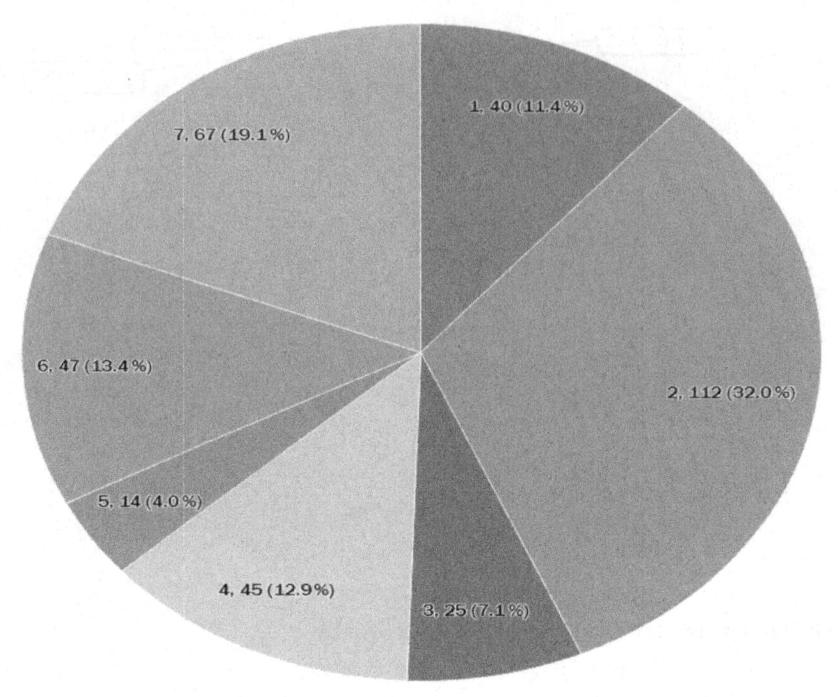

FIGURE 6.11 If they have more information regarding background of chatbot; they would be more concerned and monitor its performance more minutely.

more information regarding background of chatbot; they would be more concerned and monitor its performance more minutely.

Conclusion

The chapter empirically investigates what millennial perceive trust while interacting with the chatbots and how perception of trust can be calculated in terms of this research context. Several studies have mentioned that in the era of digitalization, to trust the machine is the only way out to survive with millennial. On the flip side of story, there are studies which mentioned that millennial still feel hesitant while conversing with the chatbots. The major findings of the study are that millennial' trust has been measured into two dimensions.

For this chapter, author analyzed trust in terms of Affect-based trust which represents the emotional affection with the chatbot. Author adapts the questions from McAllister Scale (1985) and firstly asked the respondents about the sharing relationship in terms of queries while shopping, booking movies, doctor's appointment, millennial mentioned that they are quite satisfied with the experience they have with the chatbots. Secondly, millennial trust the chatbot for talk about complexities having at job and it will listen empathetically means they are using counseling bots. Thirdly, it is also mentioned that if chatbot is not properly working at some point of time then millennial generally feel sense of loss as their work is dependent on the precondition work of that machine. Fourthly, chatbots is also used by millennial for seeking the guidance. Fifthly, it is mentioned that they have a virtual emotional bond with the machine. So, based on above-mentioned findings, it can be concluded that millennial reciprocate the affect-based trust while dealing with chatbots.

The other dimension of trust is cognitive-based trust which represented whom we will choose in our life. The sixth findings show that chatbots are dedicated towards their job irrespective of day or night. The seventh findings represent that being a machine, there is no doubt on the competence of the same. The eighth findings showed that millennial can rely on chatbots. The ninth findings represented that chatbot work with everyone without any hassles. So, all can trust the bots. The tenth findings represented that there is no chance of trust issue with any other teammates if they are working with chatbots. Last findings mentioned that there is no need for close supervision on chatbots. chatbots are considered as SOS for the human beings.

Thus, it is recommended to the software developers, IT practitioners that mostly people start working and they trust on the chatbots in INDIA. While developing the chatbots, we need to give utmost care about the privacy of personal information. Now a days, different portal, like Facebook, WhatsApp, and other social networking sites they are selling the information of people for their own benefits to other organizations which creates a sense of loss among some of the millennial while interacting with chatbots and they feel cheated and consequently not able to trust the chatbots.

So, it is suggested to secure the private information, bank details, security from theft, accurate and precise information should be delivered from the chatbot. In the coming era, chatbot technology will be considered as a landmark in all the sectors namely banking, medicine, e-commerce, tour and travel, restaurants or hotel chains etc. Consequently, increased customer attraction, engagement and loyalty which helps in increasing the effectiveness of all the sectors.

Limitations

The present study has several limitations which deserve to be mentioned. The prime limitation is small sample size of 350 millennial from a particular region namely Delhi/NCR. Only one generation cohort has been taken for the analysis in the study, other generational cohorts like generation X, generation Z can also be analyzed. Study can be enhanced through comparative analysis between the generation cohorts. Lastly, data was collected only through online survey; thus, a Quantitative Approach was used for collection of data. A more detailed analysis can be done with the help of Qualitative Approach of data collection.

References

Agarwal, S., Jindal, A., Garg, P., & Rastogi, R. (2017). The influence of quality of work life on trust: Empirical insights from a SEM application. *International Journal of Indian Culture and Business Management*, *15*(4), 506–525.

Agarwal, S., & Linh, N. T. D. (2021). *A student well-being through chatbot in higher education in higher education. Further advances in internet of things in biomedical and cyber physical systems*. Cham: Springer, ISBN 978–3–030–57834-3.

Ameen, N., Tarhini, A., Reppel, A., & Anand, A. (2020). Customer experiences in the age of artificial intelligence. *Computers in Human Behavior*, *114*, 106548.

Braganza, A., Chen, W., Canhoto, A., & Sap, S. (2020). Productive employment and decent work: The impact of AI adoption on psychological contracts, job engagement and employee trust. *Journal of Business Research*. (In press).

Carter, S. M., Rogers, W., Win, K. T., Frazer, H., Richards, B., & Houssami, N. (2020). The ethical, legal and social implications of using artificial intelligence systems in breast cancer care. *The Breast*, *49*, 25–32.

Gille, F., Jobin, A., & Ienca, M. (2020). What we talk about when we talk about trust: theory of trust for AI in healthcare. *Intelligence-Based Medicine*, *1-2*, 100001.

Lewis, J. D., & Weigert, A. (1985). Trust as a social reality. *Social Forces*, *63*(4), 967–985.

Marsh, S., Atele-Williams, T., Basu, A., Dwyer, N., Lewis, P. R., Miller-Bakewell, H., & Pitt, J. (2020). Thinking about trust: People, process, and place. *Patterns*, *1*(3), 100039.

McAllister, D. J. (1995). Affect- and cognition-based trust as foundations for interpersonal cooperation in organizations. *Academy of Management Journal*, *38*(1), 24–59.

Sharan, N. N., & Romano, D. M. (2020). The effects of personality and locus of control on trust in humans versus artificial intelligence. *Heliyon*, *6*(8), e04572.

Xu, Y., Shieh, C. H., van Esch, P., & Ling, I. L. (2020). AI customer service: Task complexity, problem-solving ability, and usage intention. *Australasian Marketing Journal (AMJ)*.

Further reading

Bunker, D. (2020). Who do you trust? The digital destruction of shared situational awareness and the COVID-19 infodemic. *International Journal of Information Management*, *55*, 102201.

Dhaggara, D., Goswami, M., & Kumar, G. (2020). Impact of trust and privacy concerns on technology acceptance in healthcare: An Indian perspective. *International Journal of Medical Informatics*, *141*, 104164.

Golembiewski, R. T., & Mcconkie, M. (1975). The centrality of interpersonal trust in group processes. In C. L. Cooper (Ed.), *Theories of group process*. New York: John Wiley and Sons.

Libai, B., Bart, Y., Gensler, S., Hofacker, C. F., Kaplan, A., Kötterheinrich, K., & Kroll, E. B. (2020). Brave new world? On AI and the management of customer relationships. *Journal of Interactive Marketing*, *51*, 41–56.

Paschen, J. (2020). *Creating market knowledge from big data: Artificial intelligence and human resources* (Doctoral dissertation, KTH Royal Institute of Technology).

Pillai, R., Sivathanu, B., & Dwivedi, Y. K. (2020). Shopping intention at AI-powered automated retail stores (AIPARS). *Journal of Retailing and Consumer Services*, *57*, 102207.

Prentice, C., & Nguyen, M. (2020). Engaging and retaining customers with AI and employee service. *Journal of Retailing and Consumer Services*, *56*, 102186.

Raats, K., Fors, V., & Pink, S. (2020). Trusting autonomous vehicles: An interdisciplinary approach. *Transportation Research Interdisciplinary Perspectives*, *7*, 100201.

Seeber, I., Bittner, E., Briggs, R. O., de Vreede, T., De Vreede, G. J., Elkins, A., & Schwabe, G. (2020). Machines as teammates: A research agenda on AI in team collaboration. *Information & Management*, *57*(2), 103174.

Chapter 7

Cognitive computing in autonomous vehicles

Atharva Sandeep Vidwans
Pune, India

Introduction

Cognitive computing is a combination of cognitive and computer science. It represents self-learning systems without human interaction based on human cognition. It is a modern computing technique that works based on neural networks (NN). Being the building block that is based on human cognition, NN is called the brain of Cognitive Computing. It mimics the functioning of the central nervous system in human beings. It generates an output depending on the input given to the NN, which is a function of the input data. The detailed working of NN is explained later in the chapter. In the past decade, Computing Hardware is advancing with the developments in cognitive computing algorithms. Thus, due to a large amount of available data to train the models, robust hardware and efficient algorithms have facilitated the application of complex algorithms for problems with ease. They have led developments in many fields, from medical and robotics to internet security.

Cognitive computing has led to the development of a new field. Unlike previous computing techniques, it is not about solving complex algorithms or doing fast calculations but finding new relations in the data. That is finding patterns in the data, the quality which was lacking in previous techniques.

So, what is cognitive computing?

It is the use of computerized models to simulate the process similar to human thoughts in complex situations in which decision-making is crucial, or tasks which require human-level expertise. Thus, cognitive computing can be defined as a problem-solving approach that uses hardware or software to approximate the form or function of the natural cognitive process.

Cognitive Computing for Human-Robot Interaction. DOI: https://doi.org/10.1016/B978-0-323-85769-7.00008-2

Natural cognition

Learning is what a system must do to improve its performance of the system, based on experience rather than hard programming. Constant learning is a critical defining feature in cognitive computing.

How the system processes or acquires the natural stimuli from the environment is called perception. This data acquired by the perception is used for the learning task.

Cognitive computing consists of two fundamental approaches:

1. Neuromorphic architecture approach: It is an approach in which there is specialized hardware that simulates neural activity that is called neuromorphic components or systems. The basic idea is to create neural network models after biological systems or components, such as neurons or synapses. They consist of hardware with a Neuro synaptic chip system that is currently developed by IBM containing upto 10 billion neurons and a hundred trillion synapses that consume around 1 kW of power called system of neuromorphic adaptive plastic scalable electronics (SyNAPSE) board.
2. Software centric approach: this method emulates neural function in software in on conventional von Neumann hardware architecture.

Many times, the term "Cognitive Computing" is used interchangeably with artificial intelligence (AI) that is the umbrella term for all the technologies related to decision-making that rely on data without direct human intervention. Several AI techniques are used to build cognitive models that mimic thought processes that include but are not limited to machine learning, deep learning, NN, neuro linguist programming (NLP).

Introduction to autonomous driving

Taxonomy for autonomous vehicle

Due to the recent advancements in cognitive computing techniques, they are used for a wide range of applications. One such prominent application is autonomous driving. By drawing the understanding from biological cognition and behavior generation in our brain, human drivers, and autonomous vehicles (AVs) coexist better. Cognitive computing in autonomous driving is based on the fact that neither the humans nor the AVs are immune to failure. Rather, it is been observed that humans and self-driving are prone to fail at different things. Thus can be used together no nullify the effects, creating safe and secure travel. Neither the machine nor the human is the best driver, but the combination of both. Therefore, fusing the strength from both sides and removing the lacunas is a viable option. It is quite natural to train the cognitive model that can estimate the risk and possibly avoid one (Nützel, 2018).

SAE Level	Name	Narrative definition	
Human driver monitors the driving environment			
0	No automation	The full-time performance by the human driver of all aspects of the dynamic driving task, even when "enhanced by warning or intervention systems"	
1	Driver assistance	The driving mode-specific execution by a driver assistance system of "either steering or acceleration/deceleration"	Using information about the driving environment and with the expectation that the human driver performs all remaining aspects of the dynamic driving task
2	Partial automation	The driving mode-specific execution by one or more driver assistance systems of *both steering and acceleration/deceleration*	
Automated driving system monitors the driving environment			
3	Conditional automation	The driving mode-specific performance by an automated driving system of all aspects of the dynamic driving task	With the expectation that the *human driver will respond appropriately to a request to intervene*
4	High automation		*Even if a human driver does not respond appropriately to a request to intervene*
5	Full automation		*Under all roadway and environmental conditions* that can be managed by a human driver

FIGURE 7.1 SAE (J3016) automation levels taxonomy.

Society of automotive engineers (SAE) has proposed different classes in AVs depending on their level of autonomy. They range from level 0, meaning no autonomy or completely manual, to level 5 which means complete and unrestricted vehicle autonomy. All six classes for autonomous autonomy are illustrated in the table below (Fig. 7.1).

Before studying the taxonomy of AVs it is necessary to go through some of the technical terms which will help to define the classes in taxonomy.

1. Driving task: driving task mainly consists of three subcategories (Nützel, 2018), namely:
   a. Perceiving the environment–This is mainly concerned with perceiving the environment in which we are driving, which includes tracking the vehicle motion and identifying the various elements around the vehicle like the road surface or the road signs and moving objects.
   b. Motion planning—motion planning task consists of how the vehicle is going to reach from point A to point B.
   c. Controlling the vehicle—After the path is defined and the waypoints are fixed then the vehicle needs to be maneuvered to follow that specific path to reach our final destination.
2. Operation design domain (ODD): It constitutes the operating conditions wherein which the vehicle is designed to perform it may include, environmental condition, road condition, weather conditions, or even time of the day. Thus for a self-driving car, the ODD needs to be defined perfectly as it is crucial for the safety of the system. ODD needs to be planned out carefully in advance based on the fact that what the environmental conditions and other parameters could be during the drive.

AVs are classified based on the amount of degree of automation and the degree of human action and human attention required. For the first three classes that is 0, 1, and 2 the human driver has to monitor the environment and then take suitable decisions. The control is with the human driver and the vehicle can just assist the driver. But for the next three classes that is for 3, 4, and 5 the vehicle takes care of the perception task for the human that is the AV monitors the driving environment. As seen from the above figure, there are six levels for an AV (Nützel, 2018).

1. Level 0: For level 0 there is no vehicle autonomy and the vehicle is completely controlled by a human. Vehicles falling into this category shows absolutely no intelligence. They are completely controlled by the human driver.
2. Level 1: For level 1, the vehicle can be programmed such that it shows either longitudinal control like acceleration and braking or, lateral control like steering. Thus the vehicle can be programmed for either the longitudinal or lateral control but not both. Still, the complete control of the vehicle is with the human. The Level 1 control is for just assisting the driver. This class is called as "Driver Assistance". For example, adaptive cruise control in which the system can keep up the speed but the driver has to steer, lane keeping assistance which can help to stay in the lane or warn in case of a drift.
3. Level 2: The second level automation is such that the AV can have more than one vehicle assistance. As against in level 1, in which the vehicle can only have either the lateral or the longitudinal control, in level 2 the AV can perform both the vehicle assistance, lateral and longitudinal control. So it can be said that the vehicle is driving itself partially based on environmental parameters. But still, the human needs to be attentive and need to give feedback to the vehicle. Thus the crucial decision-making and environment monitoring are still up to the human driver. This level is called as "Partial Automation". For example, GM super cruise, Nissan's pro pilot assist.
4. Level 3: level 3 automation is called as "Conditional Automation". In level 3 automation the system can perform object and event detection and response (OEDR) to a certain degree in addition to the level 2 control tasks. Even though the system can perform OEDR the control of the vehicle needs to be taken by the human driver in case of emergency or failure. The difference between level 2 and level 3 is that the attention of the driver is reduced and the vehicle can alert the driver in case of emergency or critical decision-making. But this creates controversy as it is not always possible for the vehicle to know that which is the critical situation. For example Audi A8 Sedan.
5. Level 4: In level 4 autonomy the system is such that it can execute a low-risk task in case the driver does not intervene during an emergency

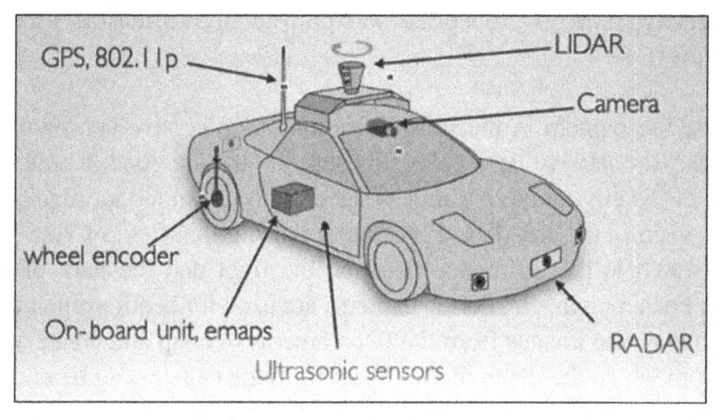

FIGURE 7.2 Components of autonomous vehicle.

situation, thus mitigating the damage. This type of control, thus, called as "High Automation". Level 4 vehicles can handle emergencies on their own but up to some extent and still requires driver intervention for some situations. This type of automation is such that it permits the driver to check the mobile phones or even watch movies. But still, the Level 4 has some lacunas, as it has a limited ODD. For example, WAYMO has achieved such a type of autonomy for limited ODD without the need for a human driver.

6. Level 5: level 5 autonomy called as "Full Automation" is a class in which the vehicle supports unlimited ODD and can operate in any conditions with full automation without human intervention. Such level 5 AVs are still not developed and require extremely high accuracy of hardware and software. Currently, every industry is jostling for this type of autonomy. And this is where cognitive computing can help to reach level 5 autonomy (Fig. 7.2).

Components of autonomous vehicle

Sensors

Even the best quality perception algorithms are restricted by their sensor data. A sensor is any device that measures a property of an environment or measures any change to that property over time. Sensors are mainly of two categories depending on what type information they record, they are as follows:

1. Exteroceptive: where "Extero" means surrounding or outside. Meaning they measure the surrounding property. For example, a camera, LIDAR, etc.
2. Proprioceptive: proprioceptive sensors are the ones that measure the property of own that is this case property of the vehicle. For example, IMU, Wheel Encoder, etc.

Different types of exteroceptive and proprioceptive sensors used in AVs are as follows:

1. Camera: the camera is the most common exteroceptive sensor used in an AV. They are passive light-collecting sensors that are good in collecting rich details of the environment. Camera selection is an important requirement for a good vision task. It is based on, camera resolution, field of view, dynamic range which is the difference between the light and the dark pixel in the Image, Focal length, etc. Stereo cameras are used for depth estimation. In stereo cameras, the images from the two cameras overlap and create a disparity map from which the depth of the object from the camera can be known.
2. LIDAR: light detection and Ranging is an important type of sensor as high accuracy vision task cannot be achieved using just the camera. In LIDAR the sensor shoots light beams in all directions and records the time taken by the light beam to return to the sensor after reflecting from the object. Thus by measuring the time we can know the distance traveled by the light as the velocity of light is known in almost all cases and varies slightly. Apparatus includes spinning element shooting light beams and gives the output in the form of a 3D point cloud. Unlike a camera, it is not affected by ambient light. LIDAR is selected based on the following factors, the number of beams in which the sensor shoots, points per second collection. The rotation rate is also an important parameter as higher the rate faster are the points updated, and also the field of view. Displayed below is a typical Image of point cloud generated using LIDAR.
3. RADAR: radio detection and ranging is the most common sensor and been there longer than the LIDAR sensor and is more robust. As they consist of radio waves they are not affected by weather conditions like precipitation, snowfall, etc. Thus making them more reliable in adverse weather. Radars are selected based on detection range, the field of view, position, and speed accuracy.
4. Ultrasonic sensors: ultrasonic sensors are also called as sound navigation and ranging (SONAR) which are low-cost sensors compared to others. They are mainly used for detecting closer objects, thus they are used predominantly during parking etc. Like RADAR they are unaffected by adverse weather conditions. SONAR sensors are selected by the parameters such as, range, the field of view, and cost.
5. Global navigation satellite system (GNSS) and inertial measurement unit (IMU): GNSS is a type of proprioceptive sensor which measures the position of the vehicle. Global positioning system is one of the GNSS which is widely used in AVs. They are used to measure the velocity and heading direction. The IMU is used to measure the angular velocity and the acceleration of the vehicle. By using these sensors together we can get a complete orientation of the vehicle.
6. Wheel encoder: Wheel encoder or wheel odometry sensor is used to track the heading direction rate and the rate of rotation of wheels (Fig. 7.3).

FIGURE 7.3 3D point cloud.

Sensor fusion: no sensor used individually provides enough information to achieve higher accuracy for the self-driving vehicle. Thus the fusion of sensor data and GPS is used that enables in determining the exact position and other objects around the self-driving vehicle with an accuracy of 10 cm. Sensor fusion can overcome certain drawbacks of the individual sensor. The most common example is the fusion of RADAR and camera. The camera can give greater accuracy when the weather is clean and clear. But during high fog or precipitation conditions where the visibility is hardly in few meters, the camera does not perform well due to unclear vision and obstructions. Thus in this case a RADAR can be used which enables the AV to see through the environment with enhanced visibility.

Actuators

Actuators of an AV are similar to what muscles are for the human body. They are responsible for moving and controlling the self-driving vehicle. Different actuation methods used for steering, braking, acceleration use an electric motor with gear reduction, or mechanical linkages or hydraulics to maneuver the vehicle.

Cognitive computing algorithms in the development of AVs

Types of learning algorithms

This part of the chapter will present general algorithms along with the cognitive algorithms used in AVs for different tasks. Cognitive computing is based on multi-core programming of CPUs, GPUs, TPUs, and Neuromorphic chips along with software development environment with intrinsic support for parallel computing and different software libraries. Before designing any algorithm it is necessary to consider certain events, a few of which are, landscape, different traffic rules in different countries, such as left-hand side and right-hand side driving, road signs, etc. Each of these parameters can affect the algorithm widely. However, as stated above the accuracy and development of any algorithm are restricted by the type of sensor used, such as camera, LIDAR, RADAR, etc. and the collected data. This data is then feed as input to the AI algorithms (Gudivada, Irfan, Fathi, & Rao, 2016).

Before diving deep into AI algorithms used in self-driving vehicles, it is necessary to know their classification. AI algorithms are classified as supervised, unsupervised, and reinforcement learning algorithms. In supervised learning, there is a set of associations between learning data and the labels, that is along with the data, the labels are fed to the AI model. The new data is repeatedly fed to the model with labels and based on that the weights are modified, to get a perfect model for prediction. For example, this process is similar to a child learning to identify new objects from the surrounding. The training dataset consists of two sets of information, first is the input data x, and expected output y. Thus (x, y) forms one example. Training dataset consists of n such pairs: $\{(x_1, y_1), (x_2, y_2), (x_3, y_3)\ldots (x_n, y_n)\}$. Some of the examples of supervised learning algorithms are decision tree, NN, Bayesian classification, etc. (Gudivada et al., 2016). In unsupervised learning algorithms, the labels are not provided thus the model draws the inference from just the input data and groups the data into different clusters based on their characteristic pattern and hidden structures. K-means clustering, genetic algorithms are different unsupervised learning algorithms. Reinforcement learning is a type of learning which is inspired from day to day learning processes of a human being. Usually, in the unknown situation or when the outcomes are not known we the humans tend to take a certain action, and then based on the output of those actions, we decide whether to repeat that action or not. Thus similarly, in reinforcement learning algorithms the theoretical system agent takes some actions and based on those actions it gets some reward. For each correct action, it is awarded some positive reward. Thus the agent performs further actions such that the notion of reward is maximized in the environment. There are basically two types of reinforcement learning based on the fact that whether the agent is rewarded or penalized called positive reinforcement and negative reinforcement learning. For cognitive analysis, unsupervised learning algorithms have a certain edge over supervised algorithms in some scenarios such as, when we do not know the pattern in the data a priori. In those cases, unsupervised algorithms are used to generate the training data which then can be used to train supervised learning algorithms. Ensemble learning is a technique in which two or more algorithms are optimally combined to give better results than using individual algorithms (Gudivada et al., 2016).

Thus any machine learning algorithm can be classified into one of the above categories. These algorithms are used in AV to perform different tasks. Let us dive deep into the cognitive computing algorithms used AVs.

Cognitive computing algorithms used in autonomous vehicle

In this section different machine learning algorithms are discussed and compared. Before directly diving into different algorithms let us understand a general view of input, output, and the AI model.

The input consists of multiple rows of data which are called features. Each example represents a data point. The number of rows of data is guided by the number of a feature used for training the model. At times there may be a large number of features to train. It is not possible every time to train the model for all the features as the process is quite time consuming, it is necessary to select the most important features. Finding good features is an art necessary to increase the accuracy of the algorithm and to reduce the time consumption. The data is denoted by X which has n × m dimensions, given by $X = \{x_1, x_2, x_3 \ldots x_n\}$. Where x is a feature vector defined as, $x_n = \{x_{1n}, x_{2n}, x_{3n} \ldots x_{mn}\}$, where m is the number of features. For supervised algorithms, the input consists of labels given by Y which is an $n \times 1$ dimension vector, where Y is defined as $\{y_1, y_2, y_3 \ldots y_n\}$, where n is a positive integer equal to the number of training examples. The most challenging part of the machine learning algorithms is to select the model for mapping feature vector to target values. There are a lot of models available, thus selecting a proper model for the required task is critical. Parameters that define the model architecture are called "Hyperparameters" and the process to select an ideal structure of the model is called "Hyperparameter tuning". It is an important step which is used as a fine tuning technique to increase the accuracy of the model. Most of the times it is a trial and error process to select proper set of Hyperparameters. During training of the model, the weight is tweaked such that the prediction matches the expected outcome. These weights are given by $\theta = \{\theta_0, \theta_1, \theta_2 \ldots \theta_n\}$.

The autonomous driving pipeline is decomposed into four different modules which can use AI or deep learning approach or traditional approach independently (Grigorescu et al., 2020):

1. Perception and localization
2. High-level planning
3. Low-level planning
4. Decision-making (Fig. 7.4)

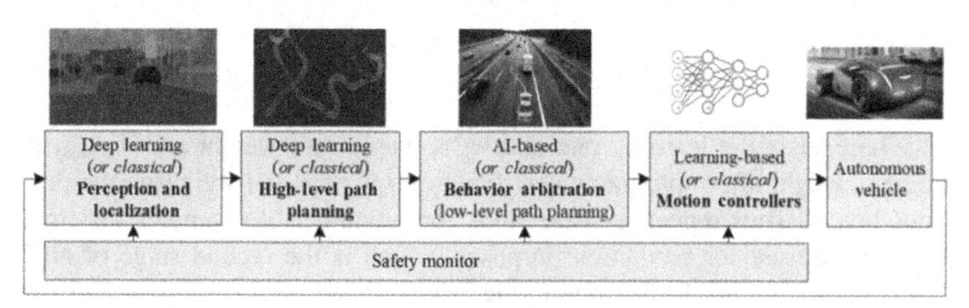

FIGURE 7.4 Architecture of deep learning based self-driving vehicle.

Let us discuss the tasks and the machine learning algorithms used in detail.

Machine learning

Neural networks

Artificial neural networks or neural nets are a type of supervised machine learning algorithm. It is biologically inspired by the neurons in the human brain. The main component of a NN is called a perceptron. It is predominantly used for the task of classification. Overall this algorithm is divided into different phases. The starting phase is known as the training phase. In this phase, the NN is fed different inputs and their corresponding labels. The learning phase itself is subdivided into two stages called feedforward and backpropagation. During the feed-forward step, the different weights are multiplied by their neuron value. Then passed through a perceptron which classifies the output into given classes depending on the activation function like the ReLU function, and then the prediction is calculated as output. After this feed-forward step the there is a backpropagation step. The cost function is shown below compares the true output value with the predicted value, given by the equation (Nützel, 2018; Gudivada et al., 2016) (Fig. 7.5). Where, $J(w)$ is the cost function which uses mean squared error to calculate the difference between the actual and predicted value. Here, n is the number of training examples. This cost function is optimized using an optimizer like Adams or gradient descent. The cost function has a bell shaped curve, thus consists of only one global minimum. In some cases, when the cost function is complex it can have multiple local minima. Thus it can happen that sometimes the optimizer can get stuck at a certain local minimum (Fig. 7.6).

The backpropagation updates the weights and biases of the NN. The Hyperparameter known as the learning rate is adjusted to control the weight change during one single iteration known as "Update Strength". Increasing the learning rate may cause higher oscillations in getting to the optimum point but reduces the time of training. Whereas, lowering the rate causes increased time consumption for the learning. Thus the value of the learning rate needs to be adjusted to suit our purpose by trial and error. The updated weights are given by the equation (Nützel, 2018),

$$\mathrm{w}_{(\mathrm{j,new})} = - \propto \frac{\partial \mathrm{J}}{\partial \mathrm{w}_{\mathrm{j}}} + \mathrm{w}_{\mathrm{j}} \tag{7.2}$$

Where, α is the learning rate and w_j is the current weight and $w_{(j,\mathrm{new})}$ is the new weight after optimizing cost function j w.r.t initial weight in the previous layers. Thus repeating these steps for n number of examples the final weights are used for prediction purposes, which is the second stage of NN, that is testing phase. In this phase, the NN is tested on unseen data and the final results are analyzed.

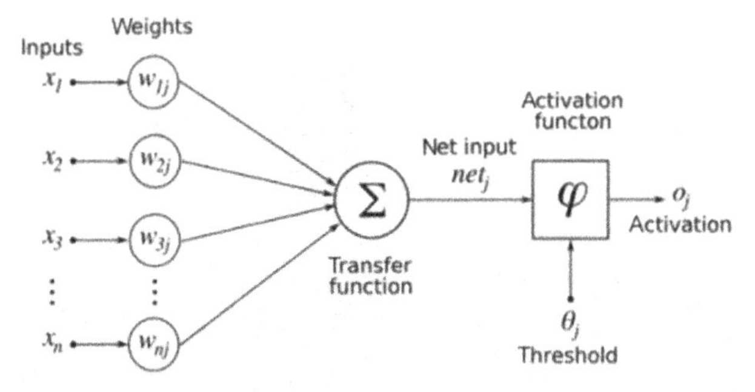

FIGURE 7.5 Diagram of artificial neural network.

$$J(w) = \left(\frac{1}{n}\right) \sum_{i}^{n} \left(\text{true}^{(i)} - \text{prediction}^{(i)}\right)^2 \tag{7.1}$$

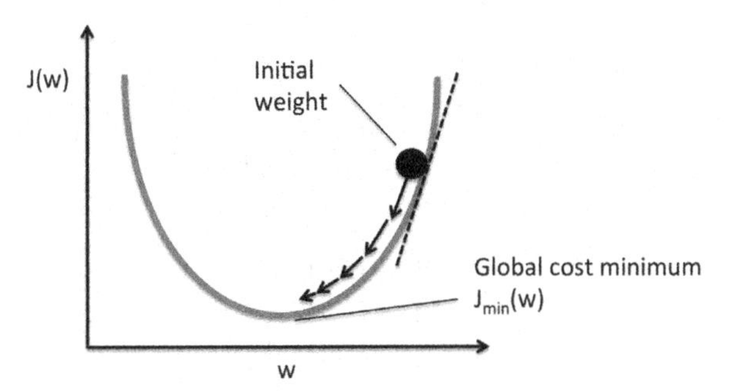

FIGURE 7.6 Cost function of a neural network having single global optimum.

Deep convolutional neural network

Traditionally, detection of objects was done in two phases, namely, feature detection which was a handcrafted approach followed by classification using ML algorithm such as SVM. But by the advent and development of NN, this traditional approach was replaced by a much accurate and automated machine approach called deep convolutional neural network (Deep CNN) (Grigorescu et al., 2020).

In the deep learning approach, the two steps of feature extraction and classification in the traditional methods are combined to form a single approach. CNN computes the feature map using the corresponding kernel by iterating the kernel over the complete image. Thus the output is the image with features whose shape and size depends on the kernel. A kernel is nothing but a square matrix that is convoluted over the complete image (Grigorescu et al., 2020) (Figs. 7.7 and 7.8).

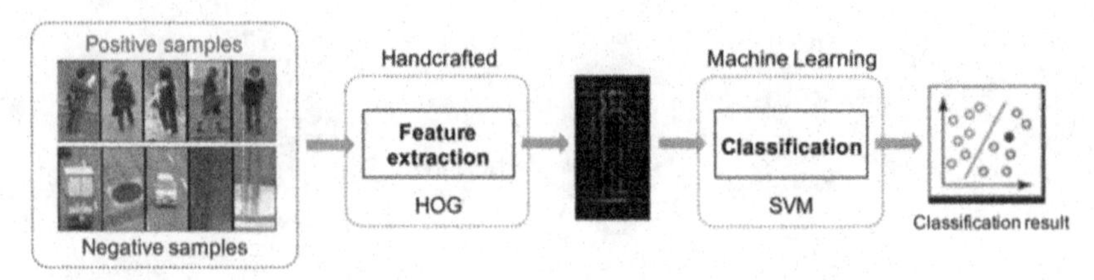

FIGURE 7.7 Traditional machine learning approach.

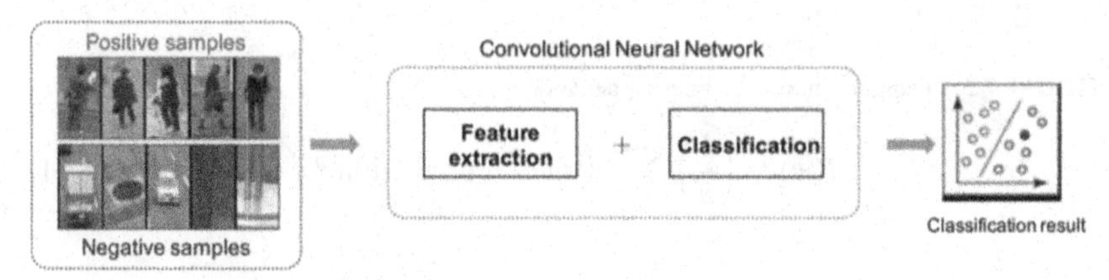

FIGURE 7.8 Modified approach using cognitive computing.

A basic example of a kernel is given by:

$$G_x = \begin{bmatrix} -1 & 0 & 1 \\ -2 & 0 & 2 \\ -1 & 0 & 1 \end{bmatrix} \tag{7.3}$$

$$G_y = \begin{bmatrix} -1 & 0 & 1 \\ -2 & 0 & 2 \\ -1 & 0 & 1 \end{bmatrix} \tag{7.4}$$

The above kernels shown are basic 3×3 kernel which detect the edges in x and y-direction. G_x is called a gradient component in x-direction and G_y is called the gradient component in the y-direction. As shown above G_x detects the vertical edge whereas, G_y detects edges in y-direction.

The result of using the above kernel on an image is given below (Fig. 7.9).

This feature extracted image is feed to the deep neural network for prediction purposes. The total CNN architecture looks like this (Fig. 7.10):

In the first layer, different kernels are applied which extract different features from the image, for example, curves, horizontal lines, vertical lines, circles, etc. Then these features are fed to the deep NN. During testing of the algorithm when the image contains those certain features, the specific neuron with that feature fires up. The detection accuracy of the deep CNN is much more than traditional handcrafted feature detection technique. Deep CNN, performs better than traditional techniques. Apart from this, deep CNN is flexible. It can be applied to various tasks just by changing the structure of the model. This is one of the advantages that cognitive computing provides over the traditional approach (Fujiyoshi, Hirakawa, Yamashita, 2019).

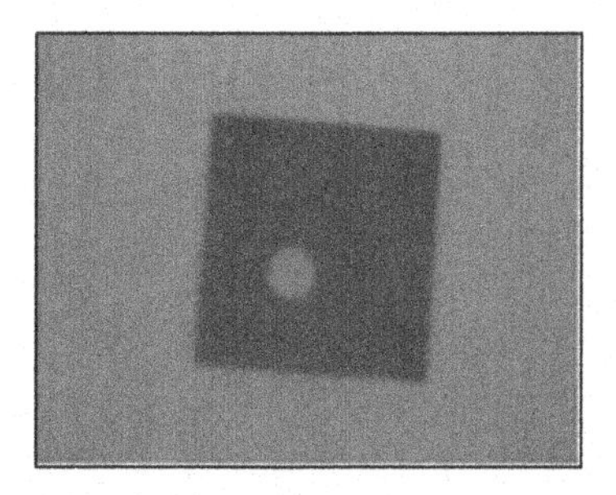

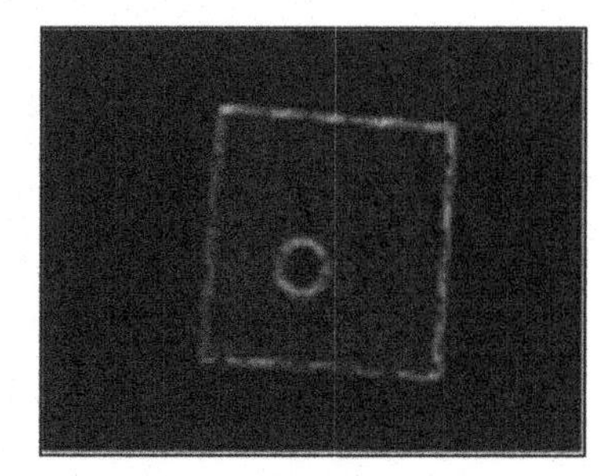

FIGURE 7.9 Before and after images of applying the kernel.

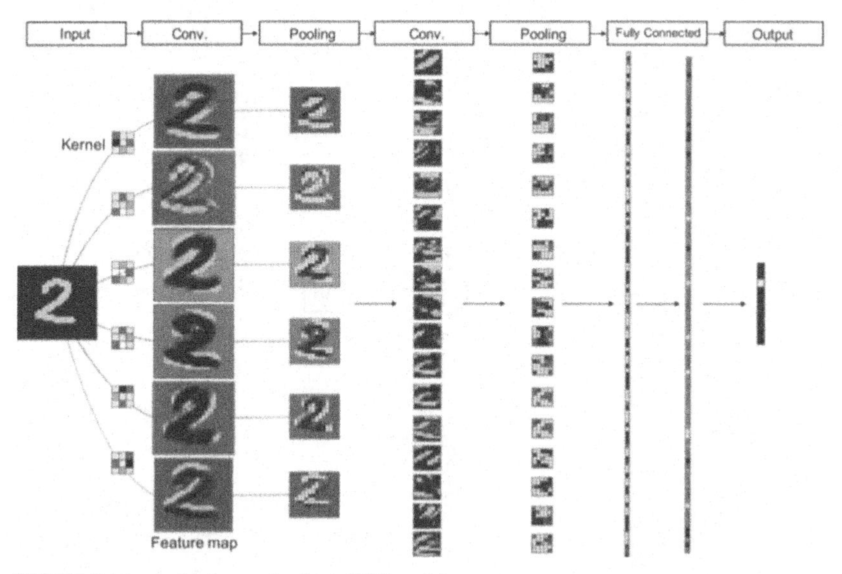

FIGURE 7.10 Architecture of a deep CNN model.

Recurrent neural network

Amongst all the types of NN, Recurrent neural networks (RNNs) are particularly good in the processing sequence of data, like music, audio, video, speech recognition, that is predicting the pattern in time series. In AV, they are mainly used to predict the motion of the object. Unlike other NN, RNNs have feedback in the hidden layer. Due to this, during the next iteration, the data from the previous layer is also included in the computation thus enabling the RNNs to consider data from not only that instance but also from the previous instances. Below is the diagram of a full RNN model (Grigorescu, Trasnea, Cocias, & Macesanu, 2020; Yao, 2018) (Figs. 7.11 and 7.12).

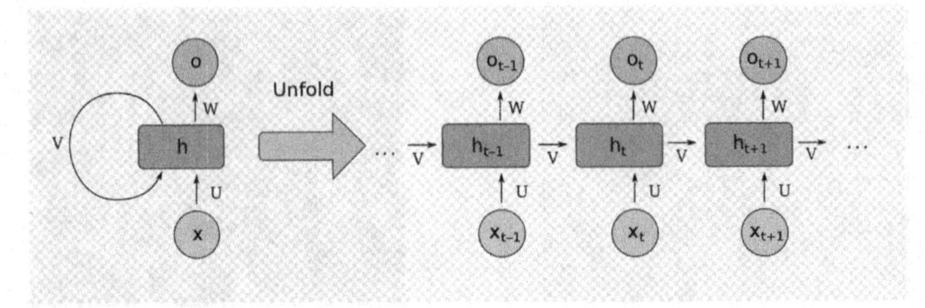

FIGURE 7.11 Folded and unfolded recurrent neural network.

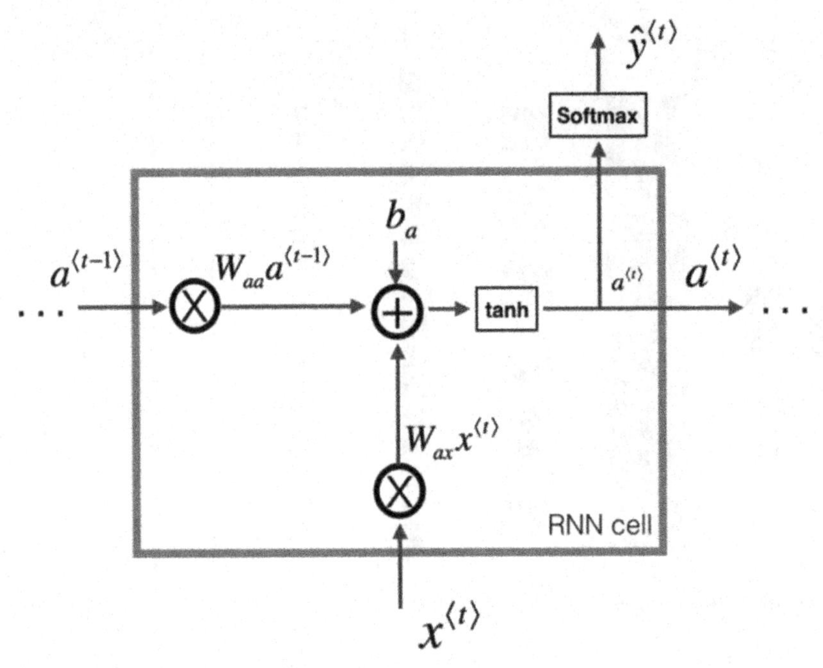

FIGURE 7.12 Recurrent neural network cell for a single time step.

In the above figure W_{ax} is the parameters that govern the connection from the x to the hidden layer. W_{aa} is the parameter that governs the horizontal connection. W_{ya} is the parameter that governs the output predictions. The value of W_{ax}, W_{aa}, and W_{ya} is the same in every time step. The value $a^{<0>}$ is considered as zero. The y and the a values for the current layer are given by the equation,

$$a<t> = g_1(W_{aa}a^{<t-1>} + W_{ax}x^{<t>} + b_a) \tag{7.5}$$

$$y<t> = g_2(W_{ya}a^{<t>} + b_y) \tag{7.6}$$

Where g_1 and g_2 are the activation functions. The most commonly used activation function tanh for g_1 and sigmoid for g_2.

By using the above equations y and a can be found out in the RNN model. Even though RNN uses the data from the previous time step it is prone to some major well-known problem. The main challenge in using RNNs is its vanishing gradient problem. As the number of time steps increases the data from the previous time steps goes on reducing to the point that during the backpropagation the change in the first layer is almost negligible. This problem can be countered by using long short term memory networks (LSTM). LSTM networks are a nonlinear function approximator for estimating the dependencies in the data. As opposed to the traditional RNN, LSTM solves the vanishing gradient problem using three gates as shown below [https://jmyao17.github.io/Machine_Learning/Sequence/RNN-1.html, Oct 2020 (accessed 4.10.2020)] (Fig. 7.13).

Forget gate: When the subject or the pattern in the video changes it is necessary to delete or forget the previously stored memory, this is where the forget gate comes into the picture. In LSTM forget gate let us do this:

$$\Gamma_f^{<t>} = \sigma(W_f\left[a^{<t-1>}, x^{<t>}\right] + b_f) \tag{7.7}$$

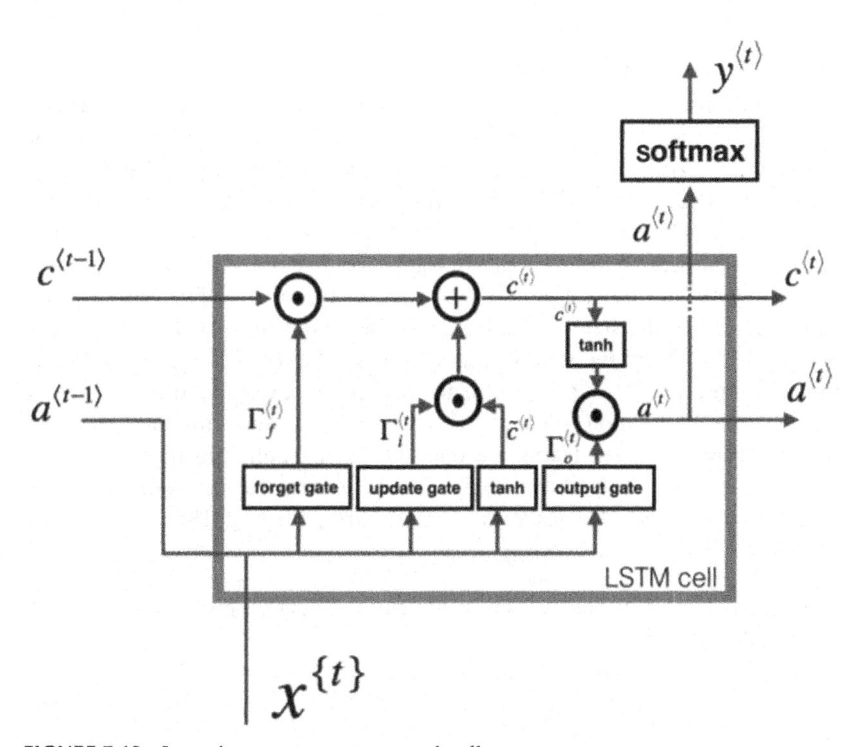

FIGURE 7.13 Long short term memory network cell.

Update gate: Called the input gate, as the name suggests, it updates the content which needs to be reflected in the new subject. To update the new subject it is necessary to create a new vector of numbers that can be added to the previous state. It is given by the equations as follows,

$$\Gamma_u^{<t>} = \tanh(W_u\left[a^{<t-1>}, x^{<t>}\right] + b_u) \tag{7.8}$$

$$\tilde{c}^{<t>} = \tanh(W_c\left[a^{<t-1>}, x^{<t>}\right] + b_c) \tag{7.9}$$

$$c^{<t>} = \Gamma_f^{<t>} * c^{<t-1>} + \Gamma_u^{<t>} * \tilde{c}^{<t>} \tag{7.10}$$

Output state: To decide which output to use the following two formulas can be applied in the model. The first equation decides what to output using a sigmoid function and the second equation multiplies that by the tanh of the previous state.

$$\Gamma_o^{<t>} = \sigma(W_o\left[a^{<t-1>}, x^{<t>}\right] + b_0) \tag{7.11}$$

$$a^{<t>} = \Gamma_o^{<t>} \times \tanh(c^{<t>}) \tag{7.12}$$

Another algorithm which is used along with LSTM is GRU. It achieves the same function as LSTM but uses only two gates instead of three.

Deep reinforcement learning

Deep reinforcement learning (DRL) have made a breakthrough in various decision-making tasks like defeating the professional human player in "DOTA", amazing victory in Atari games, deep mind victory in "Alpha Go". This section reviews the DRL in autonomous driving task, using partially observable markov decision process, which is an agent-based process. In the reinforcement learning algorithm, the agent which is the AV senses the environment with an observation $I^{<t>}$, based on the observation performs the action $a^{<t>}$ for the state $s^{<t>}$, based on the output for that state it receives reward $R^{<t+1>}$ and the state changes from $s^{<t>}$ to $s^{<t+1>}$ using a transition function $T^{s<t+1>}$ and the same process is executed till the system reaches the destination state $s_{\text{dest}}^{<t+k>}$. The agent aims to maximize the reward. Thus the agent learns the optimal driving policy in order to maximize the reward. The above logic can be expressed in model form as (Grigorescu, Trasnea, Cocias, & Macesanu, 2020; Bhatt, 2018),

$$M = (I, S, A, T, R, \Upsilon)$$

Where, I is the set of observation of the environment at any instance of time t denoted by, $I^{<t>}$.

S is the current state of the agent at time t denoted by, $s^{<t>}$

An is the action taken from the observation $I^{<t>}$ for state $s^{<t>}$ denoted by, $a^{<t>}$

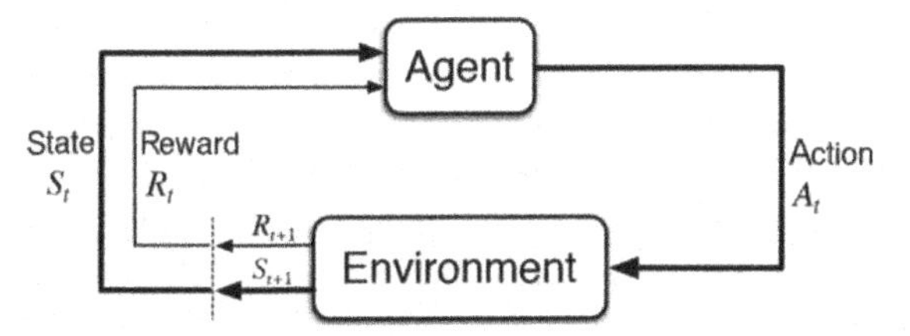

FIGURE 7.14 Block diagram of deep reinforcement learning.

R is the reward which the agent gets based on the action taken denoted by, $R^{<t+1>}$

Υ is the discount factor that takes into consideration the importance of future events to the current events (Fig. 7.14).

The agent selection of action is modeled as per the map called policy denoted by π.

$$\pi = A^{*}S \rightarrow [0, 1] \tag{7.13}$$

$$\pi(a, s) = P(a_t = a, s_t = s) \tag{7.14}$$

Which is state and action at a given time instance t. Thus the policy-map gives the probability of taking an action a in the state s. A state value function denoted by $V_\pi(s)$ follows a policy π starting from the initial state S_0. It measures how good it is to be in a state s following policy π. Thus $V_\pi(s)$ is given by,

$$V_\pi(s) = E\left[\sum_{t=0}^{\infty} \gamma^t \times r_t\right] \tag{7.15}$$

Where, $V_\pi(s)$ is the summation of all the rewards in the future multiplied by the discount factor which decides how good a state is. Thus the algorithm finds the optimum policy with maximum expected returns. The optimum action Value function is denoted by $Q \times (.,.)$ satisfies the Bellman optimality equation, which is a recursive formula given by,

$$Q^{\times}(s, a) = \sum_{s} T_{s,a}^{s'}\left(R_{s,a}^{s'} + \gamma \max Q^{*}(s', a')\right) \tag{7.16}$$

In autonomous driving applications, the observation space is mainly composed of sensor information which is made up of data from LIDAR, SONAR, camera, etc. Thus instead of the traditional approach, a non-linear Q* can be encoded in the deep learning model. Such a model is called DRL. In DRL the deep learning technique is combined with reinforcement learning to give ground-breaking results and solve the problems like never before. Although continuous control is possible with DRL, a discrete approach is used. The main problem is the training, as the agent has to learn through exploring the environment, usually through the collisions. The solution here is to use an imitation learning method called inverse reinforcement learning to learn from human driving demonstrations without the need to go to unsafe methods.

DRL suffers from a major problem known as the credit assignment problem. As the in a single episode even though all the actions are correct but due to the last few actions the agent fails, the complete reward for that episode reduces thus deducing the agent that all the actions in that single episode must be bad actions intern reducing the probability of taking those actions in the future. The solution to this is the agent should able to recognize what part of the episode is causing the less reward and eliminate that part instead of eliminating the whole episode.

End-to-end learning method

In most research in CNN and cognitive computing techniques in autonomous driving, the environment around the vehicle is recognized by using the camera or the LIDAR sensor, appropriate motion planning, and control the throttle positions. Thus they mainly use processes such as recognition of the object, segmentation, etc. But due to the progress in CNN, it is possible to directly control the valves directly from the Input Image. In this method, the network is trained using a dashboard camera or a similar method like a simulator to train the CNN model to control the vehicle under a real environment. End-to-end learning has one advantage as it does not require to completely understand its surrounding environment. The inputs of the camera are given to the CNN model and output steering angle directory and throttle angles i.e. using a driving simulator to train the CNN model and use the trained model to control the throttle and steering angles (Grigorescu et al. 2020; Fujiyoshi et al. 2019).

The structure comprises of five layers of convolutional layers along with the pooling layers and three layers of fully connected neurons. Shown below is the simple pipeline, combined pipeline with cognitive computing and combined with end-to-end approach (Figs. 7.15–7.17).

As can be seen from the above image that the end-to-end system replaces almost all the tasks in autonomous driving by the CNN model. Training of the end-to-end system is given in the following figure (Fig. 7.18).

Even with these techniques end-to-end has some drawbacks. One of the challenges in end-to-end learning is it requires a lot of driving data for training before going into the actual environment. For smaller datasets usually, the other methods are employed instead of end-to-end learning. As for smaller datasets, the accuracy of end-to-end drops drastically and thus are not

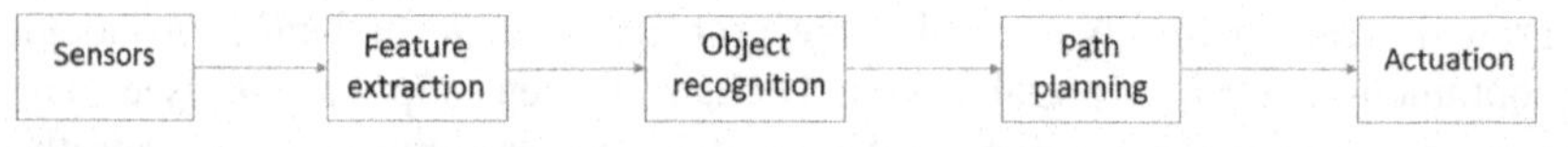

FIGURE 7.15 Simple pipeline.

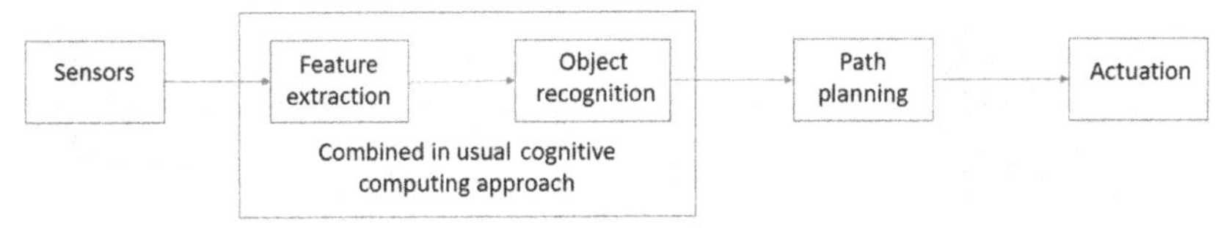

FIGURE 7.16 Combined with cognitive computing.

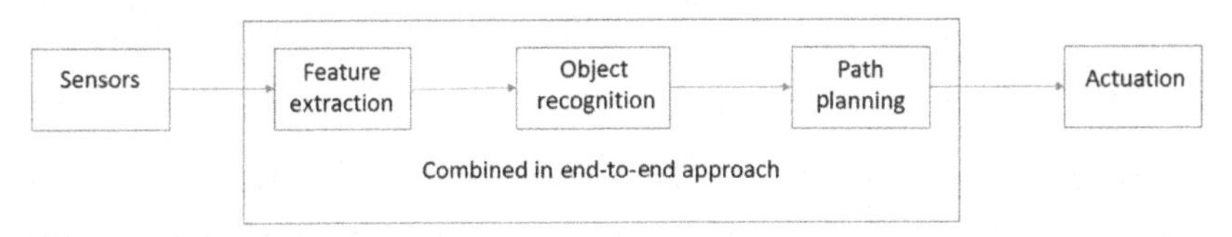

FIGURE 7.17 End-to-end learning.

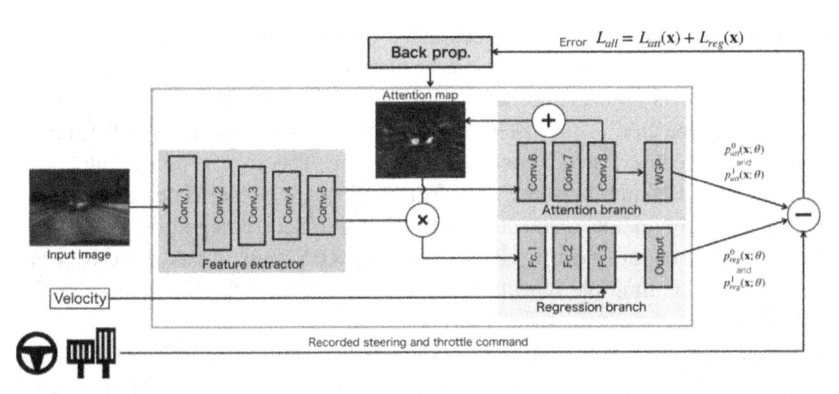

FIGURE 7.18 Regression type attention branch network (end-to-end).

preferred for smaller or medium datasets. Another disadvantage for end-to-end is that it eliminates the handcrafted ML models. When data is small the model does not get insight from only the data and hand designing a component in the model would be a way to inject manual knowledge. As a learning algorithm has two main sources of knowledge i.e. data and second is hand-design feature or components. Thus when we have large data there is less need to hand-design but when the data is small the only way to inject knowledge is by carefully hand designing the components (Fujiyoshi et al. 2019).

Deep driving

Perception and localization

CNN in object detection

In conventional machine learning approach uses a raster scan two-class classifier for detecting the object, which is based on the principle that the aspect

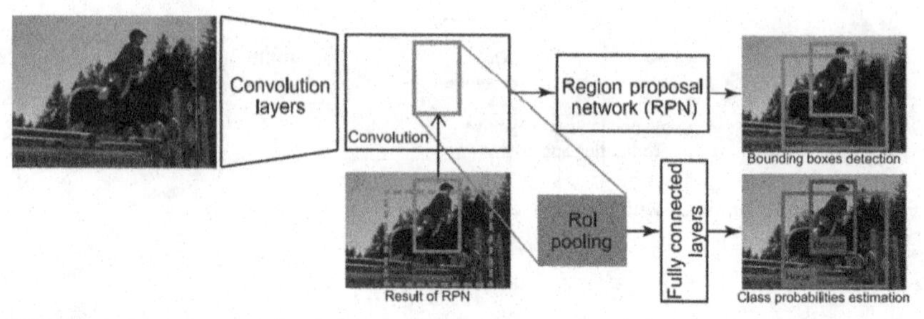

FIGURE 7.19 R-CNN structure.

ratio of the object which is to be detected does not changes, it will detect only certain categories learned as a positive sample. In some cases when there is no object in the window the algorithm searches for an object to classification. On the contrary, in the object through CNN objects with different aspects are detected, thus multiple class detection of object is possible by using region proposal network (RPN). A much faster class of CNN which is R-CNN simultaneously detects objects and detects the object class of those objects. In RPN an object is detected using raster scanning of K different windows of different sizes in the focused area known as an anchor. The region of the anchor is input to the RPN and the score of object likeliness is output. The region of the anchor is input to the connected network for object recognition when it is determined to be an object by RPN. Thus the RPN method has made it possible to recognize multiple objects in the image. Below is the block diagram for R-CNN architecture (Fujiyoshi et al.) (Fig. 7.19).

Another method was developed known as you only look once (YOLO) is a technique that detects multiple objects by giving the whole input image to Convolutional Neural Network without raster scanning. In this, the image is divided locally in 7×7 grid. Firstly a feature map is generated through CNN and then the feature map directly feed to the fully connected layers. The output obtained displays the score of the object category at each smaller grid of the two object rectangles. YOLO has an advantage as the object detection can be done in real-time. Below shows the process of object detection using YOLO (Fig. 7.20).

CNN in semantic segmentation

Semantic segmentation is defined as linking each pixel in an image to the class label. Semantic segmentation is considered a difficult task in the computer vision. But it is seen that the NN has shown much more promising results like increased accuracy, higher performance compared to traditional machine learning algorithms. A fully convolutional network (FCN) is a technique that enables end-to-end learning and can obtain segmentation using only CNN.

A typical structure of FCN is given below (Fujiyoshi et al.) (Fig. 7.21).

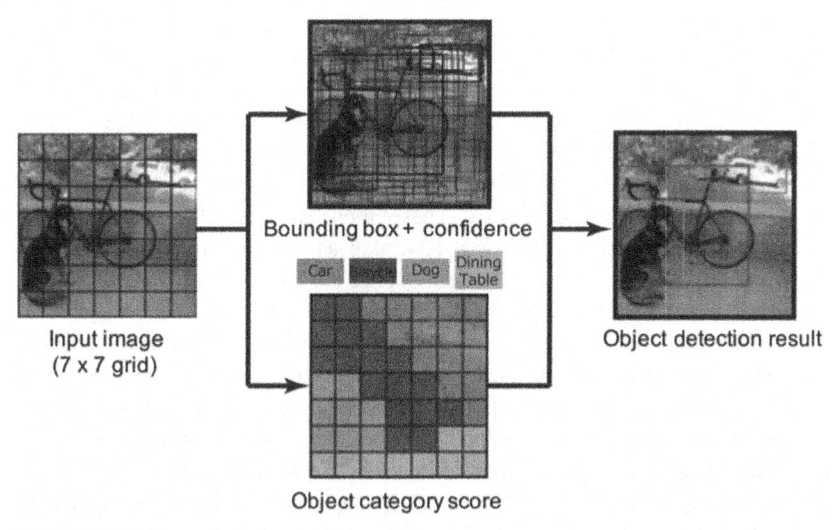

FIGURE 7.20 You only look once structure.

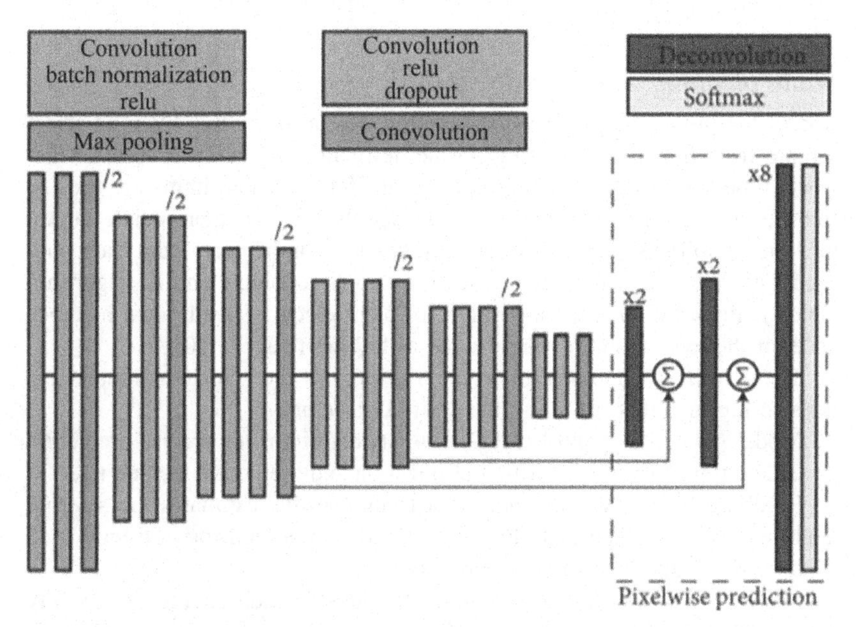

FIGURE 7.21 Fully connected network structure.

Localization

Localization algorithms are used to locate the position of the AVs as it navigates. The traditional approach to solving the localization problem is by using GPS. But GPS is not quite accurate, giving an accuracy of 2–10 m that is we cannot rely on GPS for a lane change or such tasks that require higher accuracy. Visualize localization known as visual odometry (VO) uses deep learning to localize the AV giving and accuracy of 2–10 cm. This is done by matching key-points landmarks in consecutive video frames. The key-points are input to the n-point mapping algorithm which detects the pose of the vehicle. The structure of the environment can be mapped with the computation of the camera pose. These methods are called simultaneous localization and mapping (SLAM). NN such as PosNet, VLocNet+ + uses the image data to estimate the pose 3D pose of the camera. (Bugala, 2018; Grigorescu et al., 2020).

For the driving of AVs, it is necessary to correctly estimate the motion and trajectory of the surrounding objects. This process is called as scene flow. Traditionally used methods of scene flow needs to design feature manually. This is replaced by current DL approach which automatically learns the scene flow. Even though there is a lot of improvement in DL the traditional VO algorithms are not completely supplanted by the DLs. This is because the key-point detector technique is computationally efficient and can be deployed on embedded devices without many efforts.

Route planning algorithm

One of the important and the first decision was taken by the AV is to select a route from the current location to the destination based on criteria of distance, time consumed, or fuel consumption. Heuristic algorithms are used to find the distance between two locations via the cost of intermediate nodes, they are as follows, Bellman-ford is used when non-negative edges are only used, Dijkstra's algorithms are used for a known topology, and A* algorithm which is the most common algorithm for path selection. Traditional methods used for the route planning purpose are as follows (Bugala, 2018):

The non-linear approach tracks the trajectory of the vehicle and maintains the steering such that it follows the desired trajectory.

Model predictive control (MPC) consists algorithm for precise prediction of vehicle trajectory using traffic optimization and prediction techniques.

Feedback-feed-forward control is a technique that estimates the steering angle which is required to follow the curvature in a feed-forward loop such as to minimize the feed-forward error signal.

Shared control or parallel autonomy provides an additional level of safety by taking over the task of the driver in dangerous conditions which is referred to as interleaved autonomy.

Modern algorithms are based on a machine learning model called the end-to-end learning models. These models replace the traditional models for Route Planning. These end-to-end methods apply to single components of the system (as distinguished from the end-to-end autonomy described further on), for example to keep the vehicle within a lane. Schwarting et al. (2018).

End-to-end planning is a holistic approach, that is without any breakdown into components responsible for some part of the system. There are CNN networks that learn to carry out perception and generation of routes completely based on driving sequence and LIDAR data.

Decision-making

Driving decision-making (DDM) algorithms are an important part of autonomous driving. They are responsible for making strategic decisions during driving that is whether to change the lane, to overtake, or even decision based on road signs, based on various data from the input sensors. There are a lot of decision-making algorithms available that are applied based on the capability or necessity of the algorithm. Even though it is possible to present some algorithms, which are stated below (Bugala, 2018):

1. Partially observable markov decision process
2. Support vector machine with particle swarm optimization
3. Markov decision process with reinforcement learning
4. Deep reinforcement learning

Conclusion, discussion, and further research

AVs as discussed combine different techniques and algorithms using sensor data to solve problems in self-driving such as perception, localization, planning, etc. For this cognitive computing, techniques have proved to have higher accuracy and show higher performance compared to traditional approach machine learning approach as discussed (Fig. 7.22).

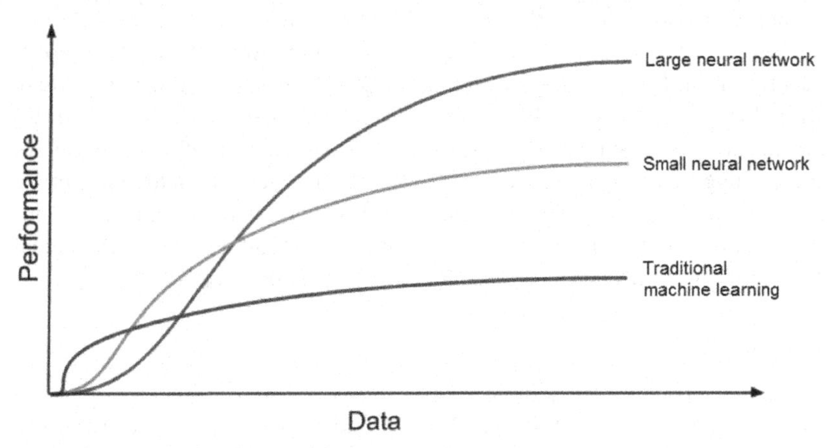

FIGURE 7.22 Comparison of different traditional machine learning algorithms and cognitive computing algorithms.

Traditional computing techniques even though it requires a small amount of data, generally shows a plateau in the performance even though a large dataset is available and is fed to the model. This is not the case with cognitive computing algorithms. For a smaller dataset, traditional algorithms prove much more useful compared to deep learning models. This is due to the fact that cognitive models require a large amount of data to function accurately. As seen from the graph, it is evident that as the data size increases the accuracy and performance of the model also increases. This useful characteristic of cognitive algorithms is used widely in AVs. For self-driving vehicles, the data collected through the different sensors is in a large amount. This data consists of data from actual driving of a human entity and also from simulators such as CARLA. In such cases, using traditional machine learning models would reduce the maximum achievable accuracy thus reducing the performance of the vehicle. As against this, if deep learning models are used, due to a higher available of data the maximum achievable accuracy can be increased to a greater extent, far beyond the conventional models.

Based on the fact in machine learning, which is "Garbage in, Garbage out," states that even though the data is available in larger quantities, it needs to be accurate and bias free, or at least the bias needs to be minimal if possible. Thus, the selection of data is a major criterion for AVs which affects the model accuracy greatly.

To summarize, cognitive computing techniques provide a greater advantage over conventional algorithms in AVs. But it is important to understand when to utilize traditional machine learning algorithms and when to opt for cognitive computing algorithms. Though deep learning provides greater advantage over other machine learning algorithms in some situations, replacing all the algorithms with cognitive computing algorithms does not seem to be a viable option. Definitely replacing most of them will prove to be a great deal of improvement. Even though using these techniques and by current technology, it is not possible to achieve full autonomy. Current developments in deep learning models like "Border Paris Method" provides an improvement over backpropagation task in deep learning by eliminating irrelevant learning patterns. Along with this, it has no problem in finding global maxima. It is called bi-propagation, quite similar to backpropagation but with the difference that is, the learning takes place separately for each layer. Thus, the hidden layers are no longer hidden. Nevertheless, technological progress will enable us to continuously change the upgrade, the choice of algorithms, and replace the traditional ones with novel methods thus reaching level 5 autonomy in AVs.

References

Bugala, M., (2018). Algorithm applied in autonomous vehicle system.

Fujiyoshi, H., Hirakawa, T., & Yamashita, T. (2019). Deep learning-based image recognition for autonomous driving. *International Association of Traffic and Safety Sciences Research, 43* (4), 244–252.

Grigorescu, S., Trasnea, B., Cocias, T., & Macesanu, G. (2020). A survey of deep learning techniques for autonomous driving. *Journal of Field Robotics*, *37*(3), 362–386. Available from https://doi.org/10.1002/rob.21918.

Gudivada, V. N., Irfan, M. T., Fathi, E., & Rao, D. L. (2016). Cognitive analytics: Going beyond big data analysis and machine learning. *Handbook of Statistics*, *35*, 169–205, https://doi.org/10.1016/bs.host.2016.07.010.

Nützel, T., (2018). *AI-based movement planning for autonomous and teleoperated vehicles including the development of a simulation environment and an intelligent agent*. Technical University of Munich.

Schwarting, W., Alonso-Mora, J., & Rus, D. (2018). Planning and decision-making for autonomous vehicles. *Annual Review of Control, Robotics, and Autonomous Systems*, *1*, 187–210. Available from https://doi.org/10.1146/annurev-control-060117-105157.

Yao, J., Research Associate Michigan State University (2018), Available at https://jmyao17.github.io/Machine_Learning/Sequence/RNN-1.html, (accessed 4.10.2020).

Bhatt, S. (2018) Reinforcement Learning, Available at https://www.kdnuggets.com/2018/03/5-things-reinforcement-learning.html, (accessed 4.10.2020).

Further reading

Available from https://towardsdatascience.com/artificial-intelligence-vs-machine-learning-vs-deep-learning-2210ba8cc4ac, Nov 2020 (accessed 20.11.2020).

Available from https://dzone.com/articles/comparison-between-deep-learning-vs-machine-learni, Nov 2020 (accessed 20.11.2020).

Available from https://www.sciencedirect.com/science/article/pii/S2095809916309432, (accessed 5.10.2020).

Available from https://www.embedded.com/the-role-of-artificial-intelligence-in-autonomous-vehicles/, (accessed 10.11.2020).

Available from https://www.robsonforensic.com/articles/autonomous-vehicles-sensors-expert/, (accessed 10.11.2020).

Bussemaker, K.J., (2014). *Sensing requirement for an automated vehicle for highway and rural environment*, Delft University of Technology.

Chen, S., Jian, Z., Huang, Y., Chen, Y., Zhou, Z., & Zheng, N. (2019). Autonomous driving: cognitive construction and situation understanding. *Science China Information Sciences*, *62*, 81101. Available from https://doi.org/10.1007/s11432-018-9850-9.

Jeffery Roderick Norman Forbes. (2002). Re*inforcement learning for autonomous vehi*cles (p. 110p) Berkley: University of California.

Kuutti, S., Bowden, R., Jin, Y., Barber, P., & Fallah, S. (2021). A survey of deep learning applications to autonomous vehicle control. *IEEE Transactions on Intelligent Transportation Systems*, *22*(2), 712–733. Available from https://doi.org/10.1109/TITS.2019.2962338.

Ni, J., Chen, Y., Chen, Y., Zhu, J., Ali, D., & Cao, W. (2020). A survey on theories and applications for self-driving cars based on deep learning methods. *Applied Sciences*, *10*(8), 2749. Available from https://doi.org/10.3390/app10082749.

O' Mahony, N., Campbell, S., Carvalho, A., Harapanahalli, S., Velasco Hernandez, G., Krpalkova, L., Riordan, D., & Walsh, J. (2019). *Deep learning vs. traditional computer vision*. Tralee: IMaR Technology Gateway, Institute of Technology Tralee.

Pendleton, S. D., Andersen, H., Du, X., Shen, X., Meghjani, M., Eng, Y. H., Rus, D., & Ang, M. H. (2017). Perception, planning, control, and coordination for autonomous vehicles. *Machines*, *5*(1), 6. Available from https://doi.org/10.3390/machines5010006.

Pu, Y., Apel, D. B., Liu, V., & Mitri, H. (2019). Machine learning methods for rockburst prediction-state-of-the-art review. *International Journal of Mining Science and Technology*, *29*(4), 565–570. Available from https://doi.org/10.1016/j.ijmst.2019.06.009.

Rao, Q., & Frtunikj, J. (2018). Deep learning for self-driving cars: chances and challenges, in: 2018 IEEE/ACM 1st International Workshop on Software Engineering for AI in Autonomous Systems (SEFAIAS), pp. 35-38.

Simhambhatla, R., Okiah, K., Kuchkula, S., & Slater, R. (2019). Self-driving cars: evaluation of deep learning techniques for object detection in different driving conditions. *SMU Data Science Review*, *2*(1), Article 23. Available from. Available from https://scholar.smu.edu/datasciencereview/vol2/iss1/23.

Chapter 8

Optimized navigation using deep learning technique for automatic guided vehicle

Anubha Parashar[1], Apoorva Parashar[2] and Lalit Mohan Goyal[3]
[1]*Department of CSE, Manipal University Jaipur, Jaipur, India,* [2]*Department of CSE, Maharshi Dayanand University, Rohtak, India,* [3]*Department of CE, J. C. Bose University of Science & Technology, YMCA, Faridabad, India*

Introduction

A smart car is generally made of two controlling units that dictate its actions, controller of low magnitude and high magnitude. Controller of high magnitude takes input from its components i.e. the driver (in this case the camera), its surroundings such as the traffic and the sensors in place, it then after deducing the correct actions to be taken sends signals to the low-level controller that controls the brakes, steering, engine, and throttle (Jiakai, 2019). To do this successfully it needs to understand driver psychology, how and when a specific maneuver is necessary, which changes with the terrain and the area the car drives in since driver temperament and driving style cannot be universally generalized (Udacity, Inc, 2019). It is also difficult for cars relying on so many sensors to be able to adapt to new surroundings and to reconfigure the system to achieve a different goal based on learning that occurs on the move, especially when noisy sensor data is received (Zheng, 2018). To understand and correctly predict such outcomes is precarious. Studies show that if the driver is given even a half a second extra before a collision, 60% of accidents can be avoided and this percentage increases to 90% if one second of warning time can be provided (Desai & Desai, 2017). Such results stress the importance of timely and correct decisions that face problems with the conventional architecture of a smart car system.

One of the breakthroughs in navigating autonomous vehicles happened because of using convolutional neural network (CNN) (Bojarski et al., 2016; Jackel, Krotkov, Perschbacher, Pippine, & Sullivan, 2006; LeCun, Cosatto, Ben, Muller, & Flepp, 2004; Pan et al., 2017; Podpora, Korbas, & Kawala-Janik,

Cognitive Computing for Human-Robot Interaction. DOI: https://doi.org/10.1016/B978-0-323-85769-7.00002-1

2014; Rajasekhar & Jaswal, 2015; Rosenzweig & Bartl, 2015; Hussain & Zeadally, 2018). In order to navigate the vehicle with complete safety of passengers, The CNN model is deployed and is trained as well as tested. The CNN model maps the complete driving model in real time by collecting information about driving behavior and taking images from the angle of steering wheel and camera (Baykal, Bulut, & Sahingoz, 2018; Billones et al., 2017; Gurghian, Koduri, Bailur, Carey, & Murali, 2016; Huang et al., 2017; LeCun, Cosatto, Ben, Muller, & Flepp, 2004; Pomerleau, 1989; Rowley, Baluja, & Kanade, 1998; Sharifara, Mohd Rahim, & Anisi, 2014). Vehicle performance is heavily dependent on the dataset, that is, if the vehicle encounters a new hazardous situation (one that the model is not trained upon) occurs, the system will not give a good analysis and an accident might happen. There are a lot of causes for the system to malfunction and crash- failure of software, or the camera sensor impairment.

The rest of the chapter follows the following pattern. In literature study, authors have explored the ultra-modern models used for self-driving vehicles. A plethora of research has been done and it has been studied carefully with more precision to find out the limitations of conventional deep neural network (DNN) based models. In section three optimization technique has been proposed which uses the CNN architecture for autonomous driving. In order to validate this methodology, we split the training data (training is done on Udacity generated training data) and it is tested with the real time data; presented in section four. At last conclusion and future work has been presented.

Literature review

Lately, computer vision is being used in autonomous vision for steering the vehicle (Mo, Kim, Kim, Mohaisen, & Lee, 2017). But computer vision alone is insufficient to provide completely accurate results so, it is being used along with sensors like radars, LIDAR, and object detection (Gong, Fan, Guo, & Cai, 2017). These sensors are then fused within a Kalman filter with the help of models. Our model is based on neural networks, has a monocular camera which is used by the detector to abstract single frames and the built system does not use any temporal features. Model has trained the network to project the depth based on labels taken from return of the radar. This network mostly learns from the annotations even though it has the information about the model and depth of the road (Parashar & Parashar, 2020).

This journey to make a smart autonomous vehicle began with the invention of modern cruise control in 1948 (Carlini & Wagner, 2017). Many dedicated steps have been taken in this direction to make self-driven cars a reality. Self-sufficient land automobile in neural network (NN) diversified the field by combining end to end learning with a NN for solving any issue. defense advanced research projects agency (DARPA) then gave perspective to what can be achieved with the technical know-how of the time (Bresson, Alsayed, Yu, & Glaser, 2017). In the recent past when CNN was trained for mapping camera pixels so they could be interpreted directly into guiding orders (Choudhury, Chattopadhyay, & Hazra, 2017), it demonstrated the reliability of this approach in contrast to the more modular alternatives (Tefft, 2016). One of the most used techniques for self-driving vehicles consists of DNN. A lot of companies that sells car like Ford, Tesla, BMW and Volvo swear by using DNN. A lot of software firms such as Uber, Lyft and Google are also deploying DNNs in their autonomous vehicles (Bertozzi, Broggi, & Fascioli, 2000). The working of such systems is based upon the acute transfer of information from Radar, Cameras and LIDAR as an input to the DNN like recurrent neural network (RNN) or CNN (Dagan, Mano, Stein, & Shashua, 2004).

The information is processed by the system in order to control the angle of steering wheel and the velocity of the vehicle being driven. For instance, the models used by Udacity make the use of both the RNN and CNN based technique whereas, the models created by NVEDIA, makes the use of only a CNN model for processing the input data provided in order to take the command of the steering wheel (Tatarek, Kronenberger & Handmann, 2018). A DNN based system can sometimes misread the given data and cause treacherous consequences (Krizhevsky, Sutskever, & Hinton, 2012). A plethora of studies has been done on the DNN based models and it is found that these systems are vulnerable to erroneous data and can malfunction easily (Yang, Zhang, Yu, Cai, & Luo, 2018). For instance, they can exploit a DNN based system by adding a small disruption in the image so that the system categorizes into a different group (Bojarski et al., 2016).

Even the objects can be morphed physically by the attackers for causing a glitch in the self-driving system based on DNN (Bhavsar, Das, Paugh, Dey, & Chowdhury, 2017). These low-cost techniques used by the attackers can cause inaccuracy in measuring distances, resolution and angles. Likewise, barricades or lumber present on the road can also lead the system to produce incorrect outputs (Yao, Fidler, & Urtasun, 2012). The foundation of the end to end approach was laid almost 14 years ago in an effort by DARPA-autonomous-vehicle (DAVE) (Pomerleau, 1989) where an autonomous car ran along an alley filled with junk. It demonstrated that this approach was viable and adequate for the functioning of a smart car. The end to end approach thus provided alternatives which were uncomplicated and easier to test (Sak et al., 2014).

With the advancement in technology, the study achieved similar functionality more skillfully and with limited resources. At the center of this end to end learning is the CNN, the driver's actions are based upon the training. The data was not hard to collect which made it possible to do rigorous experimentation as the system itself could be reproduced efficiently at a low cost. The data of a system could also be shared with another of the same kind to

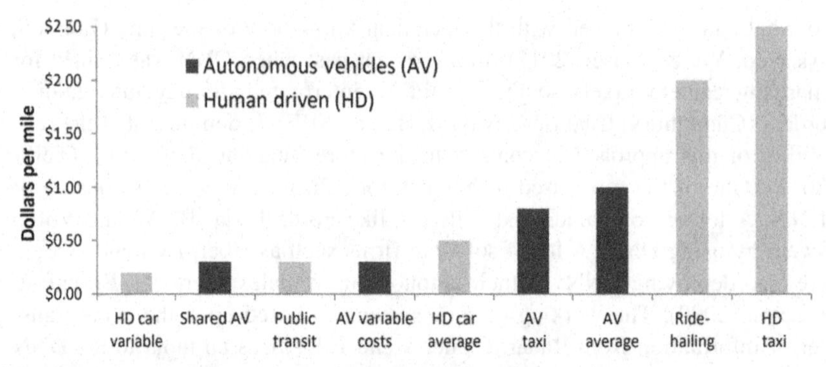

FIGURE 8.1 Cost comparison of autonomous and human driven vehicles.

improve performance. Autonomous driving technology is a fast-advancing field which has prompted a lot of questions. Some of these queries are related with the operational models suitable to the vehicles in Fig. 8.1. One of the key viable determinants of the model is hustling of the cost structures.

With the help of an extensive study of cost structures, we found that public transport will remain competitive (economically) as the demands are high for large vehicles in order to transport goods. For instance, in densely populated metropolitan areas, public transport is cheap as compared to private, pooled or autonomous cars. And in the regions where public transport is unavailable, people tend to choose pooled vehicles.

Proposed methodology

Previous methodologies discussed in literature were based on the pretrained DNN networks, but in this study an optimized navigation model has been proposed which helps the Automatic guided Vehicle more accurately with the combination of deep learning pipeline and setting the parameters of CNN network optimally.

In step one, authors have used a toy car and mounted a raspberry pi on top of it. An 8 MP camera has inserted in the raspberry pi and is attached in a way that it got a clear view of the road ahead. In step two an ultrasonic sensor is attached on the bonnet of the car for detecting any obstructions on the road in Fig. 8.2.

In step 3 the car can be driven by a remote control (the one that came along with the toy car) or could be driven autonomously (by applying DNNs). In step four, use Arduino Uno to collect the data from remote (the data here is input for Arduino) has been done.

In step five, Wi-Fi module has been used, and it is latched on the Arduino in order to connect it to the inbuilt Wi-Fi of the raspberry pi. This way the raspberry pi has the access to the data coming from the remote control (data

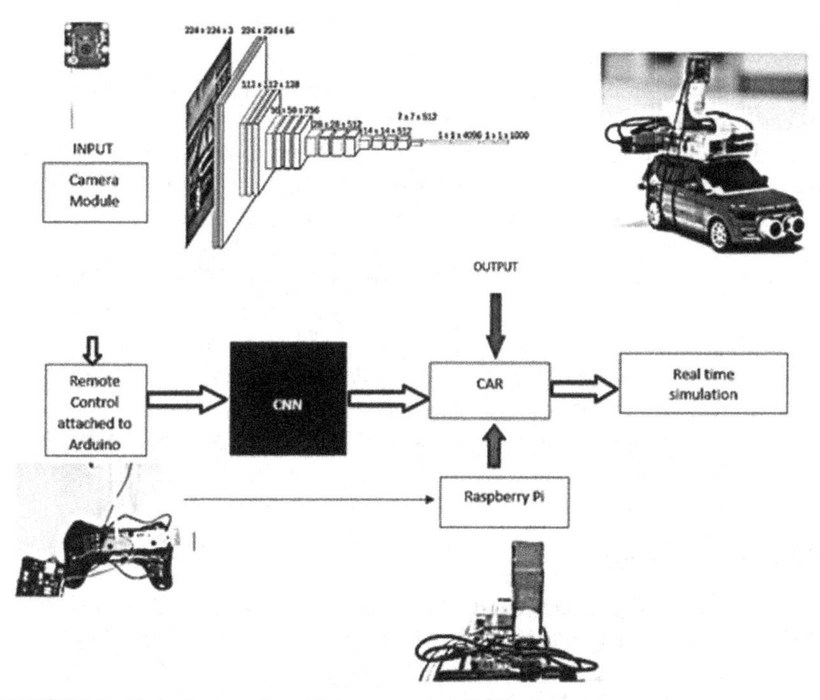

FIGURE 8.2 Trained automatic guided system controlled by the remote system.

from Arduino through Wi-Fi), ultrasonic sensor (for detecting any road lumber) and input data of the road (from camera mounted on raspberry pi).

In step six, path is formed with black sheet of paper and details are painted on it. Some barricades and debris used to create obstruction. The model is trained on various road maps- straight path, curved left, curved right and circular. In step seven, testing is also done in phases. After completion, the model is given a final test curve with a broad curvature.

Data collection

Initially for testing the equipment and the plan of action, Stimulator for autonomous car by Udacity generated training data. It gives the user the liberty to drive a car on select preset tracks and the data is collected frame by frame that will be corresponding to how you drive. The data also contains steering angle and acceleration on each frame. This data helped actualize the direction of this study and formulate what kind of data the project needs so that it could be mapped onto directions. Thus, it had to be generated independently and specific to the need of the project. This is because of two major reasons, first that the RC car is not taking directions in the form of steering angle but only left and right. The second is that the camera angle played an important role in how the car understood the road ahead which greatly affected its decision-making capabilities.

Data collection of the next set is done by dispatching direction commands to car's remote via the laptop which has transmitted to the car by the remote in Fig. 8.3. The camera view is visible on the screen of a remote device using the VNC viewer to stream the Camera's Input. The car is steered on self- laid tracks of different orientations that is similar but not the same to which the car had to be tested on. It mimicked unmarked roads and the environment where it will be tested, the car had to stop, turn and avoid a collision.

The frames have captured with the help of the pi camera staged on a RC vehicle and it also documented directions at every moment which also time-synchronized as well embedded in the final dataset. And this is repeated on diverse tracks until enough data that could be useful is collected. Thus, the dataset had frames containing the tracks as viewed by the pi camera along with the directions.

Preprocessing steps

In each image unnecessary data is cropped out that is reduced the height of the images in order to increase the contributing information in the frame and the images are resized so that less memory gets occupied. The images are then converted into YUV color-space as it allows systematic decrease in the

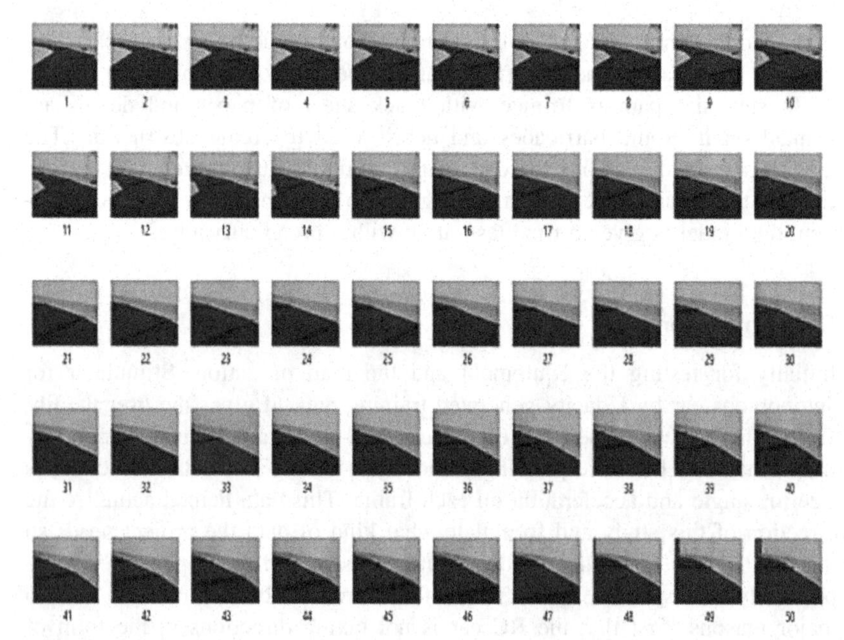

FIGURE 8.3 Sample section of the collected dataset.

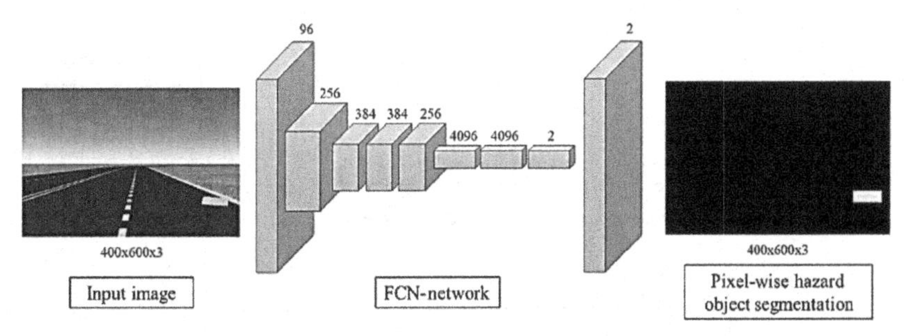

FIGURE 8.4 Feature extraction model using object detection and segmentation.

quantity of necessary information to constitute images of equivalent quality. To increase the sample data, images containing turns are flipped with the corresponding directions Fig. 8.4.

The brightness of undecided frames is increased to random degrees which are to account for the change in lighting conditions and the same procedure is followed to reduce the brightness of random images in order to make the model more robust. Furthermore, Gaussian blur is applied to each frame for smoothing the image and to reduce noise in the captured frame.

Classification

The CNN is trained with the collected data and used VGG16 model, it has 16 layers and uses soft- max function as the classification layer as shown in Fig. 8.5.

In order to train the system quickly, training is done on Udacity generated training data and tested with the real time data. For training CNN model is used and ReLU is used as activation function and softmax function is used to get the desired output.

When the autonomous vehicle turns left or right then we get a curve from training dataset to which shows the curvature of steering wheel, shown in Fig. 8.6.

Also, it is observed that unfairly large amount of the recorded data is corresponded to the car driving in a straight direction as compared to that of recorded while taking right or left turn, thus the data had to be normalized to avoid the possibility of a biased model and so that the scales could be made even.

Result analysis and discussion

This section has been divided into three subsections. The first section presents the experimental setup, in the second section detailed result analysis has been presented and the third section discusses the approach with their detailed comparison.

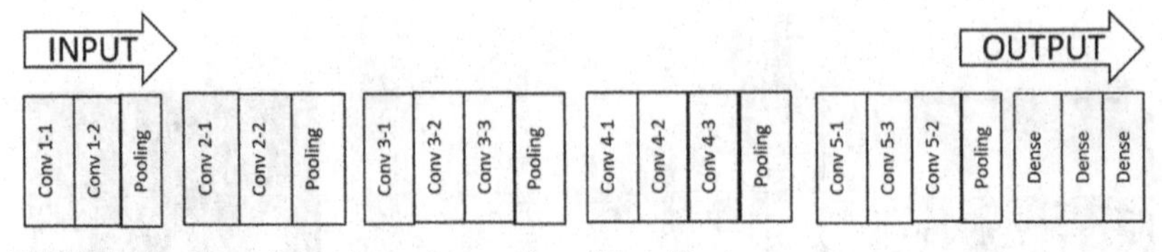

FIGURE 8.5 Convolutional neural network model architecture.

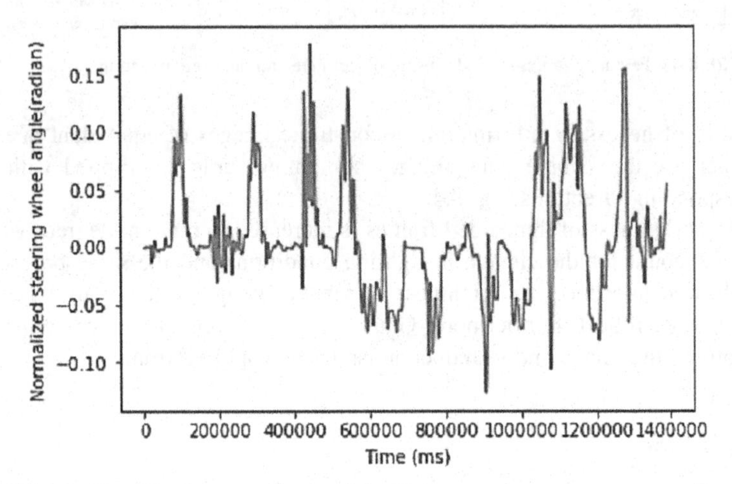

FIGURE 8.6 Normalization curve from training dataset to capture the curvature of steering wheel.

Experimental setup

Camera (as an input component): Model B + board of Raspberry Pi that draws its power from an external power bank, attached with a Pi camera is used to collect input data. The camera is fitted on top of the car in such a way that it provided vision to the car and is overlooking the road and immediate surroundings. The client program runs on Raspberry Pi for sending frames to the Neural Network which is also stored in the pi.

Processing unit (Raspberry Pi): A computer's PU manages numerous chores: input of data, training of NN and projection of data via commands and sending the resultant commands to the remote control for the final movement.

Unit of autonomous vehicle control: Controller of the car used contains a switch for turning it on and. A Raspberry Pi is used to simulate button-press actions by soldering the jumper cables to the remote and then connecting them to Four GPIO pins which were selected for collecting pin chips of the remote, communicating to actions like left, ahead, right and back. When GPIO emits a low-signal, it means that the pins have been given supply from

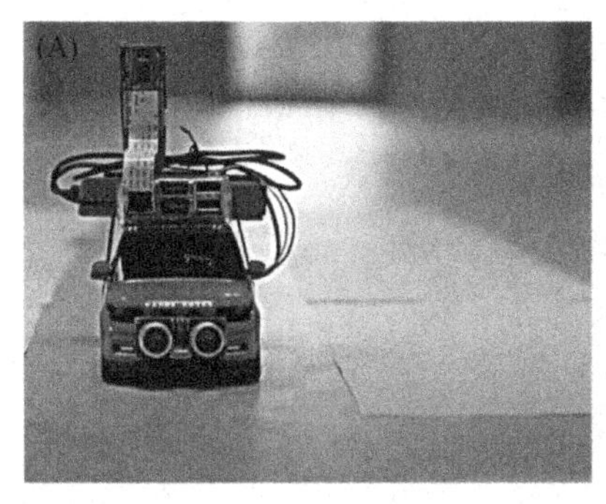

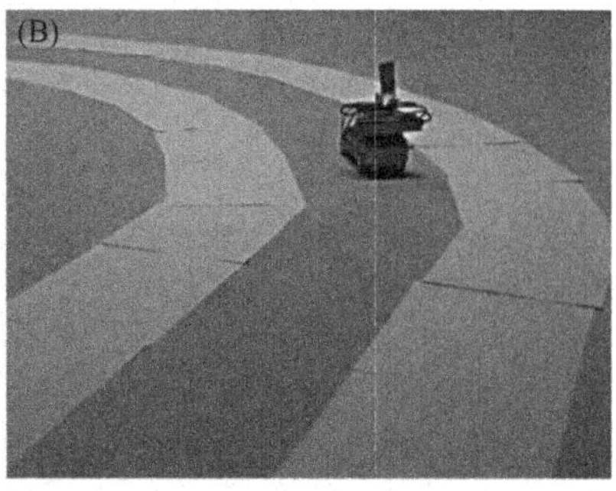

FIGURE 8.7 (A) Model training on various road maps- straight path, curved left, curved right and circular. (B) Car unable to detect road on right hand side, so taking a left turn.

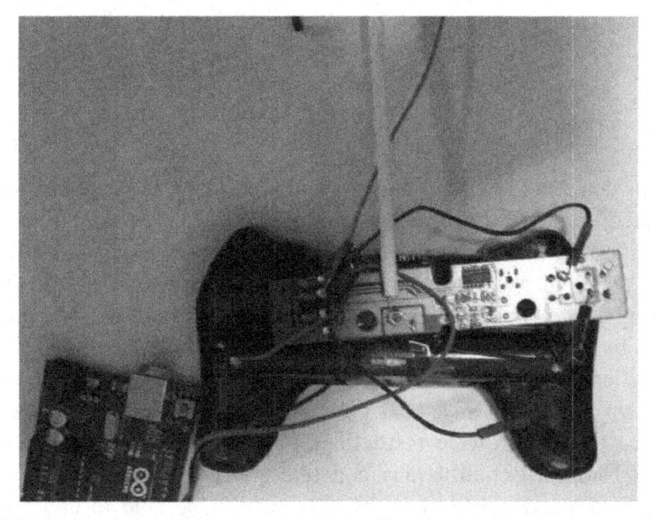

FIGURE 8.8 IR based designed remote control to collect the data from remote.

ground; whereas high-signal denotes chip resistance has no effect on the ground. Model returns the resultant commands to the pi using the serial interface, the Pi then outputs the signal to high or low after reading the input command, mimicking to drive autonomous vehicle by pressing buttons.

The model is trained on various road maps- straight path, curved left, curved right and circular. Testing is also done in phases. After completion, the model is given a final test curve with a broad curvature in Fig. 8.7A and B.

Arduino Uno is also used in this experiment to collect the data from remote (the data here is input for Arduino) in Fig. 8.8. A Wi-Fi module is also latched on the Arduino in order to connect it to the inbuilt Wi-Fi of the raspberry pi. This way the raspberry pi has the access to the data coming from the remote control (data from Arduino through Wi-Fi), ultrasonic

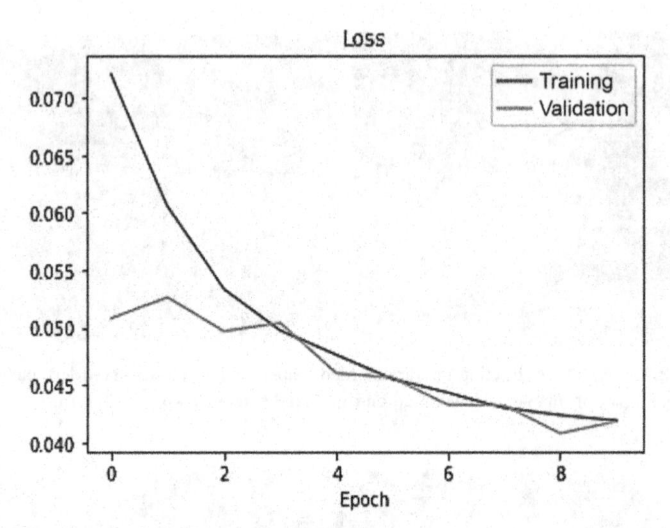

FIGURE 8.9 Shows training data and validation data.

sensor (for detecting any road lumber) and input data of the road (from camera mounted on raspberry pi).

Before testing the hardware on the actual field, the working of the model and the systems in place were to be tested. To achieve this, a video is fed to the training model frame by frame corresponding to which directions were generated by the CNN. After obtaining these results it is apparent that the basic functionality is to the point which is also confirmed by the accuracy graph shown in Fig. 8.9, results as given by the CNN for every frame.

But due to the hardware restrictions that is having only left, right, forward, and backward controls any concrete conclusions could not be drawn from this effort and field test is the only way to be able to understand the working and the shortcomings at length.

Result analysis

After rigorous test runs on tracks apart from which the car is trained on, it is concluded that the car can function competently in a controlled environment. The neural network is also working fittingly and gave better results than those obtained by previous studies as shown in Table 8.1.

The only complications faced were pertaining to the type of car used, which is later assessed could be comfortably resolved in Fig. 8.10.

A dataset of images is obtained after testing and training the data from the first trial of our self-driving vehicle. The dataset consisted of videos and images with similar structure to the tracks used by us. A test data is used to evaluate the accuracy of the model. A new curvature of road (black sheet) is

TABLE 8.1 A comparison of the accuracy of the current study with that of the others.

	Title of paper	Accuracy (%)
Muller (2006)	Off-road obstacle avoidance through end-to-end learning	75
Mori (2019)	Visual explanation by attention branch network for end-to-end learning-based self-driving	79
Xu (2017)	End-to-end learning of driving models from large-scale video datasets	84.6
Bojarski et al. (2016)	End to end learning for self-driving cars	88
Proposed method	Smart autonomous vehicle using end to end learning	98

introduced for test case. After several testing terms, we noticed that the accuracy increased and that with every try-out, the car started to make more precise turns. The car drove effortlessly and dogged the obstructions in the lane with apt precision.

Discussion

Many proposed autonomous vehicle benefits, including congestion and emission reductions, require platooning: multiple electrically connected vehicles traveling close together at relatively high speeds, preferably lead by a large truck. This requires dedicated highway lanes in Fig. 8.11. Fuel consumption by vehicle is shown according to their size along with the graph which is based on the measurements performed on a demonstrator system.

Some researchers think that the future of vehicles is electric- energy saving vehicles in Fig. 8.12. Today we have a heavy reliance on Fossil fuels for running our vehicles and these fuels are non-renewable hence, should be used judiciously. By consuming fossil fuels, we are not only exhausting our resources but also adding to the air pollution.

Autonomous vehicles are one of many factors affecting future transport demands. Whereas, with electric vehicles, we will have an one less problem to worry about, shown in Fig. 8.13.

But thinking that vehicles running on electricity will not cause any harm is a hoax as the automobile running on current produces particulate discharge from using breaks and wear and tear of tires. This emission is also hazardous

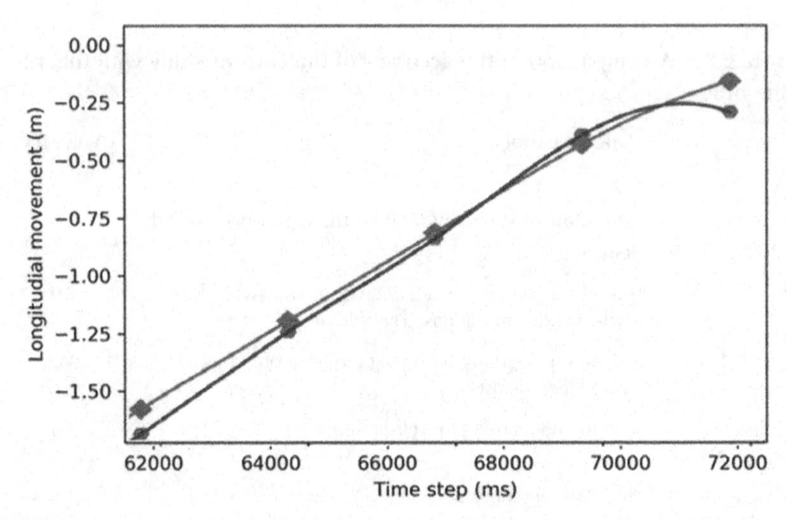

FIGURE 8.10 Accuracy comparison between convolutional neural network and deep neural network.

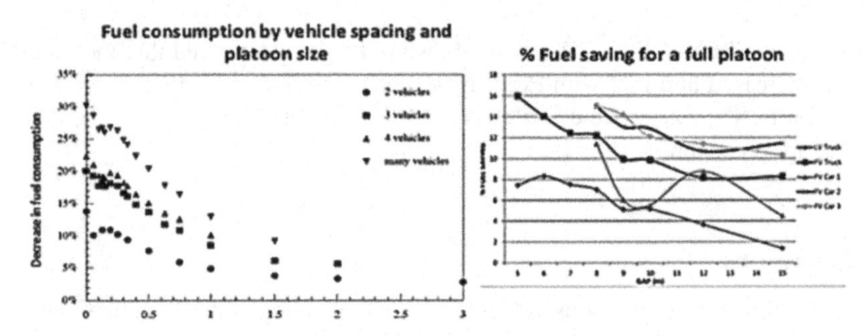

FIGURE 8.11 Pollution emissions of autonomous vehicle.

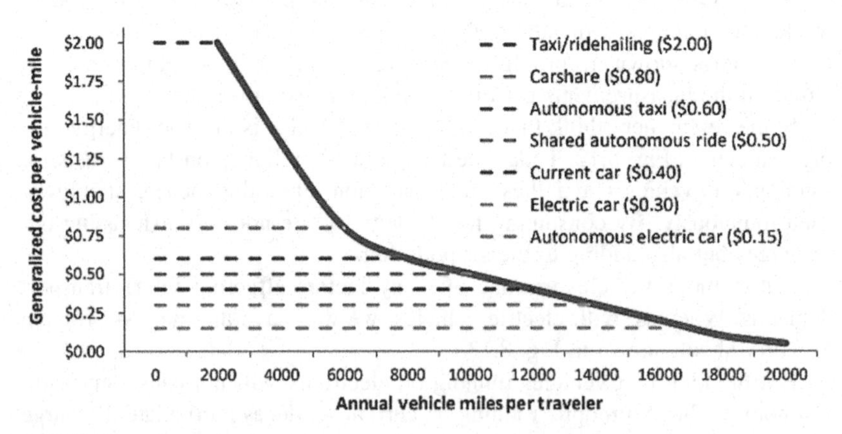

FIGURE 8.12 Generalized cost (money and time) travel demand curve.

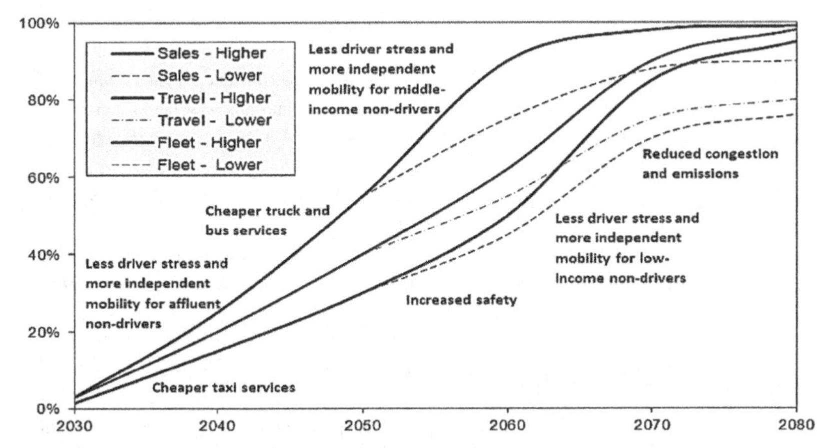

FIGURE 8.13 Smart vehicle—sales-fleet-travel-benefit projections.

Conclusion and future perspective

Learning approach in building a self-driving vehicle from beginning to end is an effective alternative to the traditional one. The car learned driving mannerisms and could detect road without the need for explicit labels, it proved to give viable results in a limited time frame which is also very cost- effective. The proposed model can be developed for industry point of view and should be viewed as independent of the vehicle like a CNG kit as the vision is to develop a setup that can be incorporated into any vehicle given that its acceleration and direction can be controlled.

A future analysis indicates that there are still about 25 years left for autonomous vehicles to take up half the automobile market and probably 40 years to be produced with autonomous fleet. In coming years, we might see the trend of discarding vehicles which are not smart and self-driving. It is obvious that at first autonomous vehicles will be very costly, and only the affluent will be able to take the luxury of using them. Nevertheless, unless autonomous vehicles become a common thing, they will not be affordable to moderate-income people/organizations.

References

S.I. Baykal, D. Bulut, O.K. Sahingoz, Comparing deep learning performance on BigData by using CPUs and GPUs. In *2018 electric electronics, computer science, biomedical engineerings' meeting (EBBT)*, 2018, 1-6, doi: 10.1109/EBBT.2018.8391429.

Bertozzi, M., Broggi, A., & Fascioli, A. (2000). Vision-based intelligent vehicles: State of the art and perspectives. *Robotics and Autonomous Systems*, *32*(1), 1−16. Available from https://doi.org/10.1016/S0921-8890(99)00125-6.

Bhavsar, P., Das, P., Paugh, M., Dey, K., & Chowdhury, M. (2017). Risk analysis of autonomous vehicles in mixed traffic streams. *Transportation Research Record: Journal of the*

Transportation Research Board, 2625(1), 51–61. Available from https://doi.org/10.3141/2625-06.

Billones, R. K. C., Bandala, A. A., Sybingco, E., Lim, L. A. G., Fillone, A. D., Dadios, E. P., Vehicle detection and tracking using corner feature points and artificial neural networks for a vision-based contactless apprehension system. In *2017 Computing conference*, 2017, 688–691. doi: 10.1109/SAI.2017.8252170.

Bojarski, M., Del Testa, D., Dworakowski, D., Firner, B., Flepp, B., Goyal, P., ... Zieba, K. (2016). End to end learning for self-driving cars, 1–9.

Bresson, G., Alsayed, Z., Yu, L., & Glaser, S. (2017). Simultaneous localization and mapping: A survey of current trends in autonomous driving. *IEEE Transactions on Intelligent Vehicles, 2*(3), 194–220.

Carlini, N., & D. Wagner. Towards evaluating the robustness of neural networks. In *Proceedings – IEEE* symposium on security and priv*acy*, 2017, 39–57. https://doi.org/10.1109/SP.2017.49.

Choudhury, S., Chattopadhyay, S. P., & Hazra, T. K., Vehicle detection and counting using haar feature-based classifier. In *2017* Eighth annual industrial automation and electromechanical engineering conferen*ce (IEMECON)*, 2017, 106–109. doi: 10.1109/IEMECON.2017.8079571.

Dagan, E., Mano, O., Stein, G. P., & Shashua, A. (2004). Forward collision warning with a single camera. *IEEE intelligent vehicles symposium*, 37–42. Available from https://doi.org/10.1109/IVS.2004.1336352.

Desai, S., & Desai, S. (2017). Smart vehicle automation. *International Journal of Computer Science and Mobile Computing, 6*(9).

Gong, T., Fan, T., Guo, J., & Cai, Z. (2017). GPU-based parallel optimization of immune convolutional neural network and embedded 2018 6th international conference on control engineering & information technology (CEIT), 25–27 October 2018, Istanbul, Turkey system. *Engineering Applications of Artificial Intelligence, 62*, 384–395.

A. Gurghian, T. Koduri, S.V. Bailur, K.J. Carey, V.N. Murali, DeepLanes: End-to-end lane position estimation using deep neural networks. In *2016 IEEE Conference on computer vision and pattern recognition workshops (CVPRW)*, 2016, pp. 38–45.

J. Huang, V. Rathod, C. Sun, M. Zhu, A. Korattikara, A. Fathi, I. Fischer, Z. Wojna, Y. Song, S. Guadarrama, K. Murphy, Speed/accuracy trade-offs for modern convolutional object detectors. In *2017 IEEE Conference on computer vision and pattern recognition (CVPR), Honolulu, HI*, 2017, pp. 3296–3297.

Hussain, R., & Zeadally, S. (2018). Autonomous cars: Research results, issues, and future challenges. *IEEE Communications Surveys & Tutorials, 21*(2), 1275–1313.

Jackel, L. D., Krotkov, E., Perschbacher, M., Pippine, J., & Sullivan, C. (2006). The DARPA LAGR program: Goals, challenges, methodology, and phase I results. *Journal of Field Robotics, 23*(11-12), 945–973.

Jiakai, Z. (2019). *End-to-end learning for autonomous driving*. New York University., Dissertation.

Krizhevsky, A., Sutskever, I., & Hinton, G. E. (2012). *ImageNet classification with deep convolutional neural networks. Advances in neural information processing systems* (pp. 1–9).

LeCun, Y., Cosatto, E., Ben, J., Muller, U., & Flepp, B. (2004). DAVE: Autonomous off-road vehicle control using end-to-end learning. Technical Report DARPA-IPTO Final Report, Courant Institute/CBLL, http://www.cs.nyu.edu/\~yann/research/dave/index.html.

Y.J. Mo, J. Kim, J.K. Kim, A. Mohaisen, and W. Lee, Performance of deep learning computation with TensorFlow software library in GPU-capable multi-core computing platforms. In

2017 ninth international conference on ubiquitous and future networks (ICUFN), 2017, 240–242. doi: 10.1109/ICUFN.2017.7993784.

Pan, Y., Cheng, C.A., Saigol, K., Lee, K., Yan, X., Theodorou, E. & Boots, B. (2017). *Agile autonomous driving using end-to-end deep imitation learning*. arXiv preprint arXiv: 1709.07174.

Parashar, A., & Parashar, A. (2020). IoT-based cloud-enabled smart electricity management system. In A. Somani, R. Shekhawat, A. Mundra, S. Srivastava, & V. Verma (Eds.), *Smart systems and IoT: Innovations in computing. smart innovation, systems and technologies* (vol. 141)). Singapore: Springer, https://doi.org/10.1007/978-981-13-8406-6_71.

Podpora, M., Korbas, G. P., & Kawala-Janik, A. (2014). YUV vs RGB-choosing a color space for human-machine interaction. *FedCSIS Position Papers*, *18*, 29–34.

Pomerleau, D. A. (1989). Alvinn: An autonomous land vehicle in a neural network. *Advances in Neural Information Processing Systems*, *1*, 305–313.

Rajasekhar, M. V., & Jaswal, A. K. (2015). *Autonomous vehicles: The future of automobiles. 2015* ieee international transportation electrification con*ference (ITEC)* (pp. 1–6).

Rosenzweig, J., & Bartl, M. (2015). A review and analysis of literature on autonomous driving. *E- Journal Making-of Innovation*, 1–57.

Rowley, H. A., Baluja, S., & Kanade, T. (1998). Neural network-based face detection. *IEEE Transactions on Pattern Analysis and Machine Intelligence*, *20*(1), 23–38.

Sak, H., Senior, A. W., & Beaufays, F. (2014). *Long short-term memory recurrent neural network architectures for large scale acoustic modeling. Interspeech* (pp. 338–342).

Sharifara, A., Mohd Rahim, M. S., & Anisi, Y. (2014). *A general review of human face detection including a study of neural networks and Haar feature-based cascade classifier in face detection. Biometrics and security technologies (ISBAST) 2014 international symposium on* (pp. 73–78).

Tatarek, T., J. Kronenberger, U. Handmann (2018). Functionality, advantages and limits of the tesla autopilot.

Tefft, B. C. (2016). *The prevalence ofmotor vehicle crashes involving road debris, United States, 2011-2014*. Washington, DC: AAA Foundation for Traffic Safety. (Technical Report).

Udacity, Inc. (2019). *Self-driving car simulator*. https://github.com/udacity/self-driving-car-sim. Accessed 05.02.2019.

Yang, Z., Y. Zhang, J. Yu, J. Cai, and J. Luo. End-to-end multi-modal multi-task vehicle control for self-driving cars with visual perceptions. In *2018 24th international conference on pattern recognition (ICPR)* 2018, 2289-2294. doi: 10.1109/ICPR.2018.8546189.

Yao, J., S. Fidler, and R. Urtasun. Describing the scene as a whole: Joint object detection, scene classification and semantic segmentation. In *Proceedings of the IEEE computer society conference on computer vision and pattern recognition*, 2012, 702–709. https://doi.org/10.1109/CVPR.2012.6247739.

Zheng W. (2018) *Self driving RC car*. https://zhengludwig.wordpress.com/projects/self-driving-rc-car/ (Accessed 22 Jan 2019).

Chapter 9

Vehicular middleware and heuristic approaches for intelligent transportation system of smart cities

Rajender Kumar[1], Ravinder Khanna[2] and Surender Kumar[3]
[1]Department of Computer Science & Engineering, IKG Punjab Technical University, Jalandhar, India, [2]Department of ECE, MM University, Sadopur, Ambala, India, [3]Department of Computer Science & Engineering, Guru Teg Bahadur College, Bhawanigarh, Sangrur, India

Introduction

Wireless communication is the transmission and reception of information between two devices that are not truly associated. Nodes can be close to each other or far apart. The nodes may be cell phones, personal digital assistants (PDAs), or other types of transceivers. Such systems have a unified structure, for example, a wireless access point (AP). These APs connect mobile wireless users to a central system. This works at the physical level of the OSI model (Kumar, Khanna, Jangra, & Verma, 2012).

Wireless telecommunications refer to the transfer of information between two or more computers that are not physically connected. Distance can be short, such as a few meters for television remote control, or as far as thousands or even millions of kilometers for deep-space radios, cellular telephones, PDAs, and wireless networking (Kumar et al., 2012). The wireless communication has seen growth in technology in multiple domains since its inception in 1880 (Kumar, Khanna, & Kumar, 2018a). Initially, wireless technology was used for personal communication. Now wireless technologies work at different frequencies to meet the requirements of corporate, personal and defense applications. There are various perspectives for wireless communications, including wireless sensor networks (WSNs), mobile ad hoc networks, wi-max, and many others. The days of flying objects are currently noticeable in military applications, so that flying planes can communicate with each other during the war or in similar cases. Wireless communication

Cognitive Computing for Human-Robot Interaction. DOI: https://doi.org/10.1016/B978-0-323-85769-7.00019-7

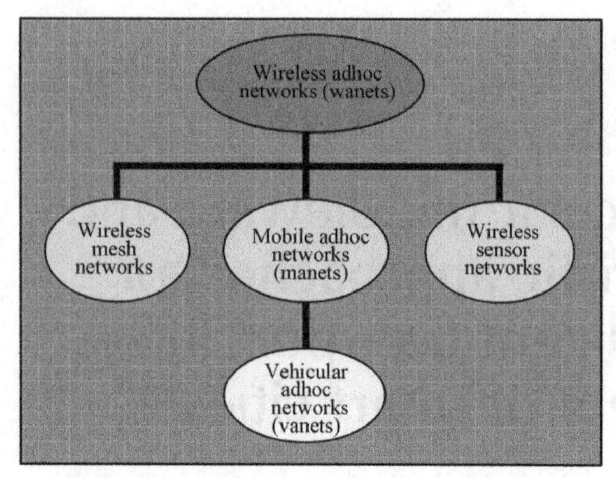

FIGURE 9.1 Hierarchy of wireless ad hoc networks.

in vehicular objects on the road is commonly known as VANET (Camp, Boleng, & Davies, 2002) (Fig. 9.1).

The implementation of vehicular ad hoc network (VANET) is widely adopted so that the high traffic regions can be controlled remotely with greater accuracy and performance (Kumar, Khanna, & Kumar, 2018b). VANETs employ mobile vehicles as network nodes. These can communicate between vehicles and/or roadside sensors. In this way cars in the network are turned into wireless routers to connect among themselves and make a broad mobile network. The term VANET was first used in 2001 (Toh, 2001) under car-to-car ad-hoc mobile communication and networking applications. It was suggested (Toh, 2001) that VANET would provide navigation, road safety, and roadside services. It was proposed that VANET would be a crucial part of the intelligent transportation system (ITS) (Kaur, Kaur, & Singh, 2012).

Development of VANET

VANETs are valuable for vehicle-to-vehicle (V2V) communications in such a special network. Vehicles can also communicate with roadside Vehicle to Infrastructure (V2I). These are two important, logical, and useful features of VANET. V2I enables the mobile node to connect to the Internet and enables global communication on the go (Majeed, Chattha, Akram, & Zafrullah, 2013).

The idea of utilizing radio communications in vehicles to enhance safety has been around for a long time, even before the appearance of digital communications. In recent years, vehicle safety has been focused on communication between vehicles. An example is the patent "Radio signaling systems for vehicles filed in 1922 and issued in 1925." Annoni and Williams (2015). In the mid-1980s, controller area network (CAN), the earliest car controller

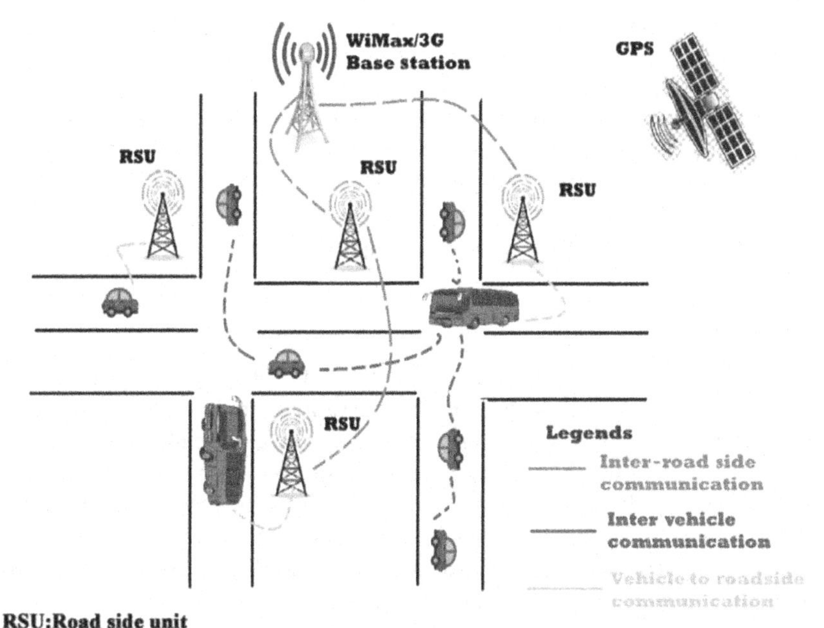

FIGURE 9.2 Vehicular ad hoc networks. Vehicle to roadside communication. Inter vehicle communication. Inter-road side communication.

network, was created by Bosch and is presently utilized in a few other automation applications (Vegni, Biagi, & Cusani, 2013).

In 1984, the radio data system (RDS) communication protocol, which included limited quantities of digital information in radio frequency-modulated transmissions to transmit more audio signals over radio waves, turned into the main digital infrastructure for a vehicle (I2V). The "RDS" turned into the European standard in 1991 (Fig. 9.2).

Around 2005, after further testing, the RDS system was refined to offer the "RDS-Traffic Message Channel." In 2000–13, ISO, functioning with IEEE, recently built up a lot of standards for conducting these communication sessions with European Telecommunications Standards Institute. Now, that everyone has smartphones in every car, it's likely that all of these cell phones support the "Ultra-High Frequency" wireless radio system, the Bluetooth, "Industrial, Scientific and Medical in the 2.4–2.485 GHz band" in the fixed and mobile devices that help in building "Personal Area Networks (PAN)". Annoni and Williams (2015).

Middleware technology

It is a technology that moves information from one program to different programs in a distributed domain and making it autonomous from the operating system, communication protocols and hardware used. Several common Middleware examples comprise enterprise application integration tools, telecommunication software and transaction monitors (Vinoski, 2004).

In the early 1960s, IBM formed "Semi-automated Business Research Environment (SABRE)" for the airline system. In the early 1970s, IBM's improved upon the system and called it transaction processing facility (TPF), which is an operating system designed for high volumes of transactions. These IBM created systems with many improvements are still in use.

Origins of middleware

In the early 1960s, IBM formed "SABRE" for the airline system. This American system handles its aircraft reservations on centralized computer PCs. The original SABRE though technically not a middleware system but rather gave the basics for the partition of transaction processing (TP), convey from equally to the operating system and the application. In the early 1970s, IBM's improved upon the system and called it TPF, which is an operating system designed for high volumes of transactions. These IBM created systems with many improvements are still in use.

Basics of middleware

Middleware like Java virtual machines and Armored Combat Earthmover (ACE) are used to form reusable event, reusable event de-multiplexing, inter-process communication and object synchronization. Distribution middleware like "Common Object Request Broker Architecture" and Java "Remote Method Invocation" form higher-level programming models that are distributed. Superior quality middleware makes consistent and scalable applications.

Communication provision

Normally middleware use distributed services over a network where applications interact throughout the network. Such distributed middleware is primarily of two types, (1) *remote procedure call* (RPC) and, (2) *messaging*.

Concurrency provision

By concurrency provision middleware applications need to be in terms of numbers of messages processed in a given period. This amounts to maximizing the concurrency in the system to enable continuous operation to multiple tasks. Middleware systems use a different technique to improve concurrency as an example using threads inside a server process permits middleware communication system to improve supervision of network links and requests, and messages which show up on these connections.

Heuristic search techniques

Heuristic technique, routinely called heuristic, is any approach to manage basic reasoning, learning, using a presence of mind system that isn't guaranteed as ideal or extraordinary, anyway satisfactory to achieve fast destinations. Circumstances where finding the ideal course of action is impossible or strange, heuristic techniques can be used to quicken the route toward finding a reasonable plan (Patterson, 2015). One case of a decent broadly useful heuristic that is helpful for a group of combinational problems is the nearest neighbor heuristic, which works by choosing the locally superior option at each progression (Rich & Knight, 2002). One of the best portrayal of the significance of heuristics in solving fascinating issues is *How to Solve It* (Polya, 1957).

Heuristic search methods

Heuristic search methods are classified into **Direct Heuristic search methods**. These aren't commonly possible since they demand a great deal of time or memory. These methods look the entire state space for an answer and use an emotional mentioning of errands. Cases of these are:

Breadth first search

It is a procedure for searching tree or graph. It begins at the root of a tree or subjective node of a graph, occasionally stated a search key, and examines the totality of the neighbor nodes at the present depth prior continuing ahead to the nodes at the subsequent depth.

Depth-first search

It is a calculation for exploring or glancing through trees or graphs. The computation starts at the root node and explores very far along each branch before backtracking. A variety of depth-first search was inspected in the "nineteenth century by French mathematician Charles Pierre Tremaux" as a technique for clarifying mazes (Rich & Knight, 2002).

Weak heuristic search methods

These are also known as "Informed Search, Heuristic Search, and Heuristic Control Strategy". This additional data is valuable to record preference amongst child nodes to examine. Individually node has a heuristic capacity related with it (Fig. 9.3).

Generate-and-test

Generate-and-test is function as (1) Generate a possible result. (2) Test to check whether this is an answer by isolating the picked point or the endpoint

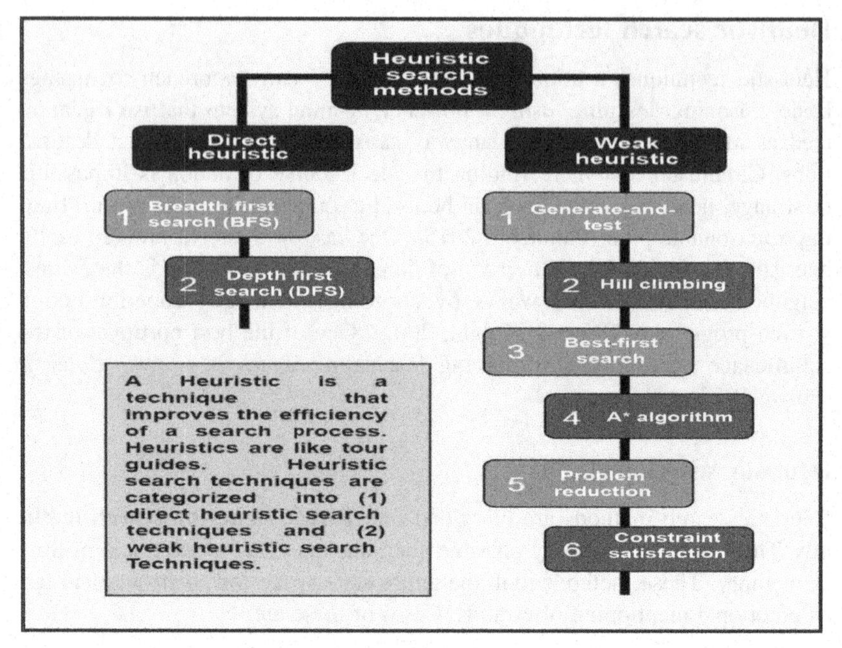

FIGURE 9.3 Heuristic search methods.

of the picked route to the plan of target states. (3) If an answer has been discovered, halted. Something different, return to step 1 (Rich & Knight, 2002).

Hill climbing

It is a variety of generate-and-test in which reaction from the test procedure is used to move in the search space. Hill climbing is as often as possible used when a better than average heuristic capacity is available for assessing states. Rich and Knight (2002).

Best-first search

At each stage of this method, we select the most likely nodes. This is done by applying an appropriate heuristic to all of them and by then build up the picked node. In case one of them is an answer, we can stop. If not, every single one of those novel nodes are supplemented to the solution of nodes delivered as of not long ago.

A Algorithm*

A* is a foundation name of numerous Artificial Intelligence frameworks and has been utilized since it was created in 1968 by Peter Hart; Nils Nilsson and Bertram Raphael. It is the blend of Dijkstra's calculation and Best-first search. It may be utilized to take care of numerous sorts of issues. A* search finds the most limited way through a search space to objective state utilizing heuristic function. A* requires heuristic function to assess the expense of path that goes through the specific state. It is characterized by the subsequent calculation.

Problem reduction

Problem reduction procedure utilizing AND-OR diagram is valuable for an issues that can be settled by disintegrating them into a set of minor issues which must be answered. Here, nodes whose successors should all be accomplished, and additionally nodes where one of the replacements must be accomplished (e.g., they are options) (Rich & Knight, 2002).

Constraint satisfaction

Numerous issues in AI can be seen as issues of *constraint satisfaction* in which the objective is to find some issue express that fulfils a given arrangement of requirements. Instances of such an issue incorporate crypt-arithmetic puzzle and some genuine perceptual naming issues. Constraint satisfaction is a two-steps process. Initially, constraints are found and circulated as far as throughout the system. In the second step, some hypothesis about an approach to support the constraints must be made (Rich & Knight, 2002).

Reviews of middleware approaches

Lee et al. (2006) proposed MobEyes for urban detecting dependent on the possibility of vehicle portability. The MobEyes exploited vehicles outfitted with camcorders and an assortment of sensors to perform occasion detecting, preparing, and message routing to different vehicles.

Riva (2007) introduced middleware that can care mobile sensing applications in Urbanets. Urbanets aid a range of administrations going from crisis and surveillance, shopping, and entertainment. They utilized vehicles to give traffic data directions to make drivers aware of coming road jams and telephones. This architecture has been implemented in Java and tried on smartphones. They test results developed in specially ad hoc network of telephones have shown their plausibility with sensible exhibition as far as dormancy, memory, and vitality utilization.

Wu et al. (2007) built up a versatile middleware platform ScudWare for smart vehicle space. In this, strategies of multiagent, context-aware, and versatile module boards were easily combined. They evaluated the ScudWare with a testbed of smart vehicle space in terms of (1) ScudWare extension; (2) coordination of the biological confirmation and AI advances; and (3) improvement applications for ITS in the smart vehicle utilizing ScudWare stage.

Bai, Ji, Han, Huang, and Zhang (2009) proposed "Middleware Enabling Context-Awareness for Smart Environment (MidCASE)" in light of remote sensor networks. A service-oriented technique was adjusted to fulfill the needs of customizing scenarios in application development. The standard-based reasoning in the process raises a decent man-machine associating condition. They employed healthcare scenario as a crucial function of middleware to validate the design principle of MidCASE.

Blair et al. (2011) introduced the CONNECT middleware architecture as new middleware that progressively created a distributed architecture foundation for the current working environment. The authors concentrated on the key part of middleware, in particular, the job of ontologies in supporting center fundamental middleware capacities identified with accomplishing interoperability. They made progressively proper availability arrangements as far as understanding a given deployment environment.

Caviglione, Ciaccio, and Gianuzzi (2011) executed a communique middleware for info motion applications on vehicular Ad hoc systems. They outlined and inspired a portion of the structural decisions hidden in the ACIS middleware.

Nour et al. (2011) introduced a middleware design for sensing and handling data produced in vehicular systems. This middleware provided the environment needed by software developers to create applications that could be used by the end-users as well as other programmers. The research solution was suited for both large and small urban areas with the difference that in the first case the data gathered were more accurate and the problem of traffic congestion becomes more stringent.

Beaubrun, Llano-Ruiz, Poirier, and Quintero (2012) recommended a middleware architecture for dispersing delay-obliged data in WSNs. The authors deployed the middleware in 03 fundamental devices: (1) the sensor nodes, (2) the gateway, and (3) the base station to assess a genuine situation. They performed 20 investigations, sending 400 notifications to the clients all through three protocols: (1) SMS, (2) email, and (3) twitter. In light of this data, two unique examinations were played out: the investigation of singular messages and the examination of the experiment. They understood that the max delay never surpasses the pre-established threshold.

Nitti, Girau, Floris, and Atzori (2014) investigated the fusion of "VANETs with the Social Internet of Things, that is the Social Internet of Vehicles (SIoV)." In the SIoV each vehicle was fit for setting up social associations with different vehicles and the roadside units (RSUs). Furthermore, the authors introduced a simulation examining sensible vehicular mobility to consider the qualities of the subsequent social organization structure.

Luo, Zhong, and Jin (2016) proposed a "SOA-based middleware (VsdOSGi)" for vehicular architecture, which enables a vehicle to impart services to others. They structured an algorithm dependent on QoS (SDQ) to guarantee offering ideal service. They demonstrated that middleware can viably decrease the quantity of message sending times, minimize bandwidth capacity while improving network use and service discovery rate with less response time.

Reviews of heuristic approaches

Caliskan, Rech, and Lubke (2005) discussed the feasibility of exchanging personal messages in urban area traffic condition. They presented a concept graph algorithm to the problem of complete area search in VANETs. The traffic was mapped to the "Traveling-Salesman-Problem (TSP)" for getting TSP Solutions. They analyzed the use of personal message to guide cars towards suitable and free parking space nearest to their destination area. The heuristic method of TSP was implemented to improve the efficiency.

Bruns and Jobmann (2006) assessed the impact of various serving systems of a fixed server for examinations the IEEE 802.11p. To upgrade its serving execution, the server was empowered to recognize and exploit mobility parameters. Improved heuristics were contrasted with existing standard methodologies. The analyzed "Enhanced Highest Density First" indicated predominant in the simulations and decreases the measure of wasted bandwidth contrasted with the well-known FCFS technique.

Chapkin, Bako, Kargl, and Schoch (2006) investigated to what degree an attacker can follow the exact area of a node, accepting an incredible assailant model where an attacker knows all neighbor connections in addition to data on node separations. They introduced another methodology which utilizes this data and heuristics to discover hub positions proficiently. The value of the outcomes was examined and contrasted.

Zhigang, Lichuan, MengChu, and Ansari (2008) proposed another clustering strategy for enormous multihop vehicular ad hoc systems. This research introduced the solidness of the projected cluster assembly, and communication expense for keeping up the construction and connectivity in an application setting. They executed CORSIM and NS-2 simulators with genuine traffic by simulating a network of one hundred cars moving in an expressway with 10 km length.

Fathian, Jafarian-Moghaddam, and Yaghini (2014) presented novel clustering algorithms based on "Ant Colony System (ACS)" and "Artificial Immune System (AIS)" as meta-heuristic algorithms. They presented results of the proposed algorithms and compare them with six VANET clustering algorithms from the literature. The results showed that the ACS-based algorithm and AIS based clustering algorithm performed the same as the Lowest-ID algorithms and highest-degree algorithms, respectively.

Vanet, Toure, Kechagia, Caire, and HadjSaid (2015) introduced two heuristic strategies for voltage constraint management through low voltage (LV) planned as a "Distribution System Operators (DSO)" subordinate help. In light of voltage sensitivity investigation, these techniques expected to assist the DSO with benefitting of the LV adaptability accessible on a given LV network to settle voltage deviations. A heuristic strategy was additionally introduced so as to refine the solution and permit the best LV adaptability dispatch utilizing low computational need.

Amuthan and Thilak (2016) explored the basic concepts in Tabu Search (TS) in application areas like VANET. They applied the meta-heuristic approach TS to find the optimized solution in search process with minimum cost function. The results focused on the benefits of integrating Tabu with other heuristics algorithm like Genetic Algorithm to give optimum solution to find neighbors for disseminating data.

Saxena, Jain, Bhadri, and Khemka (2017) approached traveling salesman problem employing GA method which more than once alters a population of distinct results, with the additional power of present day processing systems. They thought of an equal kind of GA for both multicore also many-core designs over open multi-processing and "Compute Unified Device Architecture (CUDA)" so as to make challenging issues like "Vehicle Routing Problem for google maps and DNA sequencing." They investigated both CUDA and OpenMP for different degree and structure of graphs portraying adjacent estimate to the exactness as far as most ideal path for traversal with extremely weakened execution time.

Jyothi and Jackson (2018) proposed a unique algorithm named "Ant Queue Optimization (AQO)" which was an expansion of "Ant colony optimization algorithm (ACO)." They indicated that the AQO has occupied a smaller time to progression all the vehicles inside a specified organization when contrasted with ACO. A few examinations had been completed to contrast ACO and AQO through set of vehicles, that is 10, 30, 50, 70, 100, and 200. They indicated that the AQO has occupied a smaller amount time to progression all the vehicles inside a specified organization when contrasted with ACO.

Saajid et al. (2019) analyzed V2V and V2R interchanges utilizing the heuristic methodology considered short-range protocols: ZigBee, WiFi, and Bluetooth with information rates, time, and range, different parameters included: roadway, urban streets, and blended. It worked out that, without RSUs, Bluetooth couldn't move any messages in light of its little range, the quality of ZigBee extended between 5% and 38%, and the WiFi went somewhere in the range of 30%–90%.

Conclusion and future scope

In smart cities using VANETs, the merging of middleware and heuristic approaches played a significant part. Though, Intelligent Transport System of the subsequent generation would need high mobility in transport, low latency of the system, real-time applications and networking of self-directed vehicles, which cannot be solved by old-style computing. Consequently, joining these methodologies with computing standards provided a possible solution to different issues of smart cities. Middleware technology consists of advanced wireless modules, effective algorithms, database servers, software servers, content management systems, etc. tools that ease the construction and operation of product. Nature-based algorithms is a highly active field of study which can be integrated for future scope. Many meta-heuristic-based works explain scientific observations focused on computer-driven algorithmic studies.

References

Amuthan, A. & Deepa Thilak, K., (2016). Survey on Tabu search meta-heuristic optimization. In 2016 international conference on signal processing, communication, power and embedded system (SCOPES), 1539–1543. doi:10.1109/SCOPES.2016.7955697.

Annoni, M., & Williams, B. (2015). *Vehicular ad hoc networks: Standards, solutions, and research* (pp. 3–22). Cham: Springer International Publishing.

Bai, Y., Ji, H., Han, Q., Huang, J., & Zhang, Z. (2009). Towards a service-oriented middleware enabling context awareness for smart environment. *International Journal ad Hoc and Ubiquitous Computing*, *4*(1), 24–35. Available from https://doi.org/10.1504/IJAHUC.2009.021911.

Beaubrun, R., Llano-Ruiz, J. F., Poirier, B., & Quintero, A. (2012). A middleware architecture for disseminating delay-constrained information in wireless sensor networks. *Journal of Network and Computer Applications*, *35*(1), 403–411. Available from https://doi.org/10.1016/j.jnca.2011.09.002.

Blair, G. S., Bennaceur, A., Georgantas, N., Grace, P., Issarny, V., Nundloll, V., & Paolucci, M. (2011). The role of ontologies in emergent middleware: supporting interoperability in complex distributed systems. In F. Kon, & A. M. Kermarrec (Eds.), *Middleware 2011. Lecture notes in computer science* (vol. 7049). Berlin, Heidelberg: Springer. Available from https://doi.org/10.1007/978-3-642-25821-3_21.

C. Bruns, K. Jobmann. (2006) *Evaluation of server performance in VANETs in single-hop scenarios*, 1–8.

Caliskan, M., Rech, B., & Lubke, A. (2005). *Information collection in vehicular ad hoc networks. 5th European congress on intelligent transportation systems* (pp. 1–11). Germany: Hannover.

Camp, T., Boleng, J., & Davies, V. A. (2002). A survey of mobility models for ad hoc network research. *Wireless Communications and Mobile Computing*, *2*(5), 483–502.

L. Caviglione, G. Ciaccio, V. Gianuzzi. (2011), Architecture of a communication middleware for VANET applications. In *2011 The 10th IFIP annual mediterranean ad hoc networking workshop*, 111–114. doi: 10.1109/Med-Hoc-Net.2011.5970474.

S. Chapkin, B. Bako, F. Kargl, E. Schoch., Location tracking attack in ad hoc networks based on topology information. 2006 IEEE international conference on mobile Ad Hoc and sensor systems, (2006) 870–875. doi: 10.1109/MOBHOC.2006.278667

Fathian, M., Jafarian-Moghaddam, A. R., & Yaghini, M. (2014). Improving vehicular ad hoc network stability using *meta*-heuristic algorithms. *International Journal of Automotive Engineering*, *4*(4), 891–901.

Jyothi, K., & Jackson, J. C. (2018). A time-based approach for solving the dynamic path problem in VANETs – an extension of ANT colony optimization. *Journal of Engineering Science and Technology*, *13*(3), 813–821.

Kaur, M., Kaur, S., & Singh, G. (2012). Vehicular ad hoc networks. *Journal of Global Research in Computer Science*, *3*(3), 61–64.

Kumar, R., Khanna, R., Jangra, S., & Verma, P. K. (2012). A study of diverse wireless networks. *IOSR Journal of Engineering*, *2*(11), 01–05.

Kumar, R., Khanna, R., & Kumar, S. (2018a). Deep learning integrated approach for collision avoidance in internet of things based smart vehicular networks. *Journal of Advanced Research in Dynamical and Control Systems, 10*(14), 1508–1512.

Kumar, R., Khanna, R., & Kumar, S. (2018b). An effective framework for security and performance in intelligent vehicular ad-hoc network. *Journal of Advanced Research in Dynamical and Control System, 10*(14), 1504–1507.

Lee, U., Zhou, B., Gerla, M., Magistretti, E., Bellavista, P., & Corradi, A. (2006). MobEyes: smart mobs for urban monitoring with a vehicular sensor network. *IEEE Wireless Communications, 13*(5), 52–57. Available from https://doi.org/10.1109/wc-m.2006.250358.

Luo, J., Zhong, T., & Jin, X., (2016). Service discovery middleware based on QoS in VANET. In 2016 12th International conference on natural computation, fuzzy systems and knowledge discovery (ICNC-FSKD), 2075–2080. doi: 10.1109/FSKD.2016.7603501.

Majeed, M. N., Chattha, S. P., Akram, A., & Zafrullah, M. (2013). Vehicular ad hoc networks history and future development arenas. *Information Technology & Electrical Engineering Journal, 3*(1), 1–5.

Nitti, M., Girau, R., Floris, A., & Atzori, L., (2014). On adding the social dimension to the internet of vehicles: Friendship and middleware. In 2014 IEEE International black sea conference on communications and networking (BlackSeaCom). doi:10.1109/blacksea-com.2014.6849025, 134–138.

Nour, S., Negru, R., Xhafa, F., Pop, F., Dobre, C., & Cristea, V., (2011). Middleware for data sensing and processing in VANETs. In 2011 International conference on emerging intelligent data and web technologies, doi: 10.1109/EIDWT.2011.33, 42–48.

Patterson, D. W. (2015). *Introduction to artificial intelligence and expert systems* (1st ed.). Delhi, Bangalore, Chennai: Pearson Education.

Polya, G. (1957). *How to solve it: A new aspect of mathematical method* (2nd ed.). Garden City, New York: Doubleday Anchor Books, Doubleday & Company, Inc.

Rich, E., & Knight, K. (2002). *Artificial intelligence* (2nd ed.). New Delhi: Tata McGraw-Hill Publishing Company Limited.

Riva, O. (2007). Academic Dissertation *Middleware for mobile sensing applications in urban environments*. Helsinki: University of Helsinki.

Saajid, H., Di, W., Xin, W., Memon, S., Bux, N. K., & Aljeroudi, Y. (2019). *Reliability and connectivity analysis of vehicular ad hoc networks under various protocols using a simple heuristic approach, . IEEE Access* (7, pp. 1–11).

Saxena, R., Jain, M., Bhadri, S., Khemka, S., (2017). Parallelizing GA based heuristic approach for TSP over CUDA and OPENMP. In 2017 international conference on advances in computing, communications and informatics (ICACCI), 1934–1940. doi: 10.1109/ICACCI.2017.8126128

Toh, C. K. (2001). *Ad hoc mobile wireless networks: Protocols and systems*. Prentice-Hall.

E. Vanet, S. Toure, N. Kechagia, R. Caire, N. HadjSaid, Sensitivity analysis of local flexibilities for voltage regulation in unbalanced LV distribution system. In *2015 IEEE Eindhoven PowerTech*, (2015) 1–6. doi: 10.1109/PTC.2015.7232627.

Vegni, A. M., Biagi, M., & Cusani, R. (2013). Smart vehicles, technologies and main applications in vehicular ad hoc networks. In Lorenzo Galati Giordano (Eds.), *Vehicular technologies – deployment and applications*. https://doi.org/10.5772/55492.

Vinoski, S. (2004). An overview of middleware. In A. Llamosí, & A. Strohmeier (Eds.), *Reliable software technologies – ada-Europe 2004. Lecture notes in computer science* (vol. 3063). Berlin, Heidelberg: Springer. Available from https://doi.org/10.1007/978-3-540-24841-5_3.

Wu, Z., Wu, Q., Cheng, H., Pan, G., Zhao, M., & Sun, J. (2007). ScudWare: A semantic and adaptive middleware platform for smart vehicle space. *IEEE Transactions on Intelligent Transportation Systems*, *8*(1), 121–132.

Zhigang, W., Lichuan, L., MengChu, Z., & Ansari, N. (2008). A position-based clustering technique for ad hoc inter-vehicle communication. *IEEE Transactions on Systems, Man, and Cybernetics, Part C (Applications and Reviews)*, *38*(2), 201–208. Available from https://doi.org/10.1109/tsmcc.2007.913917.

Chapter 10

Error Traceability and Error Prediction using Machine Learning Techniques to Improve the Quality of Vehicle Modeling in Computer-Aided Engineering

A. Anny Leema[1], Krishna Sai Narayana[2] and Subramani Sellamani[2]
[1]Vellore Institute of Technology, Vellore, India, [2]Renault Nissan Technology & Business Centre Pvt. Ltd.

Introduction

Geometry size and shape

To perform the software analysis, the developer needs to focus on all three dimensions outlined and the geometry is categorized into 1D, 2D, and 3D on a requirement basis. In the field of man-made brainpower, artificial intelligence (AI) methods go for distinguishing complex connections describing the systems that create a lot of yields from a lot of observational information sources, both have been considered as sources of information for the fundamental calculations. When the connections or guidelines are distinguished, expectations can be performed on new information. In the field of numerical reproduction, the adjustment and admiration procedures of CAD models to get ready reenactment models can be viewed as a fulfillment of complex undertakings including abnormal state mastery and numerous activities whose parameterization depends on profound information not regularly formalized. To address this issue, AI strategies can be a decent means to discover decisions that drive the CAD model arrangement forms. Besides, those methods can be useful to underwrite the learning implanted in a lot of adaption situations.

Seeing a lot of reproduction information requires a ton of time and human exertion. These a lot of information most ordinarily result from

Cognitive Computing for Human-Robot Interaction. DOI: https://doi.org/10.1016/B978-0-323-85769-7.00016-1

examinations, for example, robustness analysis or design space exploration, which will uncover the conduct of a structure under vulnerability. In this commitment, we will concentrate preferably on configuration space investigation over vigor examination, since this abandons us more opportunity to pick a plan later on. It is shockingly humanly impracticable to investigate numerous reproductions by hand. This pushes the longing for another computerized procedure to collect the vital data covered up in this information productively.

AI Methods can be viewed as cutting edge measurable techniques, empowering the investigation of a lot of information. The calculations endeavor to discover measurable regular examples inside the information and give techniques to explore or relate these examples. Applying machine learning in another control where computer-aided engineering (CAE) is tragically an out-of-the-case process. As an outcome, it won't work out well to just take the crude information and apply existing techniques to it. Therefore we need to propose a procedure chain for the examination of accident recreation results, using two key innovations: The field of dimensionality reduction contains numerous calculations to discover measurable examples, with the goal that one may locate a few noteworthy clasping methods of a segment when looking at a lot of reproductions. Principle Mining is an induction strategy to help relate these factual examples into circumstances and logical results. An ordinary application is to discover which structures caused explicit miss-happening modes. This information quickly helps architects to fix the undesired conduct of the structure. The general process in CAE is given in Fig. 10.1. ·

The car business today is defied which may probably the most noteworthy specialized difficulties in its history. In many districts of the world, guidelines and enactment are being instituted to implement higher efficiency and additionally lower tailpipe discharges. To meet these stricter prerequisites, car organizations are looking to enhancements in powertrain advancements, both for inward ignition motors, just as new powertrain frameworks, for example, half breed and electric powertrain frameworks. Also, diminishing vehicle weight keeps on increasing expanding significance to improve mileage.

These difficulties, alongside rapidly changing client inclinations, and the consistently expanding business sector interest for better vehicle execution and unwavering quality require car organizations to reevaluate past methodologies and to discover better approaches to configuration better vehicles at lower costs in a shorter time. Reenactment gives a solid chance and intends to realize a portion of the important cost efficiencies, while in the meantime giving the stage to help drive inventive item improvement. Besides, recreation strategies keep on progressing in advancement just as the dimension of constancy and exactness they can give. This continuous development is somewhat fueled by the determined geometric development in the limit and accessibility of superior registering.

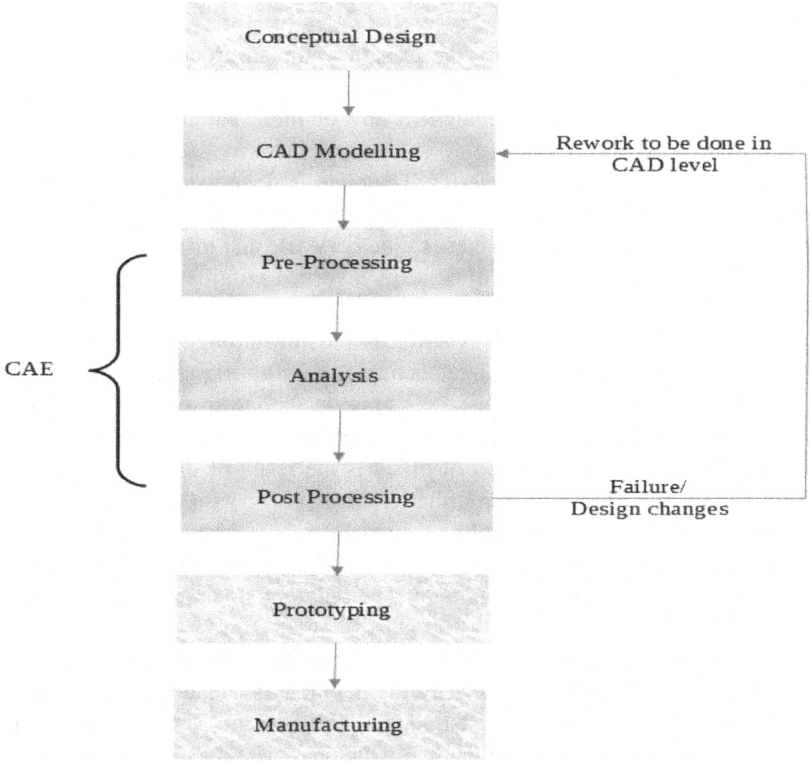

FIGURE 10.1 General process in computer-aided engineering.

Regularly propelling reproduction strategies combined with the accessibility of important processing assets laid the basis to execute a genuine reenactment based way to deal with car advancement at the framework level. As opposed to practicing recreation just to confirm the last structure or to more readily comprehend issues that were identified amid physical testing of models, a reproduction based way to deal with configuration takes into account the virtual vehicle to be tried and assessed at each period of the improvement cycle.

A recreation-based way to deal with car configuration requires more than improving reproduction innovation and across the board accessibility of elite processing to be effective. "Reenactment based" construes that recreation is omnipresent, that it saturates the whole plan process. Subsequently, the recreation client base essentially needs to extend, and it can't be normal that the present generally little network of reenactment specialists will grow comparably. Recreation in this manner must be exhibited and made accessible in such a way that is compelling for a developing network of clients that are not prepared as reenactment specialists.

Geometry check

The input CAD is checked with the Design Template to check if the thickness, material and the Property ID numbering is the same as the Design Template. If there is any mismatch in any of these parameters a query is raised to the CAD perpetrator.

This comparison with the Design Template is done using an automation tool which compares the ANSA file model by matching the Module ID of the ANSA model and its parameters to the CSV file and identifying any mismatch for the query to be raised the steps involved in finite element analysis is shown in Fig. 10.2.

The intersection is defined as surfaces passing through each other. The input CAD is checked for any Intersecting Parts after the offset and the mid surface have been taken. Intersections may occur due to error in the input CAD or due to the wrong direction of offset while moving skin surfaces to the middle. The wrong direction of offset is the result of the wrong orientation of the parts. The Intersection that arises is raised as a query to the preparators. Structure and assembling are crucial to the advancement of development. Before we dive further into CAE, let us initially comprehend the nuts and bolts of the building plan. The building structure of an item categorized into four phases.

The definition of the problem focus on the item improvement and it is important to unmistakably characterize the item that is intended to accomplish.

In the second phase, when the target of the item is recognized, it is important to orchestrate its structure. It includes meetings to generate new ideas where architects and designers meet and interface. Different structures are assessed and chose into a lot of a couple of serviceable arrangements.

The third phase is analytical process where fitting of the item into the structure is done. It requires quality and dependability investigation, cost

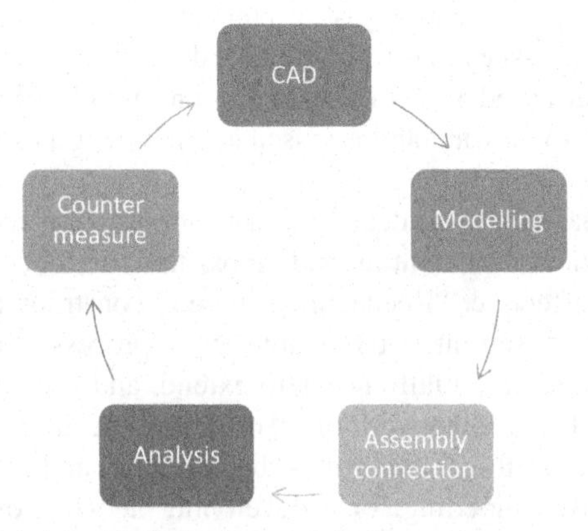

FIGURE 10.2 Steps involved in finite element analysis.

assurance, and so forth. This progression is imperative and iterative. Different answers for the item might be assessed till the ideal qualities are accomplished. This is the place CAE is mixed with CAD. Utilizing CAE programming, prepared architects can assess different parameters of the ideal item.

In the final phase prototype improvement and testing, the hypothesis is converted into useful models that are created and tried, for the most part utilizing 3D printers for quick improvement.

CAE accordingly discovers applications in building fields like liquid elements, kinematics, stress investigation, limited component examination, and so forth. Normally where item improvement is concerned. You can make 2D and 3D objects utilizing CAD, while you can investigate how that item will act utilizing CAE apparatuses. The mechanized plan instruments given by CAE have changed building examination from a "hands-on" involvement to virtual reenactment.

Literature survey

The exactness, repeatability and speed necessities of high-control laser tasks request the work of five levels of opportunity movement control arrangements that are equipped for situating and orientating the objective as for the laser(s) target connection point with high precision and accuracy. The consolidated sequential and parallel kinematic (cross breed) instrument in this paper is an appropriate possibility for this reason; in any case, various mistake sources can influence its execution. A kinematic model to dissect the blunders causing the positional and introduction deviations of the objective is portrayed thinking about two rotational degrees of the opportunity of the half breed component. Systems are sketched out to disentangle the mistake examination and to decide the blunder parameters of the instrument utilizing the blunder demonstrate and a test strategy (Karim, Piano, Leach, & Tolley, 2018).

To exploit the advanced highlights offered by CAD and CAE bundles for demonstrating and investigation amid the plan procedure, it is basic to manufacture an extension guaranteeing an intelligible connection between these apparatuses. Besides, this reconciliation system must be computerized to dispose of the monotonous costing exertion. In this paper, another computerized methodology for the CAD/CAE mix, actualized for the parametric plan and auxiliary investigation of flying machine wing structures is introduced. This technique depends on the mechanization limit accessible in present day PC supported apparatuses utilizing work in essential programming dialects just as the limit of the model information trade. The geometric and numerical models can be controlled to create an expansive assortment of conceivable structure cases through parameters presented in advance (Benaouali & Kachel, 2017).

Finding bugs in industry-estimate programming frameworks is tedious and testing. A computerized methodology for helping the way toward following from bug portrayals to pertinent source code benefits designers.

An extensive collection of past work means to address this issue and exhibits impressive accomplishments. Most existing methodologies center around the key test of improving procedures dependent on printed closeness to recognize applicable records. Be that as it may, there exists a lexical hole between the characteristic language used to plan bug reports and the formal source code and its remarks. To connect this hole, best in class approaches contain a part to break down bug history data to expand recovery execution. In this paper, we propose a novel methodology Trace Score that likewise uses ventures' necessities data and expresses reliance follow connections to additionally close the hole to relate another bug report to inadequate source code records. Our assessment of over 13,000 bug reports appears, that Trace Score fundamentally beats two best in class techniques. Further, by incorporating Trace Score into a current bug restriction calculation, we found that Trace Score altogether improves recovery execution by 49% as far as mean normal accuracy (Rath, Lo, & Mäder, 2018).

We present an exceedingly viable methodology for the alignment of vehicle models. The methodology joins the yield blunder strategy of framework recognizable proof hypothesis and the convolution indispensable arrangements of direct frameworks and stochastic math. Instead of aligning the framework differential condition straightforwardly for obscure parameters, we adjust its first fundamental. This coordinated expectation mistake minimization (IPEM) approach is worthwhile because it requires just low recurrence perceptions of the state, and creates fair-minded parameter evaluates that enhance reenactment precision for the picked time skyline. We address the adjustment of models that depict both efficient and stochastic elements, with the end goal that vulnerabilities can be figured for model expectations. We settle various usage issues in the utilization of IPEM, for example, the proficient linearization of the elements basic as for parameters, the treatment of vulnerability in beginning conditions, and the plan of stochastic estimations and estimation covariance. While the strategy can be utilized for any dynamical framework, we show its convenience for the adjustment of wheeled vehicle models utilized in charge and estimation. Explicitly we adjust models of the odometer, powertrain elements, and wheel slip as it influences body outline speed. Exploratory outcomes have accommodated an assortment of indoor and open-air stages (Seegmiller, Rogers-Marcovitz, Miller, & Kelly, 2013).

Divisions, for example, aviation, car, shipbuilding, atomic power, vast science offices or wind control need perplexing and precise segments that request close estimations and quick criticism into their assembling forms. New estimating innovations are now accessible in machine instruments, including incorporated touch tests and quick interface abilities. They give the likelihood to quantify the workpiece In-machine amid or after its production, keeping up the first setup of the workpiece and maintaining a strategic distance from the assembling procedure from being hindered to transport the workpiece to an estimating position. Notwithstanding, the detectability of the estimation procedure on a machine apparatus isn't guaranteed yet and estimation information is as yet not completely sufficiently dependable for procedure control or item approval. The logical goal is to decide the vulnerability on a machine device estimation and, accordingly, convert it into a machine incorporated discernible estimating process. For that reason, a mistake spending plan ought to consider blunder sources, for example, the machine apparatuses, segments under estimation and the collaborations between the two (Mutilba, Gomez-Acedo, Kortaberria, Olarra, & Yagüe-Fabra, 2017).

With upgrades in investigation exactness and PC execution, complex CAE models are required to steadfastly imitate a real vehicle. Be that as it may, it is hard to fulfill the need for stable quality on a constrained improvement plan if an ordinary displaying approach is utilized. In this manner, to produce top notch CAE models in a brief span, another robotization apparatus has been created utilizing ANSA. Computerization has been connected to the demonstrating steps: input parameters, fitting, association creation, and model check, for parts of taxi sheet metal and undercarriage outline. It empowers clients to create FE models for Durability, NVH and Crash examination. Associations, which had recently been especially troublesome, are presently made substantially more proficiently by perusing network data naturally in the apparatus (Yanagisawa & Kato, 2017).

Controlling the outstanding triptych costs, quality, and time amid the distinctive periods of the product development process (PDP) is an everlasting test for the business. Among the various issues that are to be tended to, the advancement of new techniques and devices to adjust to the different needs the models utilized up and down the PDP is surely a standout amongst the most testing and promising improvement territory. Today, regardless of whether strategies and devices exist, such a planning stage still requires a profound learning and a gigantic measure of time while considering digital mock-up (DMU) made out of a few a huge number of parts. Along these lines, having the capacity to evaluate from the earlier the effect of DMU adjustment situations on the recreation results would help to recognize the best situation directly from the earliest starting point. This paper tends to such a troublesome issue and uses computerized reasoning AI strategies to take in and precisely foresee practices from deliberately chosen precedents. The fundamental thought is to distinguish rules from these precedents utilized as contributions of learning calculations. When those guidelines are acquired, they can be utilized on another case to from the earlier gauge the effect of a planning procedure without performing it. To achieve this target, a technique to assemble a delegate database of precedents has been built up, the correct info (illustrative) and yield (readiness process quality criteria) factors have been recognized, at that point the learning model and its related control parameters have been tuned. One test was to recognize informative factors from geometrical key attributes and information portraying the arrangement forms. A second test was to assemble a successful learning model despite a predetermined number of precedents (Dangladea, Pernota, Vérona, & Fine, 2017).

Complex models arranged in CAD applications are regularly streamlined before utilizing them in downstream applications like CAE, shape coordinating, multigoals demonstrating, and so forth. In CAE, the meager walled models are regularly disconnected to a mid-surface for snappier investigation. Calculation of the mid-surface has been seen to be successful when the first model is DE highlighted to its gross shape. DE highlighting in this paper proposes a novel methodology for calculation such gross shape and it works in two stages. Initial, a proposed sheet metal component based characterization plot (scientific categorization) is utilized to decide the suppressibility of the highlights. Second, a technique dependent on the measure of remaining parts of the component volume is created to decide the qualification for concealment. Contextual analyses are exhibited to show the adequacy of the proposed methodology. It demonstrates that even after a significant decrease in the number of appearances the gross shape holds all the imperative highlights required for calculation of a very much associated mid-surface (Kulkarni, Kumar, Mukund, & Bernard, 2016).

Utilizing nonlinear strategies diminishes the unpredictability and time for exploring such extensive packs of gigantic numerical reproduction information. To send such a methodology for the examination of substantial scale recreation information, all things considered, conditions, it should deal with effective information stockpiling (counting pressure) which will happen on information servers later on rather than workstations as these days is the training in the industry. It will likewise require a productive exchange of the (pertinent) information among server and customer, just as proficient information handling for investigation and representation techniques, similar to the one sketched out in this work, on the server and the customer. It tends to be seen, that with the assistance of these new advances a thoughtfully extraordinary way to deal with the investigation of the vast information emerging in the virtual item improvement process ends up conceivable. The methodology permits an instinctive and intelligent treatment of the reproduction results and gives to the improvement engineer straightforward conceivable outcomes for the near and simultaneous examination of numerous reenactment results (Garcke & Teran, 2015).

How AI can be utilized to figure out how to adjust CAD models for reproduction models readiness is explored for a specific advance of defeaturing (distinguishing proof of highlight to erase or to hold). A committed system has been set up. It consolidates a few models and instruments used to help the extraction of information from the beginning and rearranged CAD models, the contributions from the specialists, just as the learning systems appropriately saying. The outcomes are promising and demonstrate that the AI systems can be a decent means to underwrite the information inserted in exact procedures like the defeaturing steps. As a matter of fact, from a database whose factors are various and heterogeneous, AI instruments have given developing guidelines except if they are formalized toward the start (Danglade, Pernot, & Véron, 2013).

Dimensionality Reduction encourages designers to rapidly recognize all significant distortion methods of a part in a huge dataset with a great many reenactments. Our geometry-based dimensionality decrease strategy has the preferred standpoint that all reproductions might be handled completely autonomous. Additionally, standardized comparability esteem between reproductions results can be processed in all respects effectively. The accompanying inquiry causes certain misshapen modes and it is identified with the assignment of discovering expansive structure subspaces that trigger the particular distortion mode. We discovered Decision—Tree-based Rule mining to be a fitting decision, since it does not just match the properties of information in the field of CAE, yet additionally addresses today's necessities of architects (Diez, 2016). The mechanical properties of composite material, for example, the carbon fiber were likewise determined amid the structure procedure (Tsirogiannis, 2015).

Techniques of AI in mechanical engineering

Today, it is very hard to set up a reenactment show while advancing the planning steps that depend on a profound learning and solid aptitude not in every case unmistakably created and therefore scarcely mechanized. In this way, various cycles are regularly required to get an ideal CAD display adjusted to a given reproduction. This adds to broaden item improvement cycles. In mechanical building, AI strategies are utilized in different applications, for example, physical conduct estimation, structure, acknowledgment, figuring out or material sciences. In those applications, classifiers are regularly used to appraise one or a few yield parameters of various natures (e.g., geometrical, measurable, and physical), or even to order shapes or 3D focuses sets. For measuring and shape structure, classifiers are regularly evaluating worldwide geometric parameters of the model. The estimation of a physical amount like a heap, a pressure, a weight, or a temperature, evacuate the need to understand complex conditions. Factual parameters don't give legitimately the physical amount. Nonetheless, they offer the chance to evaluate for instance a standard deviation, a mean, a pattern, or a physical impact likelihood. The order of shapes or digitized 3D focuses sets, is utilized for model acknowledgment and reuse. The point of our work is to from the earlier gauge the effect of the CAD show disentanglement on the aftereffects of the investigation just as the expense of the planning. Accordingly, the thought is to have the capacity to play out the estimation without doing the disentanglement itself. Here, we are not attempting to assess the investigation results but rather just the blunders because of the disentanglement. Physical amount estimation isn't valuable for our situation. A measurable parameter (e.g., level of deviation) appears to be progressively fitting in our investigation. Info factors and precedents.

When utilizing man-made brainpower procedures on CAD models, the most imperative test is to recognize the information factors to be prepared. The physical issue is commonly depicted by physical amounts vectors Geometrical information is portrayed by directions, by the histogram, or by a vector of parameters. For our situation, the intricacy of the controlled CAD models makes it hard to utilize diagrams or histograms. The abnormal state of improvement between the two designs does not permit the utilization of the focuses' directions. A vector of deliberately chosen parameters is by all accounts the sensible arrangement. The decision of the most delegate input parameters, the choice and the design of the classifiers are such huge numbers of issues that are tended to and created. Besides, in our methodology, we likewise need to recognize factors that best describe the arrangement procedure to be assessed. This is likewise a difficult issue that has been tended to in this chapter.

Information factors and precedents. When utilizing man-made brainpower methods on CAD models, the most critical test is to distinguish the info factors to be prepared. The physical issue is commonly portrayed by physical amounts vectors Geometrical information are depicted by directions, by histogram or by a vector of parameters. For our situation, the multifaceted nature of the controlled CAD models makes it hard to utilize charts or histograms. The abnormal state of improvement between two arrangements does not permit the utilization of the focuses' directions.

Understanding CAE processes

The Input CAD is checked with the Design Template to check if the thickness, material and the Property ID Numbering is the same as the Design Template. On the off chance that there is any befuddle in any of these parameters, an inquiry is raised to the CAD culprit. This Comparison with the Design Template is finished utilizing an Automation Tool. CAE apparatuses are being utilized, for instance, to dissect the strength and execution of parts and congregations. Henceforth, comprehension of devices and procedures is required for gathering information identified with blunders. The three phases of CAE tasks are preprocessing, solver, and postprocessing. The workflow of the proposed system is given in Fig. 10.3.

Implementation

Support vector machine

It is a regulated learning model with related learning calculations that break down data utilized for grouping and multivariate investigation. Given a gathering of preparing precedents, each set apart as having a place with no less

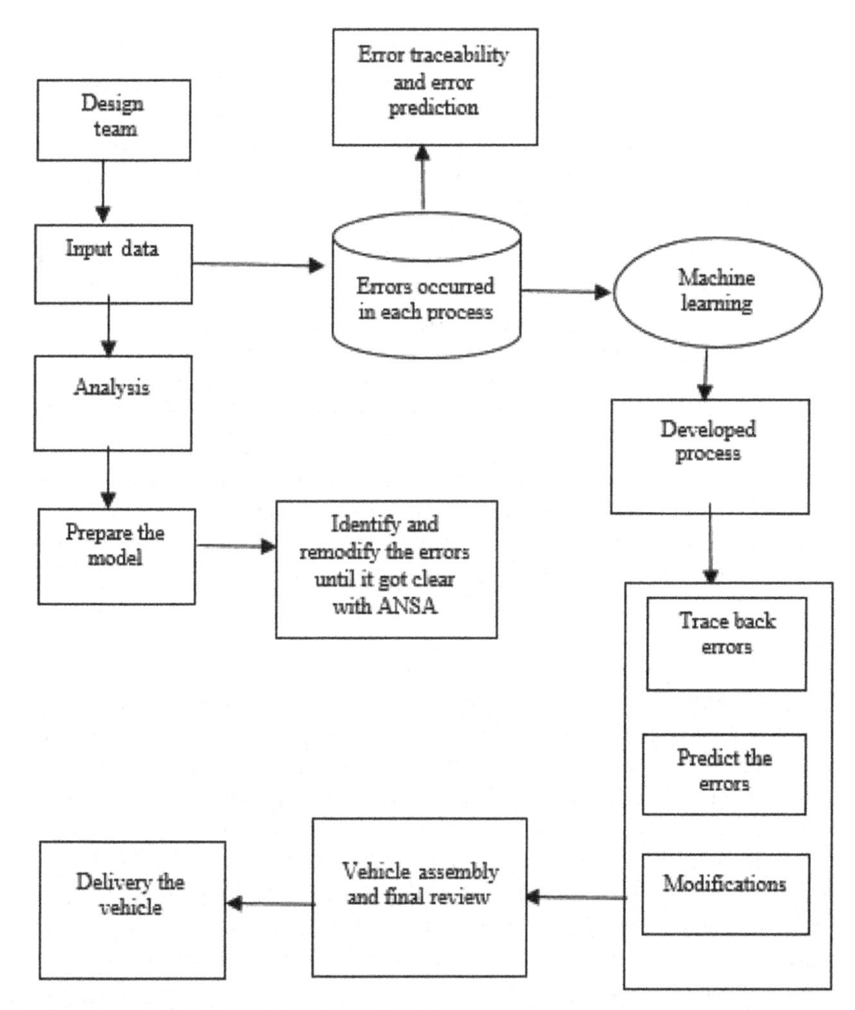

FIGURE 10.3 Workflow of the proposed system.

than one or the inverse of two classes, partner SVM preparing algorithmic principle manufactures a model that allocates new guides to somewhere around one class or the inverse, making it a nonprobabilistic parallel direct classifier even though systems like Platt scaling exist to utilize SVM in an exceptionally probabilistic order setting. partner SVM model might be a representation of the precedents as focuses in territory, mapped all together that the examples of the different classes are isolated by a straightforward hole that is as wide as potential.

Notwithstanding action direct grouping, SVMs will proficiently play out a nonstraight arrangement utilizing what's alluded to as the portion trap, certainly mapping their contributions to high-dimensional element regions.

Decision tree

The decision tree is that the principal ground-breaking far-reaching device for arrangement and forecast. A decision tree is a stream sheet-like tree structure, wherever every inside hub signifies a look on a trait, as shown in Fig. 10.6 each branch speaks to the result of the test, and each leaf hub (terminal hub) holds a class mark.

- Decision trees will generate rules. It helps to perform classification while not requiring a lot of computation. These trees are a unit ready to deal with each nonstop and all-out factors. Its offers a transparent straightforward sign of that field's area unit most significant for prediction.
- Decision trees are pricing more to do problems computationally. The method of growing a. At each hub, each competitor part field ought to be arranged before its best part will be found. In a few calculations, combos of fields are a unit utilized and a look ought to be made for best-consolidating loads. Pruning calculations may likewise be expensive since a few hopeful subtrees ought to be formed and thought about. Choice trees are all the more expensive and computational.

Naive Bayes

The issue of decision making records as having to one class or the inverse (e.g., spam or authentic, sports or governmental issues, and so forth.) because of the choices with word frequencies. With worthy prehandling, it's focused amid this space with a great deal of cutting edge techniques together with help vector machines.

Naive Bayes classifiers square measure extremely scalable, requiring variety of parameters direct inside the number of factors (highlights/indicators) in an exceedingly learning downside. Most extreme probability preparing is regularly done by assessing a shut structure articulation. Which takes straight time, rather than by extravagant reiterative estimate as utilized for a few various types of classifiers.

Backpropagation neural networks

Backpropagation is a strategy used in fake neural systems to figure an angle that is required inside the estimation of the loads to be used in the system. Backpropagation is shorthand for "the retrogressive engendering of blunders," since an erroneous conclusion is registered at the yield and conveyed in reverse all through the system's layers. It's ordinarily acclimated to train profound neural systems.

Speculation of backpropagation utilizing multi-feed forward systems, are made for potential by utilizing the chain standard to iteratively figure angles for each layer. It is intently connected with the Gauss-Newton algorithmic standard and is a part of the persistent examination in the neural back spread.

Backpropagation might be an uncommon instance of an extra broad system known as programmed separation. Inside the setting of learning, backpropagation is frequently utilized by the angle plunge enhancement algorithmic guideline to control the heap of neurons by calculative the slope of the misfortune work. The developed process uses backpropagation neural networks to trace errors.

Interacting with ANSA

The basic idea behind the scripting language is to automate many repetitive and tedious procedures with the minimum user interaction or even to perform specific tasks that are not covered by the standard ANSA commands.

Collecting entities: This function collects massively a type of entity (e.g., faces, edges, and parts) and returns it back to the user in a list.

Algorithm to collect the entities

```
import ansa
from ansa import calc
from ansa import base
from ansa import constants

def main ():
   ents = [['POINT'], ['SHELL', 'FACE']]
   nodes = base.PickEntities(constants.LSDYNA, ents[0])
   print('All selected points = ', len(nodes))
   if len(nodes) == 1 > 0:
      ret_vals = base.GetEntityCardValues(constants.LSDYNA, nodes[0], ('X', 'Y',
'Z'))
      print("point to be projected (x,y,z): ", ret_vals['X'], ", ", ret_vals['Y'], ", ",
ret_vals['Z'])
      mat = calc.ProjectPointToContainer([ret_vals['X'], ret_vals['Y'],
ret_vals['Z']],container)
   if mat:
      print("projected point (x,y,z) : ", mat[0][0], ", ", mat[0][1], ", ", mat[0][2])
      print("distance (d,dx,dy,dz) : ", mat[1], ", ", mat[2][0], ", ", mat[2][1], ", ",)
```

Deleting Entities: This function deletes an entity or a list containing entities. The function returns zero (0) if all elements were deleted, one (1) otherwise.

Algorithm to delete the entity

```
Import ansa
from ansa import base
from ansa import constants def main ():
     to_del = list ()
     ents = base.CollectEntities (constants. NASTRAN, None, 'PSHELL')
for ent in ents:
          to_del. Append(ent)
          base.DeleteEntity (to_del, True)
```

Get entity card values: This function is used to get values from an entity using its Edit Card. The labels are taken from the fields in the edit card. The respective values of these fields are then assigned the entity's parameters.

Collect boundary nodes: This function collects boundary nodes that belong outer edges of the given entities.

Measure entities: This function is used to get values from an entity using its Edit Card. The labels are taken from the fields in the edit card. The respective values of these fields are then assigned the entity's parameters.

Algorithm to get the card values

```
Import ansa
from ansa import base
from ansa import constants
def main():
vals = ('Name', 'T')
ret = base.GetEntityCardValues (constants. NASTRAN, prop, vals)
 if ret['Name']:
        print(ret['Name'])
       print(ret['T'])
```

Conceptual design phase

Requirements & specifications

Prerequisites are those which you will require to list down the motivation behind the item. For instance, while planning a fast vehicle, what ought to be the most extreme speed limit for that vehicle, goes under necessities. Particulars are the subtleties of a plan that you will require to list down the course of action of inside and outer structure.

Conceptual design views

After social occasion every one of the necessities and particulars you should begin taking a shot at producing harsh representations or formats. You ought to create 3–4 structures of a similar item and obviously with various formats. This gives an upstanding gauge study and taking a gander at the advantages and disadvantages of the considerable number of structures, the best plan can be chosen. You should incorporate 3–4 plans for each structure for better understanding. The conceptual design views are represented in Fig. 10.4.

Candidate configurations

All the generated designs undergo following two studies

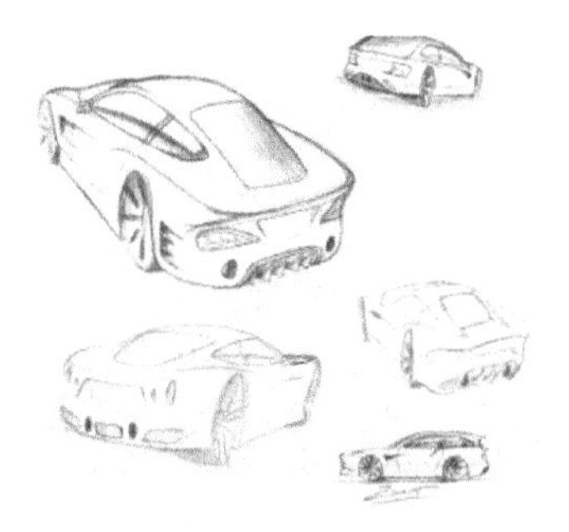
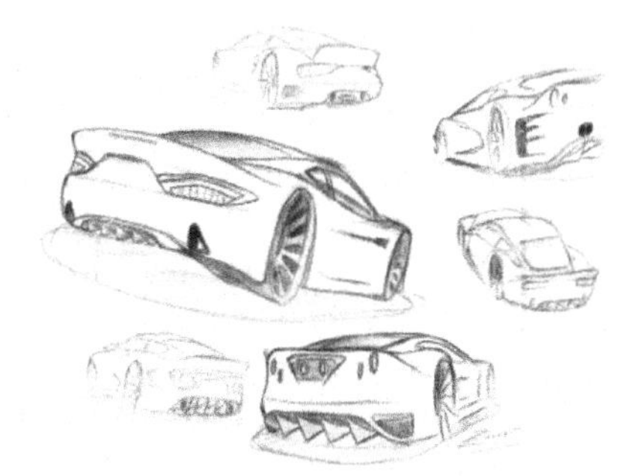

FIGURE 10.4 Conceptual design views.

Trade-off study

Here you need to make an appraisal of you structure according to cost, execution and weight. Exchange off investigation is essentially a trade off with the structure parameter. In the event that you are choosing a structure which is lighter in weight, at that point its cost will be high thus will be its execution. So fundamentally you have to bargain with anybody of the parameter. A perfect case would be a plan with minimal effort, lighter in weight and with elite.

Analysis

In this examination, you need to approve your plan on a few parameters like basic burdens, load ways, aerodynamics, weight and balance, stability, structural shape, component arrangement, aeromechanics, manufacturing, joining strategies, producibility designing and so forth.

The plan setup that passes both the above examinations is at long last chosen and is named as selected candidate design. After this, you have to make refined designs for the chose setup. This further goes to the following period of structure which is realized a preliminary design phase.

Preprocessing, solution, postprocessing

Finite element analysis facilitates to envisage potential design issues and diminish risk to your product, profits, and business. The mathematical models are used here to measure the effects of real-world conditions while assembling the products. After the simulations is performed it allows the engineers to locate budding problems in a design, including areas of tension and weak spots.

Preprocessing

The limited component work subdivides the geometry into components, whereupon are discovered hubs. The hubs, simply point to areas that are commonly situated at the component corners and maybe close to each mid-size. The observed components are 2D "twisted" marginally to fit in with a 3D surface. A model is the slight shell straight trapezoidal; flimsy shell suggests traditional shell hypothesis, direct characterizes the interjection of numerical amounts over the component, and quadrilateral portrays the geometry. For a 3D strong examination, the components have a physical thickness in each of the three measurements. Basic precedents incorporate strong straight blocks and strong illustrative tetrahedral components.

The model's degrees of opportunity (dof) is relegated at the hubs. Strong components, by and large, have three translational dof per hub. Pivots are cultivated through interpretations of gatherings of hubs concerning different hubs. Slight shell components, then again, have six dof per hub: three interpretations and three pivots. The expansion of rotational dof takes into consideration assessment of amounts through the shell, for example, twisting worries because of pivot of one hub concerning another. In this manner, for structures in which the traditional slender shell hypothesis is a substantial estimation, conveying additional dof at every hub sidesteps the need of displaying the physical thickness. The task of nodal dof additionally relies upon the class of examination. For a warm investigation, for instance, just a single temperature dof exists at every hub.

Previously, hub areas were entered in physically to estimate the geometry. The more present-day approach is to build up the work legitimately on the CAD geometry, which will be (1) wireframe, with focuses and bends speaking to edges, (2) surfaced, with surfaces characterizing limits, or (3) strong, characterizing where the material is. The strong geometry is favored, yet regularly a surfacing bundle can make an intricate mix that a solids bundle won't deal with. To the extent of geometric detail, a basic principle of FEA is to "show what is there", but then rearranging suppositions just should be connected to maintain a strategic distance from gigantic models. Examiner experience is of the pith.

The geometry is fit with a mapping calculation or a programmed free-lattice calculation. The principal maps a rectangular matrix onto a geometric district, which should in this manner have the right number of sides. Mapped lattices can utilize the exact and shabby strong direct block 3D component, however, can be very tedious, if certainly feasible, to apply to complex geometries. Free-cross section consequently subdivides coinciding districts into components, with the benefits of the quick lattice, simple work measure changing (for a denser work in locales of the expansive slope), and versatile abilities. Burdens incorporate age of tremendous models, the age of misshaped components, and, in 3D, the utilization of the fairly costly strong illustrative tetrahedral component. It is constantly essential to check basic bending before the arrangement. A seriously mutilated component will cause a network peculiarity, murdering the arrangement. A less misshaped component may fathom, yet can convey poor answers. Worthy dimensions of contortion are reliant upon the solver being utilized.

Material properties required fluctuate with the sort of arrangement. A direct statics examination, for instance, will require a versatile modulus, Poisson's proportion and maybe a thickness for every material. Warm properties are required for a warm examination. Instances of limitations are announcing a nodal interpretation or temperature. Burdens incorporate powers, weights, and warmth transition. It is desirable to apply limit conditions to the CAD geometry, with the FEA bundle exchanging them to the hidden model, to take into consideration easier use of versatile and advancement calculations. It is significant that the biggest blunder in the whole procedure is frequently in the limit conditions. Running various cases as an affectability investigation might be required.

Solution

While the prepreparing and posthandling periods of the limited component technique are intuitive and tedious for the examiner, the arrangement is frequently a clustering procedure and is requesting of PC asset. The administering conditions are gathered into lattice structure and are comprehended numerically. The get-together procedure depends not just on the kind of investigation (e.g., static or dynamic), yet additionally on the model's component types and properties, material properties and limit conditions.

On account of a direct static basic investigation, the collected condition is of the structure $Kd = r$, where K is the framework firmness grid, d is the nodal level of opportunity dof removal vector, and r is the connected nodal load vector. To value this condition, one must start with the hidden versatility hypothesis. The strain-relocation connection might be brought into the pressure strain connection to express worry as far as dislodging. Under the suspicion of similarity, the differential conditions of harmony working together with the limit conditions at that point decide an interesting removal field arrangement, which thus decides the strain and stress fields. The odds of straightforwardly tackling these conditions are probably nothing to anything besides the most minor geometries, thus the requirement for surmised numerical systems presents itself.

A limited component work is a dislodging nodal uprooting connection, which, through the component insertion conspire, decides the relocation anyplace in a component given the estimations of its nodal dof. Bringing this connection into the strain-relocation connection, we may express strain regarding the nodal dislodging, component insertion plan and differential administrator network. Reviewing that the articulation for the potential vitality of a flexible body incorporates an indispensable for strain vitality put away (subordinate upon the strain field) and integrals for work done by outside powers (subordinate upon the uprooting field), we can subsequently express framework potential vitality as far as nodal removal.

Applying the standard of least potential vitality, we may set the incomplete subsidiary of potential vitality regarding the nodal dof vector to zero, bringing about: a summation of component firmness integrals, increased by the nodal dislodging vector, breaks even with a summation of burden integrals. Every solidness vital outcomes in a component firmness network, which total to create the framework solidness lattice, and the summation of burden integrals yields the connected burden vector, bringing about $Kd = r$. By and by, coordination rules are connected to components, loads show up in the r vector, and nodal dof limit conditions may show up in the d vector or might be divided out of the condition.

Arrangement strategies for limited component lattice conditions are ample. On account of the straight static $Kd = r$, reversing K is computationally costly and numerically temperamental. A superior system is a Cholesky factorization, a type of Gauss end, and a minor departure from the "LDU" factorization topic. The K network might be productively calculated into LDU, where L is lower triangular, D is corner to corner, and U is upper triangular, bringing about $LDUd = r$. Since L and D are effectively reversed, and U is upper triangular, d might be controlled by back-substitution. Another famous methodology is the wave front technique, which collects and decreases the conditions in the meantime. Probably the best present-day arrangement strategies utilize meager network procedures. Since hub-to-hub stiffnesses are nonzero just for close-by hub combines, the firmness network has countless passages. This can be misused to decrease arrangement time and capacity by a factor of at least 10. Improved arrangement strategies are constantly being created. The key point is that the investigator must comprehend the arrangement method being connected.

Dynamic investigation for such a large number of experts implies ordinary modes. Learning of the regular frequencies and mode states of a structure might be sufficient on account of a solitary recurrence vibration of a current item or model, with FEA being utilized to explore the impacts of mass, solidness and damping changes. When exploring a future item, or a current plan with numerous modes energized, constrained reaction demonstrating ought to be utilized to apply the normal transient or recurrence condition to evaluate the relocation and even powerful worry at each time step.

This discourse has accepted h-code components, for which the request of the insertion polynomials is fixed. Another method, p-code, builds the request iteratively until assembly, with mistake gauges accessible after one investigation. At long last, the limit component technique places components just along with the geometrical limit. These systems have restrictions, yet hope to see a greater amount of them sooner rather than later.

Postprocessing

After a limited component show has been arranged and checked, limit conditions have been connected, and the model has been tackled, the time has come to research the aftereffects of the investigation. This movement is known as the posthandling period of the limited component technique.

Posthandling starts with an exhaustive check for issues that may have happened amid arrangement. Most solvers give a log document, which ought to hunt down alerts or mistakes, and which will likewise give a quantitative proportion of how respectful the numerical techniques were amid arrangement. Next, response loads at controlled hubs ought to be summed and inspected as a "once-over to make sure everything seems ok". Mistake standards, for example, strain vitality thickness and stress deviation among neighboring components may be taken a gander at next, yet for h-code examinations, these amounts are best used to target consequent versatile remeshing.

When the arrangement is checked to be free of numerical issues, the amount of intrigue might be inspected. Many presentation alternatives are accessible, the decision of which relies upon the numerical type of the amount just as its physical significance. For instance, the dislodging of a strong direct block component's hub is a three-part spatial vector, and the model's general removal is regularly shown by superposing the distorted shape over the unreformed shape. Dynamic survey and movement capacities help significantly in acquiring a comprehension of the misshaping design. Stresses, being tensor amounts, right now do not have a decent single representation procedure, and consequently determined pressure amounts are separated and showed. Chief pressure vectors might be shown as shading coded bolts, demonstrating both heading and greatness. The size of vital burdens or of a scalar disappointment stress, for example, the Von Mises pressure might be shown on the model as hued groups. At the point when this kind of showcase is treated as a 3D object exposed to light sources, the subsequent picture is known as a shaded picture pressure plot. Dislodging extent may likewise be shown by hued groups, however, this can prompt confusion as a pressure plot.

A region of postpreparing that is quickly picking up ubiquity is that of versatile remeshing. Mistake standards, for example, strain vitality thickness are utilized to remesh the model, setting a denser work in locales requiring improvement and a coarser work in regions of pointless excess. Additivity requires a cooperative connection between the model and the hidden CAD geometry, and works best if limit conditions might be connected straightforwardly to the geometry, also. Versatile remeshing is an ongoing exhibit of the iterative idea of h-code examination.

Streamlining is another territory getting a charge out of late headway. In view of the estimations of different outcomes, the model is adjusted naturally trying to fulfill certain execution criteria and is illuminated once more. The procedure repeats until some combination paradigm is met. In its scalar structure, streamlining changes shaft cross-sectional properties, flimsy shell thicknesses or potentially material properties trying to meet most extreme pressure limitations, greatest diversion requirements, as well as vibrational recurrence imperatives. Shape advancement is progressively mind-boggling, with the genuine 3D display limits being adjusted. This is best cultivated by utilizing the driving measurements as advancement parameters, however, work quality at every emphasis can be a worry.

Another bearing noticeable in the limited component field is the incorporation of FEA bundles with alleged "instrument" bundles, which break down movement and powers of vast dislodging multibody frameworks. A long haul objective would be ongoing calculation and show of relocations and worries in a multibody framework experiencing expansive uprooting movement, with frictional impacts and liquid stream considered when important. It is hard to appraise the expansion in registering power important to achieve this accomplishment, however, 2–3 requests of greatness is likely close. Calculations to coordinate these fields of investigation might be relied upon to pursue the processing power increments.

In synopsis, the limited component strategy is a general ongoing control that has rapidly turned into a developed technique, particularly for auxiliary and warm investigation. The expenses of applying this innovation to regular plan assignments have been dropping, while the abilities conveyed by the technique extend continually. With training in the strategy and in the business programming bundles ending up increasingly accessible, and people started claiming FEA has to be applied. The strategy is completely fit for conveying higher quality items in a shorter structure cycle with a decreased possibility of field disappointment if it is connected by a fit examiner. It is likewise a substantial sign of careful structure rehearses, should a sudden suit crop up. Now is the ideal time for industry to utilize this and different examination strategies.

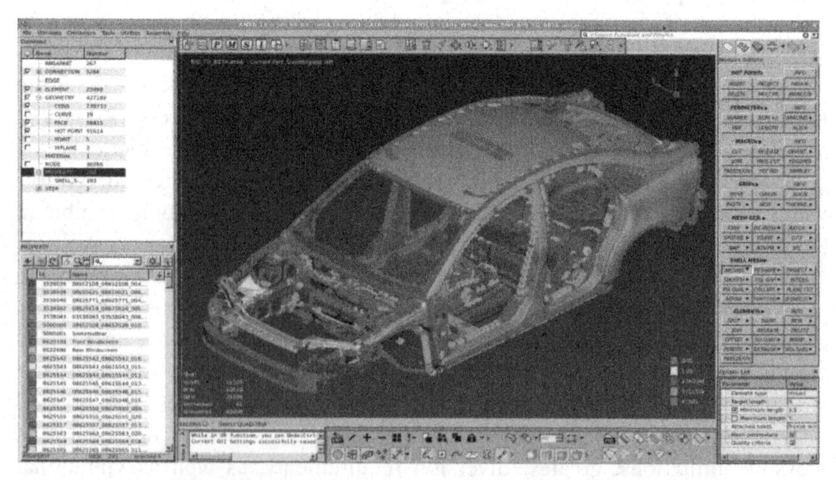

FIGURE 10.5 ANSA window.

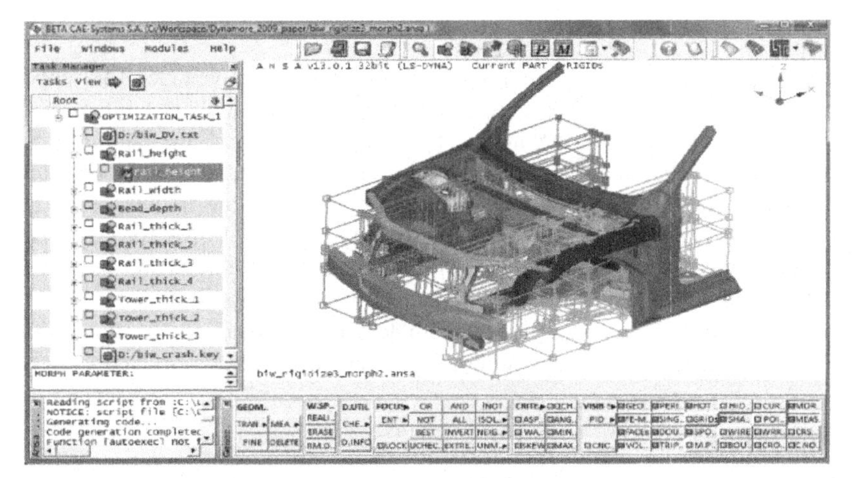

FIGURE 10.6 Task manager building.

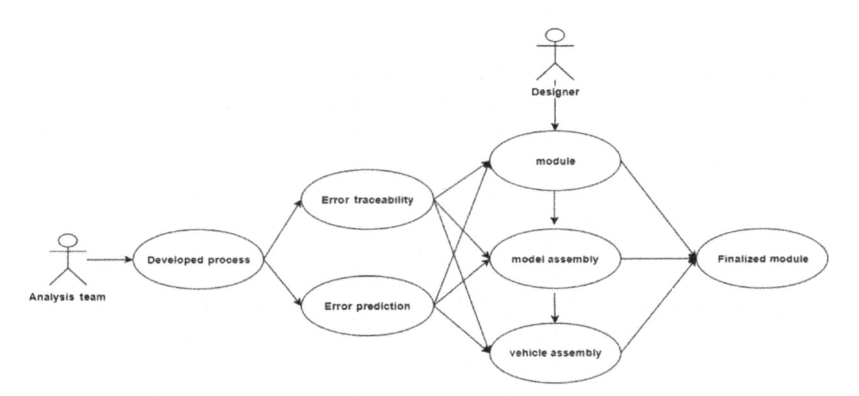

FIGURE 10.7 Task of analysis team.

Interacting with ANSA using API functions

The ANSA application programming interface (API) allows developers to access the ANSA core functionality and data. The ANSA API is an extension of the Python programming language and its window is shown in Fig. 10.5.

The basic idea behind the scripting language is to automate many repetitive and tedious procedures with the minimum user interaction or even to perform specific tasks that are not covered by the standard ANSA commands. Some of the tasks that can be performed using Python. It is known as a clear and highly readable language with a large and comprehensive standard library. Python can be utilized for both object-oriented and functional programming. It implements a dynamic typing system and automatic memory management. The task manager building is depicted in Fig. 10.6 and the task of analysis team is shown in Fig. 10.7.

Conclusion

CAE is one of the widely used approach for modeling a vehicle. The scope of this study is to automate the process by using CAE tools in-built within, for error traceability and prediction, by using this approach the rate of error occurrence in the model is reduced, and it also saves the designer's time on predicting the errors. Error tracing and error prediction are the major issues the CAD engineer faces while developing the model. Backpropagation Neural networks and Naive Bayes algorithms are used for tracing and predicting the errors respectively, by automating the process. ANSA is used in our work to automate the process with less user interaction. It allows the developers to access the ANSA core functionality and data.

References

Benaouali, A., & Kachel, S. (2017). An automated CAD/CAE integration system for the parametric design of aircraft wing. *Journal of Theoretical and Applied Mechanics, 55*(2), 447–459.

Danglade, F., Pernot, J.-P., & Véron, P. (2013). On the use of machine learning to defeature CAD models for simulation. *Computer-Aided Design and Applications, 11*(3), 358–368. Available from https://www.researchgate.net/publication/271625563.

Dangladea, F., Pernota, J.-P., Vérona, P., & Fine, L. (2017). A priori evaluation of simulation models preparation processes using artificial intelligence techniques. *Science Direct, 91*, 45–61. Available from https://doi.org/10.1016/j.compind.2017.06.001.

Diez, C. (2016). *Machine learning process to analyze big-data from crash simulations. 7th BETA CAE international conference.*

Garcke, J., & Teran, R. I. (2015). *Machine learning approaches for repositories of numerical simulation results. 10th European LS-DYNA conference 2015, Würzburg, Germany.*

Karim, S., Piano, S., Leach, R., & Tolley, M. (2018). Error modelling and validation of a high-precision five degree of freedom hybrid mechanism for high-power high-repetition-rate laser operations. *Precision Engineering, 54*, 182–197.

Kulkarni, Y., Kumar, R., Mukund, G., & Bernard, K. A. (2016). Leveraging feature information for defeaturing sheet metal feature-based CAD part model. *Computer-Aided Design and Applications, 13*(6), 885–898. Available from https://doi.org/10.1080/16864360.2016.1168238.

Mutilba, U., Gomez-Acedo, E., Kortaberria, G., Olarra, A., & Yagüe-Fabra, J. A. (2017). Traceability of on-machine tool measurement. *Sensors, 17*(7), 1605. Available from https://doi.org/10.3390/s17071605.

Rath, M., Lo, D., & Mäder, P. (2018). *Analysing requirements and traceability information to improve bug localization. MSR'18: Proceedings of the 15th international conference on mining software repositories* (pp. 442–453). New York: Association for Computing Machinery. Available from https://doi.org/10.1145/3196398.3196415.

Seegmiller, N., Rogers-Marcovitz, F., Miller, G., & Kelly, A. (2013). Vehicle model identification by integrated prediction error minimization. *The International Journal of Robotics Research, 32*(8), 912–931.

Tsirogiannis, E. C. (2015). *Design of an efficient and lightweight chassis, suitable for an electric car* (Thesis). Available from https://doi.org/10.13140/RG.2.2.30711.21920.

M. Yanagisawa, M. Kato. (2017). Automatic generation of high-quality CAE model using ANSA. 7th BETA CAE International Conference.

Chapter 11

All about human-robot interaction

Kiran Jot Singh[1], Divneet Singh Kapoor[2] and Balwinder Singh Sohi[2]
[1]ECE Department, Chandigarh University, Mohali, India, [2]Embedded Systems & Robotics Research Group, Chandigarh University, Mohali, India

Introduction

We are living in the time where robots are fulfilling various key roles in our society. We can see the presence of robots from factories to homes, from comics to movies and even The Holy Bible (Revelation 9:9, 2011). This gives rise to the question, what is a robot? The name "Robot" originated from a science fiction play by the Czech writer Karel Capek (Long, 2011). A robot can be defined as a goal-oriented machine that can sense, plan and act (Corke, 2011). In earlier days, robots were used in monotonous tasks which appeared boring to human or at the places where there is a possible threat to human life for example, mines, underwater exploration, etc. But these days robots are getting into activities which do not have a certain procedure and additionally complex.

In due course of time, robotics has advanced from fiction to reality. Software, mechanical, electrical engineering, and artificial insight (AI) is only a portion of the orders that are consolidated in modern robotics. The principle goal of robotics is the development of robots/gadgets that perform client characterized assignments. The quick development of the field in logical terms has driven the improvement of various kinds of robots such as humanoids, aerial, social, service, industrial robots and many more. The evolution of robotics over last decades is an amalgamation of different generation of robots as shown in Fig. 11.1. The advancement of robotics is affected by the innovative propels, for instance, the development of the first transistor (Kiesler & Goodrich, 2018) to integrated circuits (Asimov, 2004) further leading to the digital computer (Lehoux & Grimard, 2018). These innovative advancements further upgraded the robots and made a huge difference to develop from exclusively mechanical or pressure-driven machines to comply with programmable frameworks, which can now even know about their

Cognitive Computing for Human-Robot Interaction. DOI: https://doi.org/10.1016/B978-0-323-85769-7.00010-0

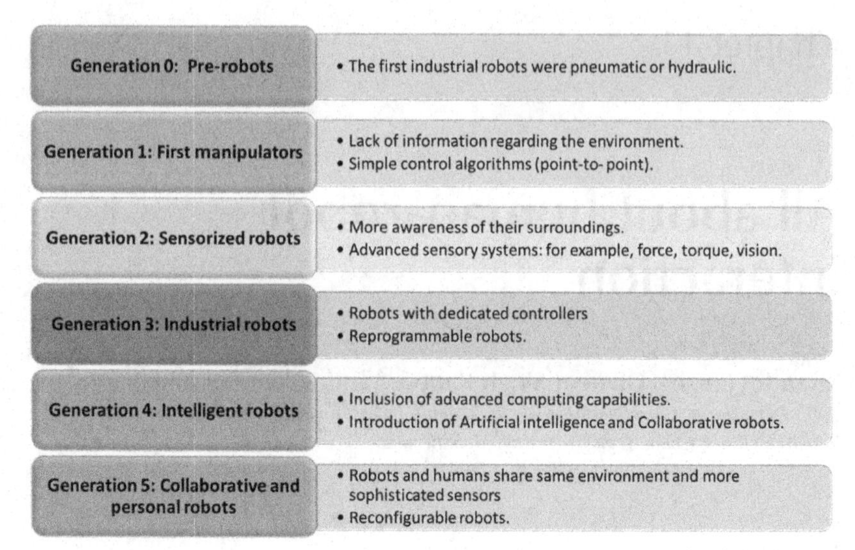

Generation 0: Pre-robots
• The first industrial robots were pneumatic or hydraulic.
Generation 1: First manipulators
• Lack of information regarding the environment.
• Simple control algorithms (point-to- point).
Generation 2: Sensorized robots
• More awareness of their surroundings.
• Advanced sensory systems: for example, force, torque, vision.
Generation 3: Industrial robots
• Robots with dedicated controllers
• Reprogrammable robots.
Generation 4: Intelligent robots
• Inclusion of advanced computing capabilities.
• Introduction of Artificial intelligence and Collaborative robots.
Generation 5: Collaborative and personal robots
• Robots and humans share same environment and more sophisticated sensors
• Reconfigurable robots.

FIGURE 11.1 Different robot generations.

surroundings. These recent advancements are bringing robots closer to human for the completion of task, which in turn requires the interaction of humans and robots. This visualization has unlocked doors for a new interdisciplinary research area called human-robot interaction (HRI).

HRI is a juncture of different areas that is computer science, humanities, social sciences, electronics, and mechanical engineering. The ultimate aim of HRI is to comprehend, recognize, and assess robotic systems by, for or with humans (Goodrich & Schultz, 2007). Communication between a robot and human is referred to as interaction. So, HRI is the science of interaction, at its core (Kiesler & Goodrich, 2018), but one cannot assume human being as known entity. Study of humans in HRI is essential to establishing norms for human–robot collaboration. One must take into consideration various cultures, groups and individuals of different ages while designing HRI. The famous three laws of robotics, originally coined by Asimov (2004) as shown in Fig. 11.2.

The research in HRI is intended to find ways by which a human and robot can interact. Various researchers are working toward developing HRI for social welfare (Lehoux & Grimard, 2018) but security and privacy are of major concern particularly in the domain of Internet of Things (IoT) (Singh & Kapoor, 2017). Advanced methods like machine learning has given the ability to plan and deploy smart robots for long term interactions (Senft, Lemaignan, Baxter, & Belpaeme, 2017). HRI can be broadly classified into two types that is physical HRI (pHRI) and cognitive HRI (cHRI) (Xing & Marwala, 2018). As the name suggests, "physical" in pHRI allows robot (equipped with sensors, control and interaction & motion planning) to

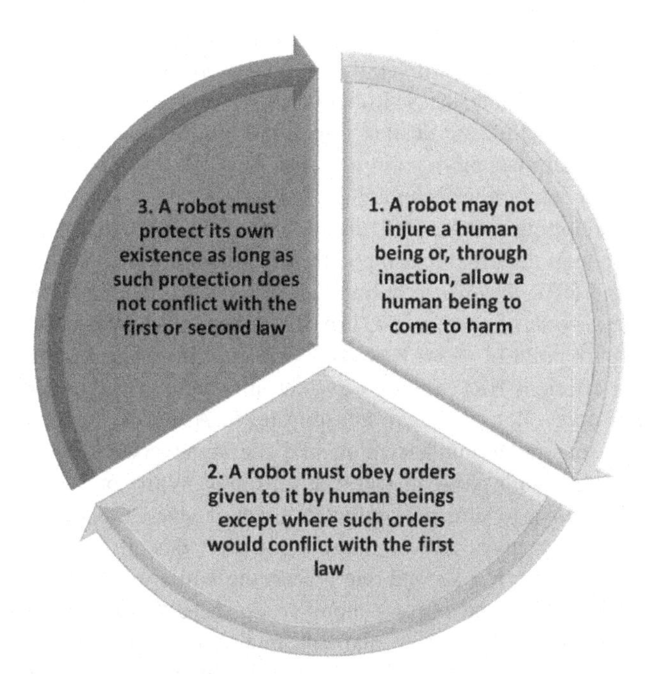

FIGURE 11.2 Asimov three laws of robotics.

closely interact with humans. Studies related to pHRI can be further subdivided into coordinated, collaborated and cooperated type of interaction. While on the other hand, cHRI facilitate the system design of robot by establishing appropriate algorithms and accurate models. The models in cHRI can be further subdivided into three types namely, human models for interaction, robot models for interaction and models for HRI.

In broader terms, HRI can be thought of as developing rules, software design for robots, testing, detecting problems and re-programming. Modern HRI emphasizes more on dynamic and efficient interactions rather than intermittent interactions in programming-based approach (Goodrich & Schultz, 2007).

Understanding human-robot interaction and human-computer interaction

The interactions study between computers and humans is known as human-computer interaction (HCI). HCI offers many vital insights but still, HRI is related yet very different domain from HCI (Dautenhahn, 2007). The idea of HRI is to make robots behave in a natural way like humans so that people do not have to act like robots while interacting with them (Reeves & Nass, 1996). HCI which used to be the interaction through keyboards and mouse, has grown to speech, character, and gesture recognition, graphical-user-interfaces, multitouch and is diversifying even more. But, HRI is not merely about software's and interfaces. Robots also have motors packed inside different types of metal/plastic designs which are mobile, look similar to living beings (human, animals etc.), so in turn, are perceived differently by humans as compared to computing technologies.

While developing robots, HCI offers a lot of support in designing technology for human needs. But computers and robots are very different things so as HCI and HRI. It is worth noting that in robotics, human and robot both are the integral part of system. So, in HRI, system is evaluated from human as well as robot point of view. Few points should always be kept in mind for a better and efficient HRI system. First, the interactions between robots and humans are more often, physical and long term. As robots are mobile, they are going to encounter humans more and the same goes for humans. But both have to deal with physical world at all times. While on the other hand, in HCI people have to turn on computers or their smart devices for interactions. So, mobility gives rise to many significant design challenges which are not there in HCI. The second one is learning while interacting. As robots have multiple sensors and possess mobility, so the data to be processed and analyzed by a robot is diverse and enormous. This gives further rise to design challenges in HRI like when to learn by themselves or from humans (e.g., Locating water pump of the garden) and how to take decisions by their own (scheduling when to water plants), which are not covered in HCI.

HCI techniques cannot be applied directly in the design and assessment of collaborating interactive robots. The varying cultural notions of "What a robot is?" and physical forms, makes it an area that is essentially different from the design of programs/software being executed by hardware platforms. Therefore, understanding of HRI and HCI techniques will be really helpful for designing better interactive robots. The theoretical study, paradigms, frameworks of HRI, can be used to apprise, guide and design improved/modified HCI techniques or develop new methods in order to develop a superior interactive robot (Fernaeus et al., 2009; Sucar, Aviles, & Miranda-Palma, 2003), as depicted in Fig. 11.3.

HRI research, encompasses of three foundations named, "aesthetic," "operational," and "social interaction." In pursuance of achieving user-friendliness/quality of interaction, social interaction is given preference above the other two parameters, catering to factors such as autonomy, interactivity and modality (Kim, Oh, Choi, Jung, & Kim, 2011). This knowledge of social interaction can guide designers to enhance existing HCI techniques while implementing intelligence. HRI can be helpful in evaluating roles and emotion of robot well in advance for robot in social interaction between robot and human.

The division of HRI can be done broadly into two approaches that is "Robot-centered HRI" and "User-centered HRI" (Kim et al., 2011).

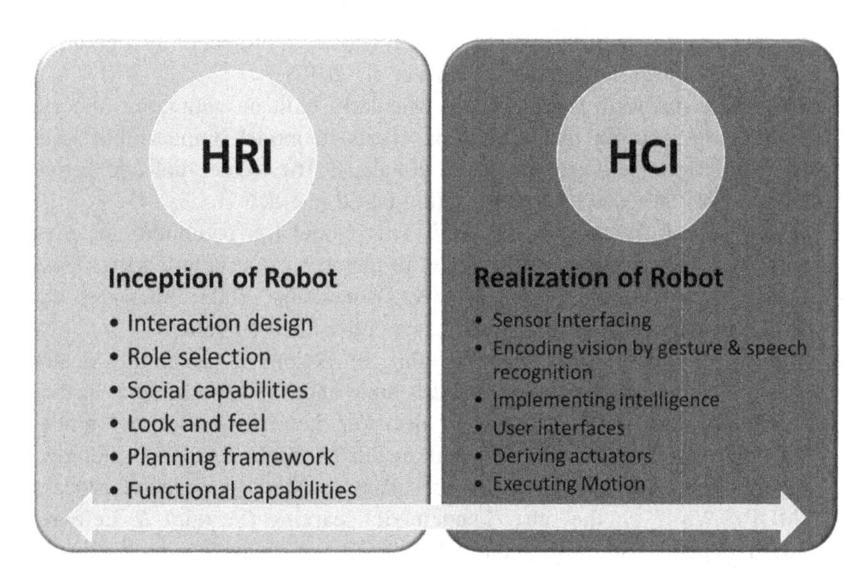

FIGURE 11.3 Linkage of human-robot interaction and human-computer interaction .

Robot-centered HRI focuses on, how robots can be made effectual to perceive, cognize, and perform in any given surroundings in order to achieve better interaction with humans. While on the other hand, User-centered HRI focuses on, how robot users observe and respond to robots, over how robots cognize and accommodate humans. So prior knowledge of these HRI factors can be assimilated through HCI techniques in interest of commendable implementation.

There is a requirement of a generalized framework for HRI encompassing various essential parameters/factors required for HRI in different domains. This framework will set a broader picture of different HRI blocks in form of general steps, so that this knowledge can applied to different problem domains. The framework discussed below, segregate complete process into various blocks and specify the flow of information. It presents an end-to-end design overview to facilitate hobbyists and researchers for creating social robots.

Framework for human-robot interaction

A framework is an organized approach for incorporating various problem domains. It provides roboticists with a platform to approach a task for a particular domain in a systematic way (Hoffman & Ju, 2014). The main spirit of any framework is to separate the complete progression into differentiable blocks so as to make the planning of the whole process organized and efficient.

1. *Dialog-based model:* In this model human and robot addresses a domain task, by engaging themselves in a dialog to share information. Such

models are very helpful in tasks like navigation, multi-robot teleoperation and collaborative exploration (Vanzo et al., 2017).

2. *Simulation-theoretic model:* It is particularly built on simulation theory, particularly inspired by observation elicits to model human-robot joint activity (Breazeal, Dautenhahn, & Kanda, 2016). Perceptual and active components are used to achieve action-based interaction.
3. *Intention and activity-based model:* This model has preconditions, executable & until conditions employed in hierarchical structure with a goal centric view (Fong, Nourbakhsh, & Dautenhahn, 2003; Sheng et al., 2015). Every action in this model has a pre-defined outcome.
4. *Models for action planning:* Planning of actions is under physically dynamic and cognitive environment is done under this model. It is further subdivided into two types that is decision theoretic model and model learning. Policies are formed and evaluated under decision theoretic model (Unhelkar et al., 2018) for example Markov decision process (MDP). While on the other hand model learning (Ferreira & Lefevre, 2015) depends on how accurately choices of transition probabilities, observational probabilities and reward have been made.
5. *Cognitive models of robot control:* It is intended to deliver better efficiency for tasks under human-robot teams (Chien et al., 2018). It addresses problems between operators and robots, by exploiting human cognition mechanism such as mental models or working memory, in order to develop control interfaces and models.

Taking cue from above listed models, a framework for HRI is shown in Fig. 11.4, which constitutes of six blocks. The first block is about analysis of

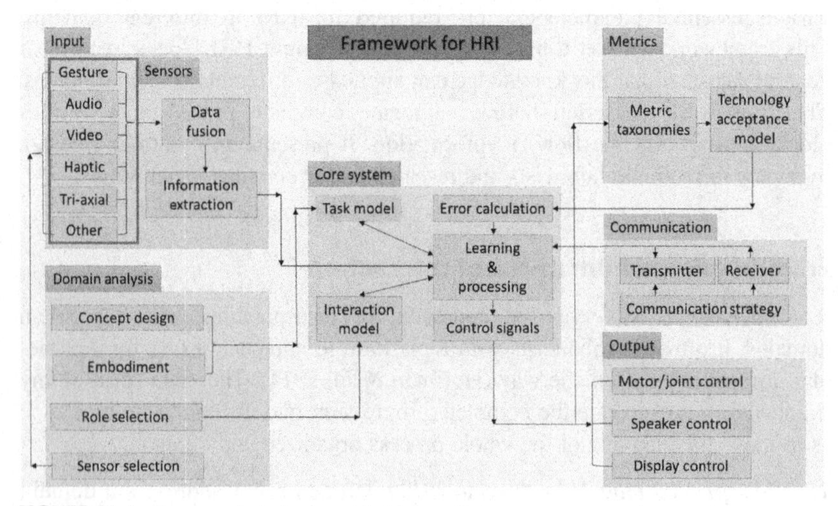

FIGURE 11.4 Framework for human-robot interaction.

domain, which defines parameters of the problem domain. The second block is input, which is resultant of data from the sensors, leading to information extraction. The third one is the core system, which is the heart of the HRI and does all the processing tasks. Fourth one is the metric taxonomies, to evaluate effectiveness of HRI. The fifth section is communication, which is responsible for transmission and reception of data. Final block is output block, where output devices are actuated by control signals. A detailed description of all blocks of framework is given below.

Domain analysis

One of the key issues of HRI is modeling of unstructured problems in different domains. With advances in robotic technology, the number of areas/domains is growing fast where robots are helpful to humans. Almost every domain involves HRI. More importantly, several of these problem domains have a wide social impact. Thus, HRI is coming up as a great multidisciplinary research area in order to address needs of society in different realms of life. There are many exciting research issues in different domains that can be explored by HRI research. Few of them are discussed below:

Concept design

The theory of making the machine for doing various tasks is leading human to make robots that can support human beings in different realms of life. As technological advances are growing, robotics is flourishing day by day which in turn is escalating HRI. Robots are being used for many purposes from domestic to defense. Nowadays, roboticists are designing application-specific robots catering to special needs for specific application only. The application areas include different types of domains (Siciliano & Khatib, 2016), for example, industrial, space, agriculture, construction, mining, surveillance, medical and competitions. Robots have to perform different tasks in different application areas and possess varied potentials.

It is very vital for designers to comprehend the concept design for various application areas. Designers must know the end users and associated tasks for developing the inspiration leading to task analysis (Myers, 1998). In order to ensure the appropriate design, designers need to build rapid and frequent prototypes which need to be tested on users with many iterations. As an initiation towards concept design a team of designers observe and evaluate the environment for patterns of actions, cultural affairs and artefacts. Diagrams and relations are established from experiences that lead to the design of storyboards, prototypes and new technology (Druin, 1999). In order to comprehend application environment and understand whether final impression of robot will be ready for use or not, designers use several

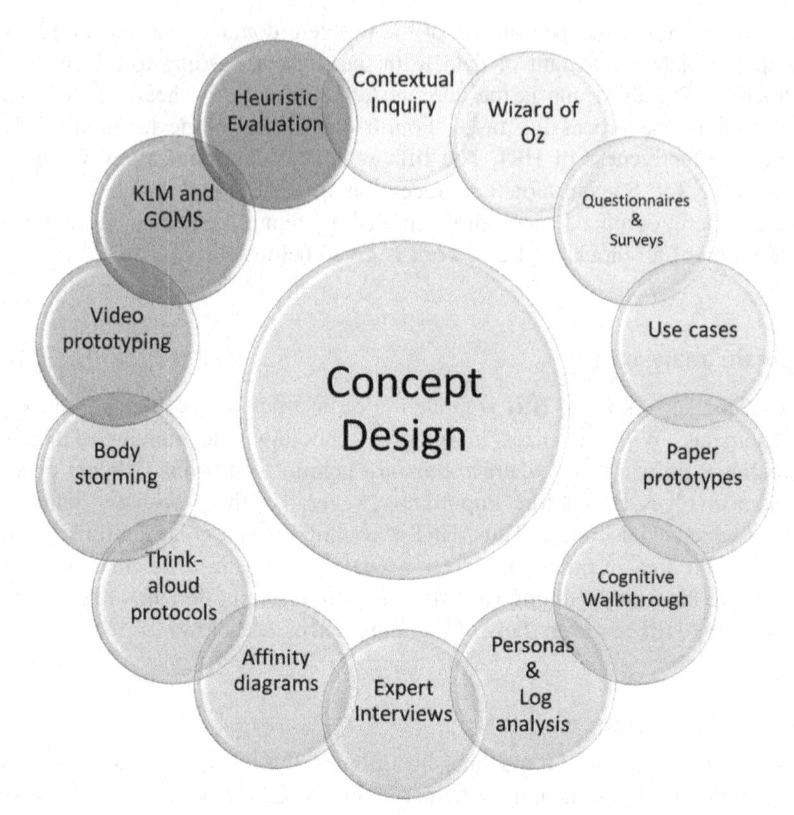

FIGURE 11.5 Various methods for concept design.

methods such as, participatory design, contextual review, improvisation with prototypes, etc. as shown in Fig. 11.5.

Embodiment

The embodiment is repeatedly conceptualized as "Intelligence requires a body and environment to progress" (Spenko et al., 2008). The morphology (structure and form) of the robot affects system behavior and helps the robot to develop social expectations (Pfeifer & Bongard, 2006). The physical embodiment of a robot is a vital aspect and has a prominent effect on the perception of robots in HRI. The core feature which is straight forwardly linked to the physical shape of that is the material and realism reality of the object in real world. Catering to same domain, a framework focusing bodily related capabilities to explain effects of varying embodiments has been proposed by Hoffmann, Bock, and Rosenthal vd Pütten (2018). The embodiment can be divided into four broad classes described below (Fong, Nourbakhsh, & Dautenhahn, 2003).

1. *Anthropomorphic*: It is the tendency to feature human physiognomies (Duffy, 2003). A robot should possess similarities to human structurally and functionally.
2. *Zoomorphic:* Designing robotic companions with designs inspired by household pets.
3. *Caricatured:* Designing characters to create anticipated interaction biases particularly to distract or focus attention on specific robot features.
4. *Functional:* Designing physical features of a robot primarily to achieve operational objectives.

Both anthropomorphic and nonanthropomorphic interface can interact with users and can be considered embodied. Another way to look at embodiment is creative embodied interfaces that caters to both of the interfaces, either present in physical or virtual world. Its major components are level of embodiment, creative context and theoretical models of creativity (Mancini & Varni, 2018). So, the morphology of a robot is intensely linked to its perceived capabilities. Recent work is being carried out on various embodiments like humanoid (Fasola & Matarić, 2013), zoomorphic (Li & Chignell, 2011), without manipulators (e.g., eMuu (Bartneck, 2003), iCat (Li & Chignell, 2011)), with manipulators (e.g., Nao (Kennedy, Baxter, & Belpaeme, 2015), Asimo (Kamide, Mae, Takubo, Ohara, & Arai, 2014)), stationary robots (e.g., Nabaztag (Hoffmann & Krämer, 2013)) and ability to move (e.g., iCub (Fischer, Lohan, & Foth, 2012), Aibo (Lee, Jung, Kim, & Kim, 2006)).

Roles

Conventionally, robots play the role of machines in the human world. Human and robots take up different roles in various situations in order to complete the tasks. Various HRI researchers came up with different roles listed below, which can be assigned to humans as well as robots in different application areas (Goodrich & Schultz, 2007; Scholtz, 2003).

1. *Supervisor:* Human monitors robot behavior but does not directly control it.
2. *Operator:* Manual supervision and interaction take place in this role.
3. *Mechanic:* Programmer or mechanic changes hardware or software physically.
4. *Peer:* Both human and robot work together to accomplish the task.
5. *Bystander:* Small understanding of robot work is required in order to ensure safety.
6. *Mentor:* Robot act as leader or teacher for human
7. *Information consumer:* Human makes use of information sent by the robot.

One of the frequently used technique by HRI researchers' is the Wizard-of-Oz (WoZ), in which a person remotely operates a robot for controlling number of things like, navigation, movement, gesture, speech, etc. (Hoffman, 2016). Researchers who employ WoZ argue that robots are not adequately advanced to interact independently with people in socially suitable or physically safe ways. So they control the robot in which robot can possess any of the above defined roles as per need of application, to test out initial phases of their design that are yet to be fully realized. Table 11.1 shows the research targeting robot role capabilities in applications domains like healthcare, education, workspaces and homes.

Sensor selection

Sensors are robots' eyes to the surroundings. Sensor data coming from different parts of robot is passed to processor extract information. Sensor selection is one of the crucial factors in robotics which primarily depends upon task requirements. Further sensor precise parameters like range, sensitivity, consumption of power, precision, etc. are also considered but the first step is selecting sensors types for the completion of task. Sensors for robotics can be classified into following categories (Deisenroth, Neumann, & Peters, 2013) discussed below.

1. *Haptic sensors:* Sensors used for detection of touch or degree of closeness. It delivers information about surface properties and forces of interaction at points of contact between the objects and robot fingers. For example capacitive, piezo-resistive, optical sensors, multimodal sensors etc.
2. *Tri-axial sensors:* Sensors used for calculating roll, pitch & yaw angles. Also used to find out acceleration, pose & orientation with respect to world frame, for example, accelerometers, magnetometers and gyroscopes.
3. *Vision & audio-based sensors:* Sensors used for capturing images, videos and audio, for example, CMOS cameras and microphone.
4. *Torque sensor*: It is used to measure exchanged contact forces and torques in order to cater to physical interaction between human and robots.
5. *Temperature & gas sensors:* Sensors used to sense the temperature of the environment including the robot and detect the presence of different type of gases in surroundings.
6. *Location and active ranging sensors:* Sensors used to find out locations and make geometric calculation, for example, global positioning system (GPS), ultrasonic sensors, laser range finders etc.

Input

Sensing the nearby environment is a necessary task for a robot. Input block incorporates all the necessary sensing mechanisms for a robot like a camera, microphone, ultrasonic sensors etc. in order to recognize modalities like

TABLE 11.1 Robot roles in different application areas.

Domain	Robot	Study design	Robot role capabilities	Adaptive features
Healthcare	NAO (Coninx et al., 2016)	Subjects: three children with grade I diabetics No. of Interactions: three interactions per child between one and two months Interaction Type: Autonomous with WoZ controlled speech Measures: User experience Method: Questionnaires and logged Data	• Switching between activities • Display gestures • Utters Speech	• User profiling • User emotions detection • Memory adaptations
Education	LightHead (de Greeff & Belpaeme, 2015)	Subjects: 41 adults Conditions: Social versus non-social No of Interactions: one-off Interaction Type: Semi- autonomous Measures: Robots learning performance and gaze behavior Method: Questionnaires and video analysis	• Playing turn taking language • Game with a human teacher in order to learn words	• Gaze-based adaptation • User-performance • based facial and verbal expressions
Workspace	Robov (Liu, Glas, Kanda, Ishiguro, & Hagita, 2017)	Subjects: 33 adults No. of Interactions: one-off Interaction Type: WoZ controlled robot Measures: Naturalness, understand-ability, perceived politeness and overall goodness of the robot's deictic Method: Questionnaire	• Displays deictic behaviors • Through gaze • casual and • precise pointing	• User pointing behaviors • based adaptations
Home	Max (Gross et al., 2015)	Subjects: nine elderly No. of Interactions: trial for three days Interaction Type: semiautonomous Measures: technology acceptance Method: Interviews	• Display emotions recognize • Detect and • track person • Give recommendations • Understand haptic input	• User-preference and emotion based adaptations

gestures, speech, distance, etc. respectively. Basically, the input block is derived from advancements and research in the field of HCI.

Multiple sensors can obtain extended coverage, multiple viewpoints both spatially and temporally in order to reduce the uncertainty. Data coming from various sensors, selected on the basis of domain analysis form a multimodal system, is passed to data fusion module. The whole drive of data fusion module is to generate useful feature sets by processing the data coming from different sensors, which is further communicated to information extraction component. Moreover, as data generated by sensors is continuous, so fusion has to be done for fixed interval of time rather than all the time, as data can be asynchronous. Various fusion models can be taken into consideration while scheming a multisensor data fusion system for example Boyd Model, intelligence cycle–based model, Thomopoulos Model, waterfall model, Omnibus model, etc. (Adla, Bazzi, & Al-Holou, 2013; Farooq, 2006). Further information can be extracted by using technique like fuzzy logic and passed to learning and processing block.

The prime advantage of fuzzy controllers in HRI is its linguistic interpretation (Shahmaleki, Mahzoon, Kazemi, & Basiri, 2010). Compensation provided by these controllers in uncertain and nonlinear conditions is decent, which makes it a good choice for HRI applications. A fuzzy controller consists of four components as shown in Fig. 11.6.

First one is fuzzification, which takes crisp inputs (multiple sensor data) and converts it into fuzzy sets based on the understanding of the problem in application domain. Second one is rule base, which consists of IF-THEN rules specifically designed by focus groups obtained by in-depth knowledge of problem. This block is directly attached to inference engine, which performs reasoning by applying different composition methods, for example, min-max, max-min etc. Finally, output of inference engine is passed to defuzzification block, to convert fuzzy sets to crisp values/control signal so

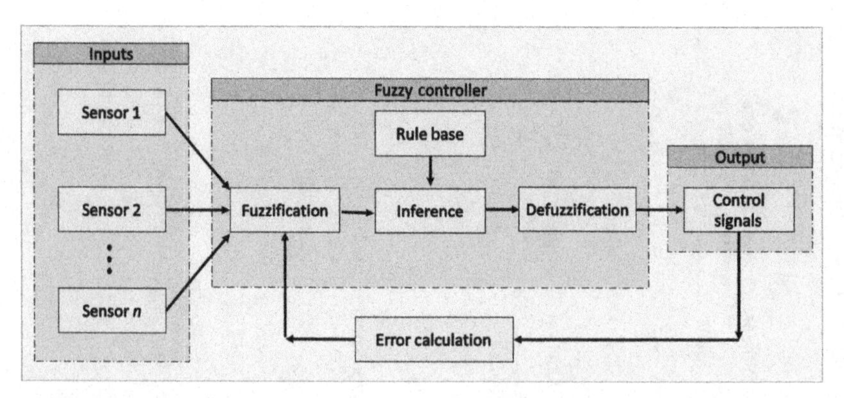

FIGURE 11.6 Fuzzy controller.

as to control output interface. A part of the output is also passed to error calculation block as feedback which is further converted to fuzzy sets in order to improve efficiency and learning.

Core system

The information coming from every block is processed by core system to generate the final output. It is further divided into five blocks which are explained below.

Task model

Making robot aware and understand about the final goal for which that robot is designed for, is one of the crucial jobs to accomplish in HRI. A robot can complete a task only if all the desired details about the domain are known to robot and human. The task model design procedures for the task which are comprehensible by the robot. There are two ways to design these procedures, first one is, from previous knowledge of circumstances provided by the expert group. Second is, it can also be based on instructions of the user in case the situation is partially or less known to a robot or human. A task model can further be divided into three sub-layers so as to organize the knowledge about task and environment (Hanheide et al., 2017). First one is *instance knowledge layer* that consists of formal information about the environment, which constitutes the knowledge about locations and categories of places belonging to different locations, where the task has to be performed. Above the first layer is the second layer, that is, *common-sense knowledge layer*, also known as default layer which consists of information about object's belongingness to a location. This layer can assume the object's location if unknown, and update the instance layer. Finally, on the top is the *diagnostic layer*, which consists of action model, that is, do's and don'ts for a robot for a particular place and has the capability to generate new knowledge. It is noteworthy that a robot can have different task models depending upon its skill level and domains in which it can work.

Interaction model

The core spirit of interaction model is communication between sender and receiver, which is subjective to the spatial distance between them means whether both are close or located at remote locations. The crucial factor is to understand that HRI is related to HCI or human-human interaction, but not similar. So, social sciences rules cannot be applied directly. Sustained interactions depend on various parameters like mood, ability, and state, etc. which change continually and dynamically. By taking the communication distance into consideration HRI can be classified into the following five types (Dautenhahn, 2007).

1. *Adaptive:* A robot should enhance its knowledge and skills by adapting to changing environment and necessities.
2. *Long/short term:* The time duration of HRI can be long or short depending on the application area.
3. *Cooperative:* Robots and humans work collaboratively on a single task and take help of each other to accomplish the same.
4. *Multimodal:* Robots are given human-like features with an emphasis on gestures, postures, emotions, expressions.
5. *Indulgent:* A robot should detect and understand human activity and act accordingly.

The interaction can be done by various means like audio, gestures, animated character(s), and send formatted information in case of large distances. The queries that rise during interaction and suggestions given to the user are managed by interaction model. It lay down the borderline of robot behavior. In order to have a broader idea of type of interactions in various domains, different application areas with respect to different parameters are presented in Table 11.2.

Diverse emotional features have been examined in the HCI studies in the HRI field recently. Few user experiences (UXs) related characteristics such as engagement, safety, teamwork, acceptance, and amiability, have been widely investigated in HRI but still not well associated with UX studies. UX design can be helpful in directing the relevance of assessing socially interactive robots and UX design which can give insights to develop new tools, in order to address several issues related to HRI. Investigations have already started (Khan & Germak, 2018; Tonkin et al., 2018), presenting the importance of crafting meaningful UX for social robots.

Based on the type of interaction explained above, interaction model can be divided into two layers (Lee, Kim, Yoon, Yoon, & Kwon, 2005). The first layer is a *reactive layer*, which is responsible for immediate response to the inputs decoded by the system. In other words, the robot evaluates the behavior of the user as a part of the input block and interaction model uses this information based on the situation to activate anticipated emotive action. On the top is the *deliberative layer*, which manages secondary emotions related to task and supports social relation with the user. This layer works in conjunction with task model and may suppress actions of reactive layer if found inappropriate with respect to task or in a state of emergency.

Learning and processing

The heart of HRI resides in learning and processing block. It takes valuable info from task model, interaction model and input block for processing and perform desired task be generating control signals. Further, it also makes utilization of diverse calculation approaches like optimization techniques, neural systems, deep learning (DL), etc. in order to enlarge and improve knowledge and learning of robot for accomplishing forthcoming tasks. The

TABLE 11.2 Different application areas and parameters.

Application area	Type of interaction	Roles			Functionality of robot	Sensors & interfaces	Contact with humans	Embodiment	Tasks of robot
			Proximate	Remote					
Surveillance	Short term adaptive	Human	Peers	Supervisor operator information consumer	• Well-defined • Can adapt for unfamiliar situations	Camera Proximity Gas Temperature Tri-axial GPS	Negligible	Functional	Search and rescue operations at inaccessible or dangerous places
		Robot	Peers						
Education	Long term adaptive multimodal	Human	Information consumer		• Initially-defined • Adapt to become more robust	Camera Speech recognition Tactile Display	Interactive	Anthropomorphic Zoomorphic	Robots as teacher, guide, companion with social skills
		Robot	mentor						
Home	Long termadaptive multimodal indulgent	Human	Supervisor Bystander	Supervisor Information Consumer	• Well-defined	Speech recognition TactileDisplay	Occasional	Zoomorphic Functional	Robots providing services/ assistance in dull & repetitive tasks
		Robot							
Industry	Long term cooperative	Human	Supervisor Peers Bystander Operator	Supervisor information Consumer Operator	• Well-defined • Interfaces to support humans	Camera Proximity GasTemperature Tri-axial	Less but ensuring user safety	Functional	Robots providing services/ assistance at manufacturing or construction units for heavy duty or automating tasks
		Robot	PeersTool						

(*Continued*)

TABLE 11.2 (Continued)

Application area	Type of interaction	Roles			Functionality of robot	Sensors & interfaces	Contact with humans	Embodiment	Tasks of robot
			Proximate	Remote					
Space	Adaptive cooperative	Human	Peers Mechanic	Supervisor Operator Information Consumer	• Well-defined	Camera Tri-axial	Very less	Functional	Robots for exploration Robots as astronaut assistant
		Robot	Peers						
Defense	Short term adaptive cooperative	Human	Supervisor Peers Mechanic Bystander Operator	Supervisor Information Consumer	• Well-defined • Interfaces to support humans	Camera Gas Chemical RADAR GPS	Less but ensuring user safety	Functional	Robots for patrol support, surveillance, transport, demining
		Robot	Peers						
Medical	Short term adaptive cooperative	Human	Supervisor Operator Information Consumer		• Well-defined	Camera Tactile Optical Temperature Pressure	Quite close	Functional	Robots doing scans, giving therapy's and performing complex procedures
		Robot	Tool						
Entertainment	Long term adaptive multimodal	Human	Supervisor Information Consumer Operator	Supervisor Information Consumer	• Initially-defined • Adapt to become more robust	Camera Speech recognition Tactile Display	Interactive	Anthropomorphic Zoomorphic Caricatured	Robot as social companion that knows individual with social skills
		Robot							

biggest challenge of this block is to maintain system firmness and dynamic behavior of the robot, which can be applicable to machine learning as well.

DL in robotics deals with very specific problems like robot learning, reasoning, embodiment challenges and many more which are typically not addressed by machine learning and computer vision. It takes immediate outputs (object segmentation, detection, depth estimation etc.) into account in order to generate actions in real world. Because of advent of DL many new features have been introduces in HRI. Features of HRI with or without involvement of DL are shown in Table 11.3 (Zhang, Qin, Cao, & Dou, 2018).

Deep reinforcement learning (DRL) techniques are widely used for achieving state-of-art results for learning, navigation and manipulation tasks in robotics (Tai, Zhang, Liu, Boedecker, & Burgard, 2016). MDP is used for formalizing robotic task in DRL by making a series of observations, leading to actions and generating rewards. MDP has five key parameters namely states, actions, transition dynamics, rewards and discount factor. The MDP can be described as a process which is in a state at each time step. The decision maker can take any action in present state. As decision maker moves to next state it receives a corresponding reward which is evaluation of action-

TABLE 11.3 Human-robot interaction (HRI) contrast with or without deep learning (DL).

Features	HRI without DL	HRI with DL
Autonomy	Low	High
Information input	Complete, accurate command	Incomplete, vague, and noisy dialog
Information type	Inherent information	Inherent information, Impromptu information (based on knowledge graph etc.)
Information environment	Physical space (low dimension)	Physical space (low dimension) Cyberspace (high dimension)
Interaction medium	Point-to-point, end-to-end mapping Contact-based user interface	Peer-to-multi or multi-to-peer Multimodal interfaces
Information processing	Sequential processing	Parallel processing
Tolerance	Low	High
Affordance	Low	High
Output and feedback	Near-time	In-time real-time

based on performance. To extract better reward form environment the agent, learn to make decisions by making use of trial and error paradigm.

Control signals & error calculation

Control signals are the output of learning and processing block, which are sent to different output interfaces to produce desired output in the real physical world. These signals can derive various output devices like motors, speakers, and displays etc. as per requirement of task & robot. A part of the output is also fed back to the metric framework so as to calculate error. Error calculation is performed by evaluating results of benchmark and summation block, and passed to learning and processing block for making effective changes in rule base to increase the efficiency of the system.

Metrics and acceptance for human–robot interaction

To evaluate the overall performance of the system comprising of human and robot as a team, diverse set of metrics is required. Usually, metrics are taken/chosen subjectively which are mostly specific to application. Considering robots as costly resources in terms of money, computation and learning involved, multiple metrics can be applied to widespread tasks in order to elevate the performance level and reduce production time. Furthermore, technology is only appreciated and valuable when it is acknowledged and used. So, an understanding of acceptance of technology is as important as acceptance. In this article, a comprehensive metric taxonomy has been shown in Fig. 11.7, derived from various studies in the literature which is frequently used by technology acceptance models.

Metric taxonomies

Metric selection for any application plays a very crucial role in the evaluation system. A designer should select metrics in such a way that all the aspects of problem domain and interaction are covered, while avoiding the selection of too many metrics which may lead to false results or correlations. These results of these need to be combined with quantitative measures in order to maintain the integrity of the system. This helps in developing superior understanding of HRI. In order to deal with this confusion, we have divided metrics into eleven categories, so as to cater wide diasporas of domains in different realms.

Navigation

The capability of a robot to move in an environment from a known location to the desired location is called as navigation. A robot can encounter various problems like route planning, and obstacle avoidance during navigation. In order to segregate the whole process navigation can be divided into three categories, which are global, local and obstacle encounter (Steinfeld et al.,

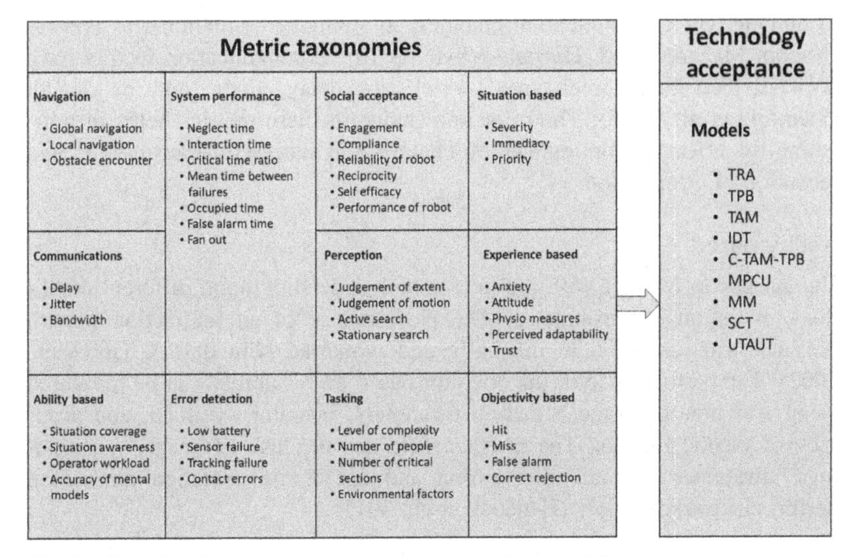

FIGURE 11.7 Metric taxonomies and technology acceptance models.

2006). These three types help in calculation of effective measures like deviation from scheduled route, hazards avoided effectively, time for accomplishment of task, operator intervention time, generation of map for unknown terrain and total area covered.

System performance

The score of success in terms of completeness and precision to achieve desired goals is known as system performance (Neerincx, 2011). With advancement in technology and growing needs, system performance can be looked upon by different perspectives like scenarios, autonomy and real-time performance while deciding the resultant metrics (Singer & Akin, 2011). Keeping the above-stated issues into concern system performance has been further broken down into seven parts, as neglect time (Goodrich & Olsen, 2003), interaction time (Goodrich & Olsen, 2003), critical time ratio (Glas, Kanda, Ishiguro, & Hagita, 2009), mean time between failures (Glas et al., 2009), occupied time (Wang, Wang, & Lewis, 2008), false alarm time (Elara, Calderon, Zhou, & Wijesoma, 2009), and fan out (Goodrich & Schultz, 2007), depicted in Fig. 11.4. These metrics help in analyzing the system performance by looking into the time taken for a successful interaction towards completion of a task.

Communications

The prime objective of communication in HRI is to forward significant information when required and avoid overloading situation to maintain effectiveness of the system. In a robot-centric environment, robots pass information to humans via communication channels to strengthen situational awareness (Kaupp, Makarenko, & Durrant-Whyte, 2010). Communication factors have been divided into three types, which are delay, jitter, and bandwidth (Steinfeld et al., 2006). The prior knowledge of these metrics helps in mitigating the effect of communication channel and successful transmission and reception of information.

Ability based

Human has to interact with multiple robots depending upon different conditions in various environments. The performance of an interaction pattern deviates with variation in difficulty and workload (Crandall & Goodrich, 2002). The performance of the operator based on his ability can be measured using situation coverage, situation awareness, operator overload, and accuracy of mental models. The metrics measures the task coverage, surroundings' awareness, human intervention and operator expectation on human performance respectively (Hancock et al., 2011).

Error detection

In order to ensure safety, a robot must report errors to operator. If there is detection of any kind of hazard, a robot must halt to a safe state when it does not receive any directions from operator. The errors that can be reported are low battery, sensor failure, tracking failure in case of non-updating of location data, and contact error which reports the unintentional collisions in the environment (Glas et al., 2009; Steinfeld et al., 2006).

Perception

The Illustration of interpretations based upon the data received from proximity sensors is called as perception. In HRI, perception is performed by variety of sensors like camera, tactile, tri-axial, proximity sensors, etc. Perception can be performed by human, robot or by both (Casper & Murphy, 2003). It can be categorized as judgment of extent, judgement of motion, active and stationary search (Tittle, Woods, Roesler, Howard, & Phillips, 2001).

Social acceptance

The utmost objective of social acceptance is to employ robots in human daily lives for the jobs it is intended to support (Dillon, 2001). Roboticists are designing biologically inspired robots in order to cater social needs of human. It is a very extensive domain and affected by many factors which can be, but not limited to, engagement (Schulte, Rosenberg, & Thrun, 1999), compliance (Goetz & Kiesler, 2002), reliability (Hancock et al., 2011), reciprocity (Bolt, 1980), and self-efficacy (Weiss, Bernhaupt, Lankes, & Tscheligi, 2009). These factors decide the trust towards and acceptability of a robot in human lives.

Tasking

The analysis of diverse environments while designing a robot for same application comes under tasking. Roboticists are meant to formulate generic metrics for same application in different situations for example a surveillance robot can be used in space, remote areas or buildings. Tasking can be parameterized using complexity of a task (Hancock et al., 2011), number of people interacting with robot (Glas et al., 2009), number of critical sections (Asimov, 2004), and environmental factors like temperature, pressure, visibility, etc. (Glas et al., 2009).

Situation based

During the execution of tasks, depending on the level of autonomy involved, various errors are reported by robot which may require user attention (Glas et al., 2009). Some errors can be severe, which is a measure of probable dangerous consequence that can be caused by a hazard (ISO E. 14971). Some of them need to be addressed on urgent basis which is called immediacy. On the whole, all the errors need to be tackled on a significance basis in order to maintain effectiveness of the system by prioritizing.

Objectivity based

The study of HRI in target recognition tasks can be done on the basis of four parameters, derived from signal detection theory, are hit, miss, rejection and false alarm (Macmillan, 2002). These parameters are objective in nature and are usually employed in applications where human–robot collaboration is required to a greater extent to achieve the objective (Oren, 2008).

Experience based

Measuring the performance of a complete system from user point of view, based on the experiences and expectations, is the key essence of experience based metric evaluation. It is quite a tedious job to quantify metrics based on experience, since it refers to a subjective process. Based on a broad and thorough analysis, it can be divided into five categories which are anxiety, attitude, physio measures, perceived adaptability, trust (Heerink, Krose, Evers, & Wielinga, 2009). These techniques help to quantify a metric's importance in designing an HRI system.

Technology acceptance models

Describing user acceptance in robotics is often considered as one of the mature research area in literature. It has roots in various domains like psychology, computer systems, sociology etc. based on intention of use and individual perception towards robots (Sepasgozar, Hawken, Sargolzaei, & Foroozanfa, 2019). Technology acceptance models helps to evaluate how users accept and use robots based on its usefulness, ease to use, performance and many other factors discussed in metric taxonomies. Various models have been proposed till date (de Boer & Åström, 2017) and being used for robotics as discussed below.

1. *Theory of reasoned action (TRA)*: This model establishes a connection between behavior intention and attitude. It is based on the theory that people decide to take action depending upon external and internal information in their mind.
2. *Theory of planned behavior (TPB):* This model is a modification over TRA model. It also includes perceived behavioral control to compliment attitude toward subjective norm and behavior.
3. *Technology acceptance model (TAM)*: Perceived ease of use and usefulness are two major components catered by TA. It describes the degree of to which user believes that using a particular system will be free from mental and physical effort.
4. *Innovation diffusion theory:* This model gives the basis for how innovation is rejected or adopted based on characteristics like acceptance; relative advantage, compatibility, difficulty, trialability, and perceivability.
5. *Combined TAM and TPB (C-TAM-TPB):* As name suggests this model is created by combining two models (TAM and TPB) to inculcate two components that is society and control which were missing in TAM.
6. *Model of PC utilization:* This model is based on theory of human behavior and makes a distinction between affective and cognitive elements of attitudes. It majorly concentrates on six parameters, which influence intention of behavior that is job-fit, difficulty, long-term consequences, affect in the direction of use, social factors and facilitating settings.
7. *Motivational model (MM):* It is based on psychological aspects while addressing extrinsic and intrinsic motivations parallel, which is used at individual level.
8. *Social cognitive theory:* This model is used in sociological field by reproduction of the observed conduct as an interaction between personal, behavioral, and environmental components.
9. *Unified theory of acceptance and use of technology (UTAUT):* This model targets to understand the usage and acceptance level of a new technology. Three main components of UTAUT are effort expectancy, performance expectancy, and social influence which have direct impact on behavioral intention.

Communication

The primary drive of communication in the domain of robotics is to obtain directions via receiver and send valuable data via transmitter. The luxury of mobility and freedom from tangled wires makes wireless communication more deployable technology in robotics. There is three type of communication in robotics. (1) Between base station and robot, (2) communication between two or more robots that is interrobot communication, and (3) communication within individuals blocks of robot that is intra robot communication (Knoll & Prasad, 2012). Different technologies can be positioned to satisfy the requirement of task like near field communication, Wi-Fi, Bluetooth, cellular, ZigBee, radio frequency, and many more protocols. While designing network links for communication technologies, roboticists take care of different parameters like latency, compression losses, bandwidth, encryption, reliability etc.

Output

Robots make connection to real world with the help of output devices. Output devices are essential for any robot, to make it capable of interaction in different environments. Based on the domains discussed above, output is divided into three categories. First one is the motion where numerous types of accessories are installed at moving joints of the robots to achieve different degrees of freedom in motion and perform various gestures. These accessories can be motors (DC, AC, stepper, servo, etc.), hydraulic or pneumatics devices and many more, depending upon the requirement of application. Audio is the second category, which is utilized by the robots to interact with humans by running audio data on speakers. Speakers are also used by robots to play alarm or warning signals. Finally, display is the third category, which is utilized by robots to show information, emotions by making use of animated faces, etc. Robots nowadays are equipped with IPS-LCD, OLED, LEDs, segmented display, etc. for displaying outputs.

Discussion

HRI and its components have been discussed in the above sections. Every component is crucial for HRI implementation in robotic products. The development team needs to follow a certain process in order to achieve HRI and other engineering aspects of a robotic product.

A good process helps in breaking up the problem into phases, and can be treated as a guide while designing robots for a particular domain as presented in Fig. 11.8. The process is broken down into four phases that is ideation, design, engineering, and validation which are discussed below.

1. *Ideation:* This phase is further broken down into two parts that is problem research and concept prototyping. In problem research, concept design (as discussed earlier) is formulated by defining problem through reviewing prior work extensively. Once problem is defined designers need to build motivation by selecting role of robot, in order to develop

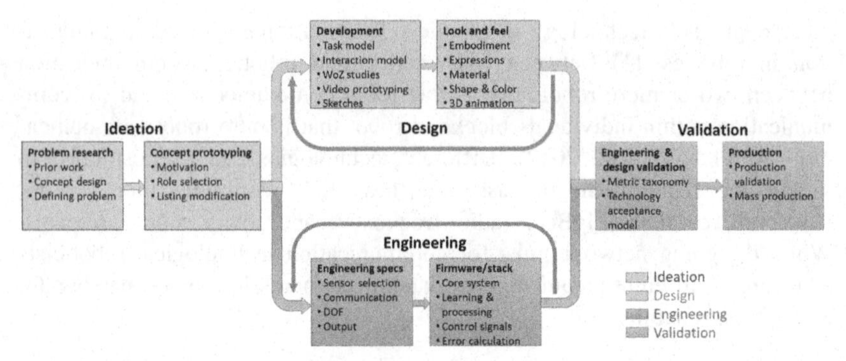

FIGURE 11.8 Phases of robotic product development.

thoughts for determining outcomes, as a part of concept prototyping. Designers also list down modification in prior design if required. At the end of ideation phase everyone in development team is clear about purpose and outcome of robot.

2. *Design:* This phase is divided into two parts, that is, development and look & feel catering to HRI aspects of robot. In development process, designers come up with task and interaction model (as discussed earlier) by conducting video and WoZ studies. Finally, sketches are prepared which lays down foundation for look and feel. Robot embodiment and expressions are designed in this process. Designers put thought on shape and color of robot leading to material selection, for example, wood, plastic, matte plastic etc. Finally, 3D renders in form of animation are generated which need to be agreed upon and forwarded to engineering team. Design phase is a reiterative process which is conducted till outcomes are met from ideation phase.
3. *Engineering:* This phase consists of two parts, that is, engineering specifications and firmware/stack. In engineering specifications, engineering team make a documentation of specification form technological point of view. They list down sensors required for robot for sensing outer world for better HRI, communication technology required to transmit and receive information and output devices for making robot capable for having interaction. Once specifications are finalized engineering teams start writing firmware for the same. Adequate emphasis is laid on learning and processing part which is heart of any robot and responsible for robot actions and interaction. Finally, a part of firmware generates control signal to activate output modalities and error calculation is done via. feedback mechanism. Similar to design phase, engineering phase is also is also reiterative process which is conducted till design specifications are met.
4. *Validation:* This phase is subdivided into two parts, that is, engineering & design validation and production. In engineering & design validation, metric taxonomy is listed and passed through technology acceptance

model in order to analyze the final output and generate feedback for learning and control process for further improvement. When robot performance matches a certain agreed level than robot products are mass produced and goes through quality checks.

In a nutshell, these four phases are very dependent on each other. A small change in one activity of one phase leads to modifications in every phase. It is to note that design and engineering phase are executed parallel and in a reiterative manner in order to maintain constant feedback for both phases. Designers and engineers conduct meetings on daily basis to achieve desired outcome.

Concluding remarks

In the near future, it looks very evident that very important roles will be played by robots in numerous areas by working together with humans. Success of these robotic system lies in the efficient interaction between robot and humans. HRI is an interdisciplinary problem that needs policy makers, researchers, engineers and users, who understand technology, sociology, & psychology to come together and develop. In order to use robotic technology in a responsible and ethical way, systematic structures and a set of rules need to be developed.

This article is a result of detailed literature review, which presents framework and metric taxonomies for HRI. The main goal of the presented framework is to divide the complete process of HRI into differentiable blocks in a generalized way, so that it can be applied to various problem domains. In a nutshell, this article is a step toward developing a platform to enable systematic approach for HRI.

References

Adla, R., Bazzi, Y., & Al-Holou, N. (2013). Multi sensor data fusion, methods and problems. In *Proceedings of the international conference on parallel and distributed processing techniques and applications (PDPTA). The steering committee of the world congress in computer science, computer engineering and applied computing (WorldComp).* (p. 105).

Bartneck, C. (2003). Interacting with an embodied emotional character. In *Proceedings of the 2003 international conference on designing pleasurable products and interfaces* (pp. 55–60). ACM.

Bolt, R. A. (1980). "Put-that-there": Voice and gesture at the graphics interface. *ACM, 14*(3), 262–270.

Breazeal, C., Dautenhahn, K., & Kanda, T. (2016). *Social robotics. Springer handbook of robotics* (pp. 1935–1972). Cham: Springer.

Casper, J., & Murphy, R. R. (2003). Human-robot interactions during the robot-assisted urban search and rescue response at the world trade center. *IEEE Transactions on Systems, Man, and Cybernetics, 33*(3), 367–385.

Chien, S. Y., Lin, Y. L., Lee, P. J., Han, S., Lewis, M., & Sycara, K. (2018). Attention allocation for human multi-robot control: Cognitive analysis based on behavior data and hidden states. *International Journal of Human-Computer Studies*, *117*, 30–44.

Coninx, A., Baxter, P., Oleari, E., Bellini, S., Bierman, B., Henkemans, O. B., & Hiolle, A. (2016). Towards long-term social child-robot interaction: using multi-activity switching to engage young users. *Journal of Human-Robot Interaction*, *5*(1), 32–67.

Corke, P. (2011). *Robotics, vision and control: Fundamental algorithms in MATLAB* (Vol. 73). Springer.

Crandall, J. W., & Goodrich, M. A. (2002) Characterizing efficiency of human robot interaction: A case study of shared-control teleoperation. In *IEEE/RSJ International Conference on Intelligent Robots and Systems* (vol. 2, pp. 1290–1295). https://doi.org/10.1109/IRDS.2002.1043932

Dautenhahn, K. (2007). Methodology & themes of human-robot interaction: A growing research field. *International Journal of Advanced Robotic Systems*, *4*(1), 15.

de Boer, W., & Åström, J. M. (2017). *Robots of the future are coming, are you ready? A study investigating consumers' acceptance of robotics*. Digitala Vetenskapliga Arkivet (DiVA).

de Greeff, J., & Belpaeme, T. (2015). Why robots should be social: Enhancing machine learning through social human-robot interaction. *PLoS One*, *10*(9), e0138061.

Deisenroth, M. P., Neumann, G., & Peters, J. (2013). A survey on policy search for robotics. *Foundations and Trends® in Robotics*, *2*(1–2), 1–142.

Dillon, A. (2001). *User acceptance of information technology. Encyclopedia of human factors and ergonomics*. Taylor and Francis.

Druin, A. (1999). Cooperative inquiry: Developing new technologies for children with children. In *Proceedings of the SIGCHI conference* on human factors in computing *systems* (pp. 592–599). ACM.

Duffy, B. R. (2003). Anthropomorphism and the social robot. *Robotics and Autonomous Systems*, *42*(3), 177–190.

Elara M.R., Calderon C.A.A., Zhou C., & Wijesoma W.S. (2009) False alarm demand: A new metric for measuring robot performance in human robot teams. In *2009 4th* international conference on autonomous robots and age*nts* (pp. 436–441). https://doi.org/10.1109/ICARA.2000.4803988.

Ellison, H., & Asimov, I. (2004). *I, Robot: The illustrated screenplay*. Ibooks, 271 p.

Farooq, M. (2006). Sensor data fusion and integration with applications to target tracking and robotics. In *2006 49th IEEE* international midwest symposium on circuits and sy*stems* (vol. 2, pp. xli-xlii). https://doi.org/10.1109/MWSCAS.2006.382190

Fasola, J., & Matarić, M. J. (2013). A socially assistive robot exercise coach for the elderly. *Journal of Human-Robot Interaction*, 2(2), 3–32.

Fernaeus, Y., Ljungblad, S., Jacobsson, M., & Taylor, A. (2009). Where third wave HCI meets HRI: report from a workshop on user-centred design of robots. In *2009 4th ACM/IEEE* international conference on human-robot interact*ion (HRI)* (pp. 293–294). https://doi.org/10.1145/1514095.1514182.

Ferreira, E., & Lefevre, F. (2015). Reinforcement-learning based dialogue system for human–robot interactions with socially-inspired rewards. *Computer Speech & Language*, *34*(1), 256–274.

Fischer, K., Lohan, K.S., & Foth, K. (2012). Levels of embodiment: Linguistic analyses of factors influencing HRI. In *2012 7th ACM/IEEE International Conference on Human-Robot Interaction (HRI)* (pp. 463–470). https://doi.org/10.1145/2157689.2157839.

Fong, T., Nourbakhsh, I., & Dautenhahn, K. (2003). A survey of socially interactive robots. *Robotics and Autonomous Systems*, *42*(3–4), 143–166. Available from https://doi.org/10.1016/S0921-8890(02)00372-X.

Glas D.F., Kanda T., Ishiguro H., & Hagita N. (2009) Field trial for simultaneous teleoperation of mobile social robots. In *2009 4th ACM/IEEE* international conference on human-robot inter*action (HRI)* (pp. 149–156). https://doi.org/10.1145/1514095.1514123

Goetz J., & Kiesler S. (2002). Cooperation with a robotic assistant. In *Extended abstracts of the 2002* conference on human factors in computing sy*stems, CHI 2002, Minneapolis, MI, USA, April 20-25, 2002*. https://doi.org/10.1145/506443.506492

Goodrich M.A., & Olsen D.R. (2003) Seven principles of efficient human robot interaction. In *SMC'03* conference pro*ceedings. 2003 IEEE* international conference on systems, man and cyberne*tics. Conference* theme – system security and assu*rance (Cat. No.03CH37483)* (vol. 4, pp. 3942–3948). https://doi.org/10.1109/ICSMC.2003.1244504.

Goodrich, M. A., & Schultz, A. C. (2007). Human-robot interaction: A survey. *Foundations and Trends in Human-Computer Interaction*, *1*(3), 203–275.

Gross, H.M., Mueller, S., Schroeter, C., Volkhardt, M., Scheidig, A., Debes, K., & Doering, N. (2015). Robot companion for domestic health assistance: Implementation, test and case study under everyday conditions in private apartments. In *2015 IEEE/RSJ International Conference on Intelligent Robots and Systems (IROS)* (pp. 5992–5999). https://doi.org/10.1109/IROS.2015.7354230

Hancock, P. A., Billings, D. R., Schaefer, K. E., Chen, J. Y., De Visser, E. J., & Parasuraman, R. (2011). A meta-analysis of factors affecting trust in human-robot interaction. *Human Factors*, *53*(5), 517–527.

Hanheide, M., Gobelbecker, M., Horn, G. S., Pronobis, A., Sjoo, K., Aydemir, A., & Zender, H. (2017). Robot task planning and explanation in open and uncertain worlds. *Artificial Intelligence*, *247*, 119–150.

Heerink, M., Krose, B., Evers, V., & Wielinga, B. (2009). Measuring acceptance of an assistive social robot: a suggested toolkit. In *RO-MAN 2009 – The 18th IEEE* international symposium on robot and human interactive communicati*on* (pp. 528–533). https://doi.org/10.1109/ROMAN.2009.5326320.

Hoffman, G. (2016). Openwoz: A runtime-configurable Wizard-of-oz framework for human-robot interaction. In *AAAI* spring symposium on enabling computing research in socially intelligent human-robot interacti*on, Palo Alto, CA*.

Hoffman, G., & Ju, W. (2014). Designing robots with movement in mind. *Journal of Human-Robot Interaction*, *3*(1), 91–122.

Hoffmann, L., & Krämer, N. C. (2013). Investigating the effects of physical and virtual embodiment in task-oriented and conversational contexts. *International Journal of Human-Computer Studies*, *71*(7–8), 763–774.

Hoffmann, L., Bock, N., & Rosenthal vd Pütten, A.M. (2018). The peculiarities of robot embodiment (EmCorp-Scale): Development, validation and initial test of the embodiment and corporeality of artificial agents scale. In *HRI'18: Proceedings of the 2018 ACM/IEEE* international conference on human-robot interaction (pp. 370–378). https://doi.org/10.1145/3171221.3171242

ISO E. 14971 (2000-12) EN ISO 14971/A1 (2003-03) EN ISO 14971/EC (2002-02). Medical devices-Application of risk management to medical devices. European Standard.

Kamide, H., Mae, Y., Takubo, T., Ohara, K., & Arai, T. (2014). Direct comparison of psychological evaluation between virtual and real humanoids: Personal space and subjective impressions. *International Journal of Human-Computer Studies*, *72*, 451–459. Available from https://doi.org/10.1016/j.ijhcs.2014.01.004.

Kaupp, T., Makarenko, A., & Durrant-Whyte, H. (2010). Human–robot communication for collaborative decision making—A probabilistic approach. *Robotics and Autonomous Systems*, *58*(5), 444–456.

Kennedy, J., Baxter, P., & Belpaeme, T. (2015). Comparing robot embodiments in a guided discovery learning interaction with children. *International Journal of Social Robotics*, *7*(2), 293–308.

Khan, S., & Germak, C. (2018). Reframing HRI design opportunities for social robots: lessons learnt from a service robotics case study approach using UX for HRI. *Future Internet*, *10*(10), 101.

Kiesler, S., & Goodrich, M. A. (2018). The science of human-robot interaction. *ACM Transactions on Human-Robot Interaction (THRI)*, *7*(1), 9.

Kim, M., Oh, K., Choi, J., Jung, J., & Kim, Y. (2011). *User-centered HRI: HRI research methodology for designers. Mixed reality and human-robot interaction* (pp. 13–33). Dordrecht: Springer.

Knoll, A., & Prasad, R. (2012). Wireless robotics: A highly promising case for standardization. *Wireless Personal Communications*, *64*(3), 611–617.

Lee K.W., Kim H.R., Yoon W.C., Yoon Y.S., & Kwon D.S. (2005) Designing a human-robot interaction framework for home service robot. In *ROMAN 2005. IEEE International Workshop on Robot and Human Interactive Communication, 2005* (pp. 286–293). https://doi.org/10.1109/ROMAN.2005.1513793

Lee, K. M., Jung, Y., Kim, J., & Kim, S. R. (2006). Are physically embodied social agents better than disembodied social agents? The effects of physical embodiment, tactile interaction, and 'people's loneliness in human–robot interaction. *International Journal of Human-Computer Studies*, *64*(10), 962–973.

Lehoux, P., & Grimard, D. (2018). When robots care: Public deliberations on how technology and humans may support independent living for older adults. *Social Science & Medicine*, *211*, 330–337.

Li, J., & Chignell, M. (2011). Communication of emotion in social robots through simple head and arm movements. *International Journal of Social Robotics*, *3*(2), 125–142.

Liu, P., Glas, D. F., Kanda, T., Ishiguro, H., & Hagita, N. (2017). A model for generating socially-appropriate deictic behaviors towards people. *International Journal of Social Robotics*, *9*(1), 33–49.

Long, T. (2011). *Robots first czech*. Wired. Available from https://www.wired.com/2011/01/0125robot-cometh-capek-rur-debut/.

Macmillan, N. A. (2002). Signal detection theory. In H. Pashler, & J. Wixted (Eds.), *Stevens' handbook of experimental psychology: Methodology in experimental psychology* (pp. 43–90). John Wiley & Sons Inc.

Mancini, M., & Varni, G. (2018). A framework for creative embodied interfaces. In *AVI'18: Proceedings of the 2018* international conference on advanced visual interfaces (Article No. 71, pp. 1–3). https://doi.org/10.1145/3206505.3206577.

Myers, B. A. (1998). A brief history of human-computer interaction technology. *Interactions*, *5* (2), 44–54.

Neerincx, M. A. (2011). Situated cognitive engineering for crew support in space. *Personal and Ubiquitous Computing*, *15*(5), 445–456.

Oren, Y. (2008). *Performance analysis of human-robot collaboration in target recognition tasks*. Ben-Gurion University of the Negev.

Pfeifer, R., & Bongard, J. (2006). *How the body shapes the way we think: A new view of intelligence*. MIT press.

Reeves, B., & Nass, C. (1996). *How people treat computers, television, and new media like real people and places* (pp. 3–18). CSLI Publications and Cambridge University Press.

Revelation 9:9. (2011). *Holy Bible new international version*. Biblica Inc. http://biblehub.com/niv/revelation/9.html.

Scholtz J. (2003) Theory and evaluation of human robot interactions. In 36th annual Hawaii international conference on system sciences, 2003. Proceedings of the. (10 pp.). https://doi.org/10.1109/HICSS.2003.1174284

Schulte J., Rosenberg C., & Thrun S. (1999) Spontaneous, short-term interaction with mobile robots. In *Proceedings 1999 IEEE International Conference on Robotics and Automation (Cat. No.99CH36288C)* (vol. 1, pp. 658–663). https://doi.org/10.1109/ROBOT.1999.770050

Senft, E., Lemaignan, S., Baxter, P.E., & Belpaeme, T. (2017). Leveraging human inputs in interactive machine learning for human robot interaction. In *Proceedings of the* companion *of the 2017 ACM/IEEE* international conference on human-robot int*eraction* (pp. 281–282). ACM.

Sepasgozar, S. M., Hawken, S., Sargolzaei, S., & Foroozanfa, M. (2019). Implementing citizen centric technology in developing smart cities: A model for predicting the acceptance of urban technologies. *Technological Forecasting and Social Change*, *142*, 105–116. Available from https://doi.org/10.1016/j.techfore.2018.09.012.

Shahmaleki, P., Mahzoon, M., Kazemi, A., & Basiri, M. (2010). *Vision-based hierarchical fuzzy controller and real time results for a wheeled autonomous robot. Motion control.* InTech. Available from http://doi.org/10.5772/6959.

Sheng, W., Du, J., Cheng, Q., Li, G., Zhu, C., Liu, M., & Xu, G. (2015). Robot semantic mapping through human activity recognition: A wearable sensing and computing approach. *Robotics and Autonomous Systems*, *68*, 47–58.

Siciliano, B., & Khatib, O. (Eds.), (2016). *Springer handbook of robotics.* Springer.

Singer S.M., & Akin D.L. (2011) A survey of quantitative team performance metrics for human-robot collaboration. In *41st International Conference on Environmental Systems.*

Singh, K. J., & Kapoor, D. S. (2017). Create your own internet of things: A survey of IoT platforms. *IEEE Consumer Electronics Magazine*, *6*(2), 57–68.

Spenko, M. J., Haynes, G. C., Saunders, J. A., Cutkosky, M. R., Rizzi, A. A., Full, R. J., & Koditschek, D. E. (2008). Biologically inspired climbing with a hexapedal robot. *Journal of Field Robotics*, *25*(4–5), 223–242.

Steinfeld A., Fong T., Kaber D., Lewis M., Scholtz J., Schultz A., & Goodrich M. (2006) Common metrics for human-robot interaction. In *HRI'06:* p*roceedings of the 1st ACM SIGCHI/SIGART conference on* h*uman-robot interaction* (pp. 33–40). https://doi.org/10.1145/1121241.1121249

Sucar, O.I., Aviles, S.H., & Miranda-Palma, C. (2003). From HCI to HRI – usability inspection in multimodal human – robot interactions. In *The 12th IEEE* international workshop on robot and human interactive communi*cation, 2003. Proceedings. ROMAN 2003* (pp. 37–41). https://doi.org/10.1109/ROMAN.2003.1251773.

Tai, L., Zhang, J., Liu, M., Boedecker, J., & Burgard, W. (2016). A survey of deep network solutions for learning control in robotics: From reinforcement to imitation. *Journal Of Latex Class Files*, *14*(8), 1, arXiv preprint 1612.07139.

Tittle, J. S., Woods, D. D., Roesler, A., Howard, M., & Phillips, F. (2001). The role of 2-D and 3-D task performance in the design and use of visual displays. *Proceedings of the Human Factors and Ergonomics Society Annual Meeting*, *45*(4), 331–335.

Tonkin, M., Vitale, J., Herse, S., Williams, M.A., Judge, W., & Wang, X. (2018). Design Methodology for the UX of HRI: A Field Study of a Commercial Social Robot at an Airport. In *Proceedings of the 2018 ACM/IEEE* international conference on human-robot in*teraction* (pp. 407–415). ACM.

Unhelkar, V. V., Lasota, P. A., Tyroller, Q., Buhai, R. D., Marceau, L., Deml, B., & Shah, J. A. (2018). Human-aware robotic assistant for collaborative assembly: Integrating human motion prediction with planning in time. *IEEE Robotics and Automation Letters*, *3*(3), 2394–2401.

Vanzo, A., Croce, D., Bastianelli, E., Gemignani, G., Basili, R., & Nardi, D. (2017). *Dialogue with Robots to Support Symbiotic Autonomy. Dialogues with social robots* (pp. 331–342). Singapore: Springer.

Wang J., Wang H., & Lewis M. (2008) Assessing cooperation in human control of heterogeneous robots. In *2008 3rd ACM/IEEE international conference on human-robot interaction (HRI)* (pp. 9–15).

Weiss A., Bernhaupt R., Lankes M., & Tscheligi M. (2009) The USUS evaluation framework for human-robot interaction. In P*roceedings of the symposium on new frontiers in human-robot interaction* (vol. 4, pp. 11–26).

Xing, B., & Marwala, T. (2018). Introduction to human robot interaction. In *Sm*art maintenance for human–robot intera*ction* (pp. 3–19). Cham: Springer.

Zhang, S., Qin, J., Cao, S., & Dou, J. (2018). HRI design research for intelligent household service robots: Teler as a case study. In *Int*ernational conference of design, user experience, and usa*bility* (pp. 513–524). Cham: Springer.

Further reading

Antona, M., & Stephanidis, C. (Eds.), (2018). *Universal access in human-computer interaction: Methods, technologies, and users: 12th international conference, UAHCI 2018, Held as Part of HCI International 2018, Las Vegas, NV, USA, July 15–20, 2018, Proceedings* (Vol. 10907). Springer.

Cheng, Y. W., Sun, P. C., & Chen, N. S. (2018). The essential applications of educational robot: Requirement analysis from the perspectives of experts, researchers and instructors. *Computers & Education*, *126*, 399–416.

Edwards, C., Edwards, A., & Omilion-Hodges, L. (2018). Receiving medical treatment plans from a robot: evaluations of presence, credibility, and attraction. In *Companion of the 2018 ACM/IEEE* international conference on human-robot interaction (pp. 101–102). ACM.

Goguey, A., Casiez, G., Cockburn, A., & Gutwin, C. (2018). Storyboard-based empirical modeling of touch interface performance. In *Proceedings of the 2018 CHI* conference on human factors in computing syst*ems* (p. 445). ACM.

Jensen, S.Q., Fender, A., & Müller, J. (2018). Inpher: Inferring physical properties of virtual objects from mid-air interaction. In *Proceedings of the 2018 CHI* conference on human factors in computing system*s* (p. 530). ACM.

Lee, K.H., Chua, K.W.L., Koh, D.S.M., & Tan, A.L.S. (2018). Team cognitive walkthrough: fusing creativity and effectiveness for a novel operation. In *Congress of the international ergonomics association* (pp. 117–126). Cham: Springer.

Lin, X., & Chen, T. (2018). A qualitative approach for the 'elderly's needs in service robots design. In *Proceedings of the 2018* international conference on service robotics technolog*ies* (pp. 67–72). ACM.

Luria, M. (2018). Designing robot personality based on fictional sidekick characters. In *Companion of the 2018 ACM/IEEE* international conference on human-robot intera*ction* (pp. 307–308). ACM.

Ramachandran, A., Huang, C.M., Gartland, E., & Scassellati, B. (2018). Thinking aloud with a tutoring robot to enhance learning. In *Proceedings of the 2018 ACM/IEEE* international conference on human-robot interact*ion* (pp. 59–68). ACM.

Reig, S., Norman, S., Morales, C.G., Das, S., Steinfeld, A., & Forlizzi, J. (2018). A field study of pedestrians and autonomous vehicles. In *Proceedings of the 10th* international conference on automotive user interfaces and interactive vehicular ap*plications* (pp. 198–209). ACM.

Rozo, L., Amor, H. B., Calinon, S., Dragan, A., & Lee, D. (2018). Special issue on learning for human–robot collaboration. *Autonomous Robots*, 1–4.

Tennent, H., Lee, W.Y., Hou, Y.T.Y., Mandel, I., & Jung, M. (2018). PAPERINO: Remote wizard-of-Oz puppeteering for social robot behaviour design. In *Companion of the 2018 ACM* conference on computer supported cooperative work and social comput*ing* (pp. 29–32). ACM.

Chapter 12

Teleportation of human body kinematics for a tangible humanoid robot control

Tutan Nama[1] and Suman Deb[2]
[1]*Department of Computer Science and Engineering, Indian Institute of Technology Kharagpur, Kharagpur, India,* [2]*Department of Computer Science and Engineering, National Institute of Technology Agartala, Agartala, India*

Introduction

The natural human body is a complex structure balanced on knee joints and can adjust itself almost in every environmental situation. As technology evolved, many humanoid robots are developed and being used in various fields instead of humans. Humanoid robots are trained by programming or human gestures, but these programmings are very complex. There are 244 degrees of freedom (DOFs) in a human body. Each DOF consists of three directional movements. Even the teleportation of social kinematics movements into a humanoid robot is quite complicated. It is infeasible to teleport all these DOFs into a single robot by transferring coordinates. As of now there is no such kind of humanoid robot that has all 244 DOFs. This work is an attempt for human gesture teleportation into a humanoid robot. This work is a Human-Machine-Machine Interaction (HMMI) (Chang, Lee, Chao, Wang, & Chen, 2010; Guo & Sharlin, 2008; Nikolaidis, Lasota, Ramakrishnan, & Julie, 2015) model where human hand gestures are captured by a depth sensing device and then transferred the kinematics into a humanoid robot after necessary transformations. For capturing human hand gestures (Shamsuddin et al., 2011; Shang, Jong, Lee, & Lee, 2007; Sinha & Deb, 2017; Taheri, Alemi, Meghdari, PourEtemad, & Basiri, 2016; Tan & Biswas, 2007), the device used in this work is a leap motion controller (LMC). The LMC is a tiny gesture recognition device with three IR LED (Infra Red Light Emitting Diode)s and two IR (Infra Red) cameras. IR LEDs are responsible for creating an IR field above the LMC, and IR cameras are accountable for capturing hand gestures acting on that IR field. Human hand gestures collected by the LMC have teleported into

Cognitive Computing for Human-Robot Interaction. DOI: https://doi.org/10.1016/B978-0-323-85769-7.00011-2

a humanoid robot platform NAO (from SoftBank Robotics Corp) in this work. Aldebaran nao robot has used as a target machine for teleporting (Jamet, Masson, Stilgenbauer, & Baratgin, 2018; Shamsuddin et al., 2011) human body kinematics movements. Nao robot has 26 DOFs, but only 10 DOFs consider mapping through human hand gestures in this work. Human hand gestures control these 10 DOFs (HeadPitch, HeadYaw, RHand, LHand, RShoulderRoll, LShoulderRoll, RShoulderPitch, LShoulderPitch, RElbowRoll, and LElbowRoll) (Hand pitch direction, Hand yaw direction, Palm position X, Palm position Y, Palm position Z, and Palm normal). The human hand pitch direction and yaw direction are used to control Nao HeadPitch and HeadYaw movement, respectively. Nao robot shoulder roll, shoulder pitch, and elbow roll are mapped here by palm position x, palm position y, and palm position z, respectively. The head and hand movement speed of the nao robot is controlled by palm normal.

Gesture recognition and teleportation are vastly applicable in the field of animation, avatar creation, and control. Technology nowadays takes roles to create virtual environments with virtual elements and work with real-world objects collaboratively. A tangible robot or machine can be controlled intangibly through gestures with the help of technology. The idea of teleportation of human gestures into a humanoid robot is to make heuristics based robot kinematics movement and experiment on training human with elementary percept action coordination. In this work Leap motion is used as a medium to sense gestures, and the nao robot as a target machine to verify teleportation and both devices are very much useful in elementary learning purposes.

Translating human body gestures to programmatic inputs to communicate and control a humanoid robot is too complicated. As a solution to this teleportation complexity in this work, the LMC is used for capturing gestures and passing features of these gestures for controlling a nao robot.

Literature survey and related works

Gesture recognition and teleportation are vastly applicable in the field of animation, avatar creation, and control. Technology nowadays takes roles to create virtual environments with virtual elements and able to work collaboratively with real-world objects. A tangible robot or machine can be controlled intangibly through gestures with the help of technology. The idea of teleportation human gestures into a humanoid robot to train these robots and find out the learning outcomes comes from applying a LMC and nao robot. Leap motion is a medium to teleport gestures, and the nao robot is a target machine to verify teleportation (Staretu and Moldovan, 2016). Both devices are very much useful in elementary learning purposes (Mubin & Ahmad, 2016). Controlling a humanoid robot through gesture is a practical approach, although machine learning concepts used vastly to do this (Rodriguez et al., 2014). Generally, teleportation of gesture is a coordinate to coordinate mapping (Shamsuddin et al., 2012). Coordinate sets of human body joints are mapped to a robot body joints for teleporting gestures, and then the robot reacts according to human action. To simplify this coordinate to coordinate mapping, the LMC can quickly transfer human hand gestures into a humanoid robot and control it. The classified objective of the work was to teleport the human hand gesture (Naidu and Ghotkar, 2016) into a humanoid robot called nao through a LMC (Guo & Sharlin, 2008; Nikolaidis et al., 2015). Some theoretical concept is described in this section as a literature review, and a few papers related to this work are included (Jamet et al., 2018).

Human body kinematics and Nao robot body kinematics: A human body has several joints, and each joint has different DOF. A human body has 244 DOFs with various movements, and a nao robot has 26 DOFs. Fig. 12.1 shows the nao robot body joints comparing with human body joints. These joints, head joint, shoulder joint, and elbow joint from human body joints are used to teleport on humanoid robot joints.

Gesture recognition from joint movements: The gesture recognition process is done in this work through leap motion (Chan, Halevi, & Memon, 2015; General, Silva, Esteves, Halleran, & Liut, 2016; Nifal, Logine, Sopitha, & Kiruthika, 2017). Various approaches have been made using cameras and computer vision algorithms to interpret sign language (Mohandes, Deriche, & Oladimeji, 2014; Savur & Sahin, 2015) but LMC, which quickly takes skeleton movements of human body joints. In this work, ten movements of human joints have teleported into humanoid robots. Leap motion captures human hand gestures as frames and passes to system scripts for feature extraction.

FIGURE 12.1 (Hussein, Torki, Gowayyed, & Saban, 2013; Nikos, 2013) Description of Body joints of NAO robot (NAO Documentation—Aldebaran 2.1.4.13 documentation, 2019).

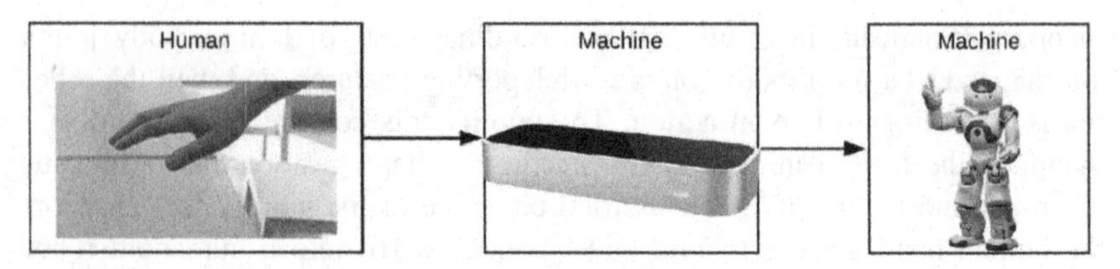

FIGURE 12.2 Human-machine-machine interaction.

Teleportation of gesture: As the word, teleportation means to transfer something from one end to another, the teleportation of gesture is to transfer human gesture into a machine (Gouda & Gomaa, 2014; Yu, Xu, Wang, Yang, & Liu, 2015) or computing devices. This work mainly focuses on the teleportation of hand gestures into a humanoid robot nao through leap motion.

Human-machine interaction: Human-Machine Interaction (HMI) (Guo & Sharlin, 2008; Nikolaidis et al., 2015) is all about interacting and communicating between human and automated systems. Communication between people and machines requires interfaces: The place where or action by which a user engages with the machine. In this work, accessing human hand gestures through leap motion is a HMI. The teleportation of gestures from leap motion to a nao robot is also a HMI.

Human-machine-machine interaction: The concept of Human-Machine-Machine Interaction (HMMI) (Chang et al., 2010; Guo & Sharlin, 2008; Nikolaidis et al., 2015) is a two-layer interaction among human and machines. When a machine instructs by a human can instruct another machine, then the interaction among them is called HMMI. In this work, the HMMI concept used as an interaction model with a closed-loop interaction model (Fig. 12.2).

Fundamentals

Apparatus

The hardware requirements for the system development are LMC, Aldebaran nao Robot, Router, PC. Software components are core software (Python 2.7 - 32 bit, Choregraphe, Leap Motion Community Setup) and SDKs and Packages (Leap SDK, naoqi SDK, PyQt5).

Leap motion controller

A miniscule gesture recognition device called leap motion has various applications in the interaction design field, and we are trying to achieve and explore those for a depth sensing and gestral percept device (Avola et al., 2014).

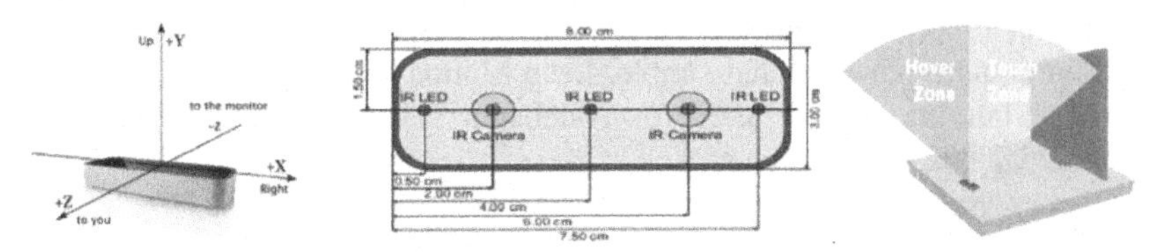

FIGURE 12.3 Axis view (left), Internal view (center), and layers view (right) of the leap motion controller (Python SDK Documentation—Leap Motion Python SDK v3.2 Beta documentation, 2019).

The SDK extends an interface to the Leap Motion with two layers, Hover zone and Touch zone. Fig. 12.3 given below shows the external, internal, and layers view of a LMC.

NAO robot platform

SoftBank Robotics formally known as Aldeberan Robotics developed an integrated, programmable, medium-sized humanoid robot called NAO. NAO (version V6) is a 58 cm, 5 kg robot, communicating with remote computers via an IEEE 802.11 g wireless or a wired Ethernet link. The nao has 26 DOFs and features a variety of sensors and joints. The bipadel locomotion of the human body like structure is created with several geared joints with functional torso, a head, two arms, and two legs. The head joint has two types of motion HeadYaw and HeadPitch. Each of those motions is purely manipulated through the LMC in this project work. Each arm of a nao robot has a total of six joints, namely, Shoulder Pitch, Shoulder Roll, Elbow Yaw, Elbow Roll, Wrist Yaw, RHand, or LHand, shown in Fig. 12.1. All these arm joints are successfully manipulated in this project work through a LMC.

Router and PC

A network device router is required to establish a connection between the Nao web server and the computer. Nao needs to connect with a server to operate through a computer. After turned on, nao automatically connects to the network of the active router. Nao has a sensor/button on his chest. Nao says its IP address when that button is pressed. A PC needs to establish this project. The system must have an OS, WiFi driver, 4GB RAM, and other basic specifications.

Python 2.7 - 32 bit

In this work, the selected scripting language is python, although the LMC and Aldebaran Nao support many programming languages. It is found that python is a more convenient programming language in both the devices, and python is quite powerful and simple. Leap Motion can support every python version, but the python SDK for Aldebaran Nao 2.1.4.13 only supports python 2.7 - 32-bit interpreter.

Choregraphe

Choregraphe is a cross platform desktop IDE that allows you to create animations, behaviors, dialog and test them on a simulated robot or directly deploy on a real one. Choregraphe allows producing applications comprising dialogues, services, and authoritative behaviors, such as gestural interaction, complex body postures in a node based developing approach.

Leap motion community setup

Leap motion community setup is the platform provided by the leap motion community used to connect, control, and access leap motion features through a computer. It also provided an application store where developers can upload or download leap motion-based applications. Two types of leap motion community setup are available for VR application, and another one is for Desktop application. The desktop setup is used in this project work.

Leap SDK

LeapSDK is a leap motion software development kit for the developers of leap motion applications. This SDK contains different libraries and dll files for the various programming languages supported by the leap motion. In this work, the LeapSDK version Orion 3.2.1 + 45911 is used. naoqi SDK The naoqi SDK is the interacting software development kit for nao robot and python 2.7 - 32 bit. This SDK provides Nao robot-related library files to the python interpreter to access nao features and develop nao robot applications.

PyQt5

PyQt5 is a set of Python bindings for Digia's Qt cross-platform Graphical user interface (GUI) toolkit. PyQt5 supports Python v2 and v3. PyQt5 is extensively used for evolving GUIs that can be deployed on various operating systems. It's one of the most general GUI selections for Python programming. For connecting nao robot and leap motion easily with the system, a python GUI is used, and that interface is developed through the PyQt5 python GUI tool.

Setting up apparatus and extracting features

This section describes how to set up a leap motion device and Aldebaran nao robot for further use. This section also shows how to extract basic features from leap motion and nao robot.

```
import Leap
import sys

class LeapMotionListener(Leap.Listener):
    def on_frame(self, controller):
        frame = controller.frame()
        for hand in frame.hands:
            timestamp = frame.timestamp
            direction = hand.direction
            normal = hand.palm_normal
            Pitch = direction.pitch * Leap.RAD_TO_DEG
            Yaw = direction.yaw * Leap.RAD_TO_DEG

def main():
    listener = LeapMotionListener()
    controller = Leap.Controller()
    controller.add_listener(listener)
    print("Press ENTER to quit")
    try:
        sys.stdin.readline()
    except KeyboardInterrupt:
        pass
    finally:
        controller.remove_listener(listener)

if __name__ == "__main__":
    main()
```

Steps to setup leap motion

Leap motion community provides different documentation for a different programming language. They also provide different lib file and dll files for different programming languages in a single SDK folder. In this project, python is chosen as a programming language.

Steps to setting up a simple leap motion project in python are shown below:

1. Download and install the leap motion community setup.
2. Run diagnostic visualizer and check leap motion working correctly or not.
3. If leap motion working properly, then download LeapSDK.

4. Create a project folder and copy Leap.py file from the LeapSDK-lib folder and paste it into that project folder.
5. Copy Leap.dll, Leap.lib, and LeapPython.pyd from the LeapSDK-lib-x86 folder and paste it into that project folder.
6. Create a new python file into the project folder and add your code to it.

Steps to setup nao robot

Aldebaran Nao is developed by Softbank robotics company for research and education purposes. Nao applications can be developed with C, C\#, or python. For this project work, python is chosen.

Steps to setting up a simple Nao project with python is shown below:

1. Download and install manually pynaoqi to the python installation folder.
2. Turn on Nao and connect with a network.
3. Connect the computer to the same network.
4. For work with Choregraphe nao simulator, open choreograph and connect to the nao through the IP address and port number.
5. Create a project folder.
6. Create a python file on that folder.
7. Add your Nao code to that file with a proper IP address and port number.

Extracting features from leap motion

Some DOF of a human hand are used in this work, and this section describes how to extract those features through a LMC using python scripts. The code segment (Python SDK Documentation—Leap Motion Python SDK v3.2 Beta documentation, 2019) given below shows how to access leap motion frame data for palm position, palm normal and direction.

Extracting features from nao robot

The code segments given below show for control a nao robot using python commands.

Accessing nao robot features using ALProxy object: ALProxy is an object that gives you access to all the methods or the module for nao robot through python commands (NAO Documentation—Aldebaran 2.1.4.13 documentation, 2019). The basic syntax of this is given below.

```
from naoqi import ALProxy
ttsProxy = ALProxy("ALTextToSpeech", robotIP, port)
memoryProxy = ALProxy("ALMemory", robotIP, port)
motionProxy = ALProxy("ALMotion", robotIP, port)
postureProxy = ALProxy("ALRobotPosture", robotIP, port)
behaviourProxy = ALProxy("ALBehaviorManager", robotIP, port)
```

class **ALProxy**(*name, IP, port*)

name - The name of the module

IP - The IP of your robot
port - The port on which NAOqi listens (9559 by default)

```
from naoqi import ALProxy

tts = ALProxy("ALTextToSpeech", robotIP, port)
tts.say("Hello, world!")
```

Making a nao robot speak: The *ALTextToSpeech* proxy provides the service to make nao speak at the text written on **say** function.

Making nao robot move: To make NAO walk, you have to use *ALMotionProxy::moveInit()* (to set the robot in a correct position), and then *ALMotionProxy::moveTo()*. The *moveTo()* function has three arguments first one is the distance in X axis(+ve value for forward and − ve value for backword), the second argument is for distance in Y axis (+ve value for left and − ve value for right) and the third one is for the angle (θ).

```
from naoqi import ALProxy

motion = ALProxy("ALMotion", robotIP, port)
motion.moveInit()
motion.moveTo(0.5, 0, 0) #forward movement
motion.moveTo(-0.5, 0, 0) #backward movement
motion.moveTo(0, 0.5, 0) #left side movement
motion.moveTo(0, -0.5, 0) #right side movement
```

In the *moveTo* method, we can also use a fourth parameter called frequency for controlling the robot speed. The value of the frequency ranges from 0.0 to 1.0, and 0.5 is the default value.

```
from naoqi import ALProxy

motion = ALProxy("ALMotion", robotIP, port)
motion.openHand('RHand')
motion.openHand('LHand')
motion.closeHand('RHand')
motion.closeHand('LHand')
```

Opening and closing nao robot fingers: This is an example of the opening and closing of nao robot fingers. The concept of opening and closing fingers can help to grip or release any object by nao.

Methods to get and set nao robot hand positions: This example shows two important methods from naoqi used in this project work. Here *getPosition()* function is used to get all joints value of a Nao robot arm (an example is given for left arm), and *setPosition()* function is used to put manual coordinates for arm joints. In this project work, palm position and palm normal of a human hand is mapped to the nao robot through the *setPosition()* function by passing them as arguments.

```
from naoqi import ALProxy

motion = ALProxy("ALMotion", robotIP, port)
chainName = "<Arm name>"
frame = motion.FRAME_TORSO
useSensor = False
current = motionProxy.getPosition(chainName, frame, useSensor)
target = [<target position>]
fractionMaxSpeed = 0.5
axisMask = 7
motionProxy.setPositions(chainName, frame, target, fractionMaxSpeed, axisMask)
```

Method

System design

In this work, an interacting model with a graphical interface was designed to integrate the LMC and Aldebaran nao robot for gesture teleportation. The model was used as a learning tool. Fig. 12.4, given below, shows the fundamental system architecture of this work.

The actions performed by a human on the leap motion tracking zone is mapped to the robot through different working phases, and those actions are reflected on the robot. Fig. 12.5 given below shows the human hand action mapped to the nao robot.

This work learning perspective is that it can help children determine how accurately they perform their tasks. In this work, the main motto is to teleport the robot gesture through leap motion and use the model as a learning tool for alphabet learning, number learning identifying body parts.

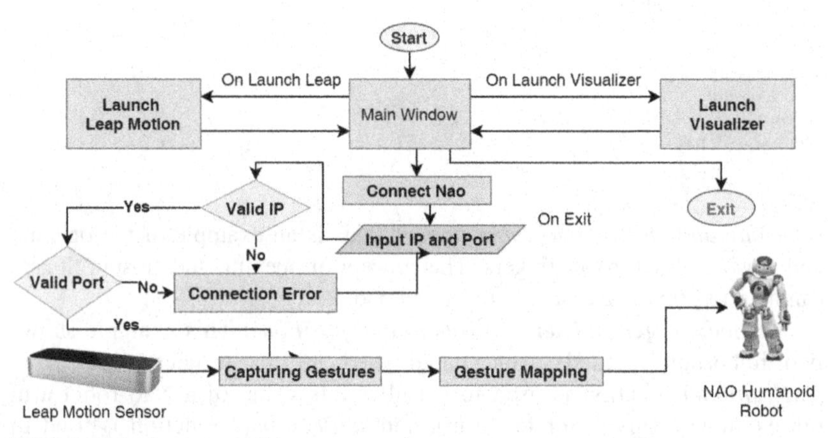

FIGURE 12.4 System workflow.

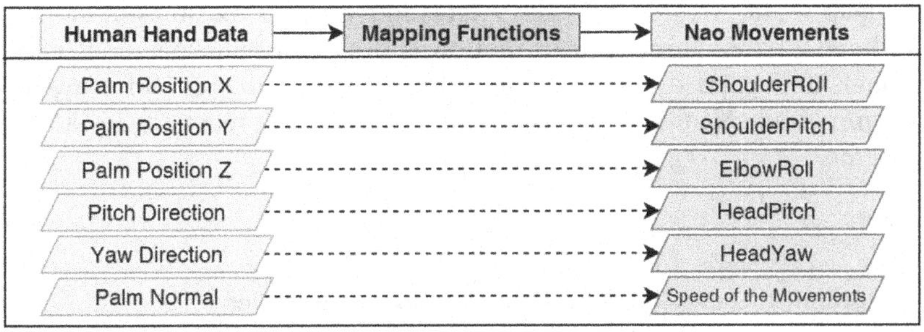

Human Hand Data	Mapping Functions	Nao Movements
Palm Position X		ShoulderRoll
Palm Position Y		ShoulderPitch
Palm Position Z		ElbowRoll
Pitch Direction		HeadPitch
Yaw Direction		HeadYaw
Palm Normal		Speed of the Movements

FIGURE 12.5 Percept-action mapping.

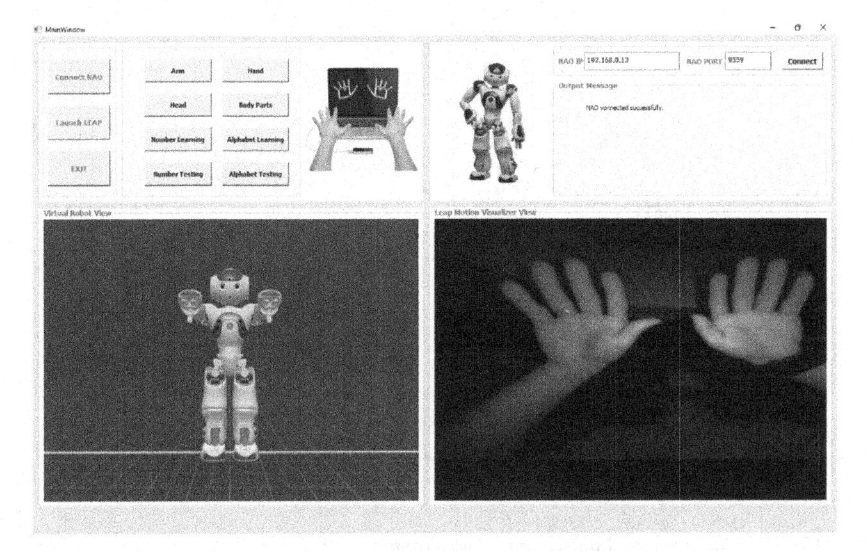

FIGURE 12.6 Graphical user interface for the proposed model.

A GUI is designed for connecting the nao robot to the system and to switch between operations. The user interface shows three different operational modes, and we can easily switch among them by clicking through the mouse. Fig. 12.6 shows the main window content that is useful to connect the nao robot, launch the LMC, launch visualizer to see correct calibration of hand gesture, and switch operations.

System modules

In this section, detailed algorithmic steps are described. Here three modules are designed to collect gestures, extract features from gestures, and map these gestures into nao. Algorithm-1 shows the steps of capturing hand gesture through leap motion and extract palm position, palm normal, and palm direction. Algorithm-2 shows the steps of mapping gestures into Nao and Algorithm-3, shows nao reaction to acting gestures.

Algorithm 1: Data acquisition from leap motion controller.

Step 1: START
Step 2: **FOR** valid frame in leap motion *ON_FRAME* **DO**
    Step 2.1: Create two instances *Frame1*and *Frame2* for previous frame and current frame.
    Step 2.2: **IF** Hands on *Frame1* is **NOT** empty **THEN**
        Step 2.2.1: **IF** Gesture is valid **THEN**
            Get first hand from *Frame1*
            Get fingers from first hand
            Get hands vector data from first hand:
                Read hand normal
                Read hand direction
                Read hand palm position X, Y, Z coordinates
            Passes all collected data for mapping into nao robot
        Step 2.2.2: **ELSE**
            Hand position is out of range
    Step 2.3: **ELSE**
        No hand detected empty frame
Step 3: STOP

Algorithm 2: Mapping collected leap motion data into nao robot.

Step 1: START
Step 2: **IF** Nao is connected **THEN**
    Step 2.1: **IF** Leap Motion is connected **THEN**
        Load leap motion data(as shown in algorithm 1) and pass to nao robot(as shown in algorithm 3)
    Step 2.2: **ELSE**
        Leap motion connection error
Step 3: Nao robot connection error
Step 4: STOP

Algorithm 3: Reaction of the nao robot on leap motion data.

Step 1: START
Step 2: **IF** Nao is connected **THEN**
  Step 2.1: Read head and hand angles:
    Head_Yaw and Head_Pitch
    LShoulderRoll, LShoulderPitch, LElbowRoll, LElbowYaw
    RShoulderRoll, RShoulderPitch, RElbowRoll, RElbowYaw
  Step 2.2: Steering robot head and arm with collected head and hand angles combining with leap motion data
Step 3: **ELSE**
  Nao connection error
Step 4: STOP

Results and discussion

This teleportation model had reflected some learning outcomes like head movements, hand movements, opening and closing of fingers, arm type detection, identifying body parts, alphabet learning, and number learning. This section shows all targeted learning outcomes.

Head movement

On head operation selection, nao can move his head according to human hand direction (direction.pitch, direction.yaw) acting on the leap motion tracking zone. This operational mode pitch direction has used to navigate the robot head forward-backward movements and yaw direction for left-right movements. The sequence of images given in Fig. 12.7 shows the nao head movement through the LMC.

FIGURE 12.7 Head movements of the NAO robot are controlled by hand gestures (head move towards the right side, the head moves towards the left side, the head moves downwards and head moves upwards shows by images from left to the right, respectively).

Hand open-close

On opening and closing of the human hand on the leap motion tracking zone, the robot opens and closes its hand as a reaction. This operational mode can be used to grab any object, followed by the hand movement operating mode (Fig. 12.8).

Hand movement

Nao hand movements were controlled by hand palm position and palm normal. Left-hand, right-hand, or both hands can move simultaneously—the set of figures given below shows the entire hand gesture teleportation (Figs. 12.9 and 12.10).

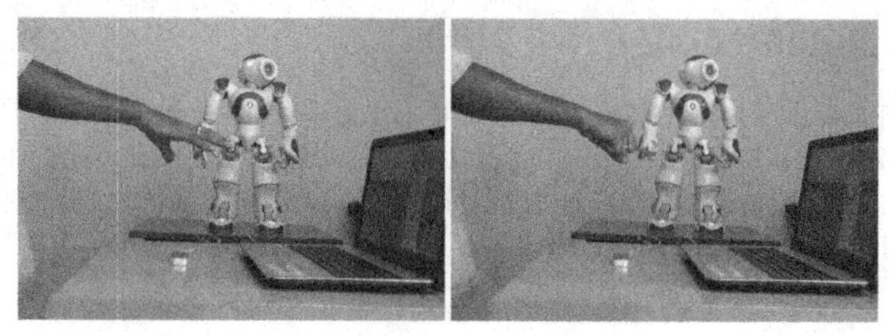

FIGURE 12.8 Hand open close of NAO robot (left: all fingers extended, right: all fingers closed).

FIGURE 12.9 Hand movements of NAO robot (single hand movements).

FIGURE 12.10 Hand movements of the NAO robot (two hand movements).

Numbers and alphabets learning

The number of fingers extended on the leap motion tracking zone can be count by extracting the LMC features. In this operational mode, the feature of counting fingers has been used to learning numbers from one to ten. The number also provides feedback to the user through audio by extracting the nao robot voice feature and visual feedback on the user interface (Fig. 12.11).

The operational mode of alphabet learning is a simple addition of the NAO robot speech recognition feature. In this operational mode, NAO read all the alphabets in order with a time delay to memorize, and after learning trial, a testing mode is on for testing the knowledge of the user. In the testing mode, any alphabet randomly arises on a window in the interface; the user needs to spell the alphabet. The feedback, correct or wrong, is provided by the robot through its voice feature.

Body parts identification

There is another operational mode, body parts identification, and learning powered by the nao robot speech recognition feature. In this mode, the robot hears for some predefined name of body parts as speech, namely, head, nose, eye, hand, etc. and points those parts by the right-hand movement.

The set of figures given below shows the outcomes of the mode called body parts identification (Figs. 12.12–12.17).

To identify each body part robot, have to point out by its right hand and point out the different body parts DOF of right hand to be set with

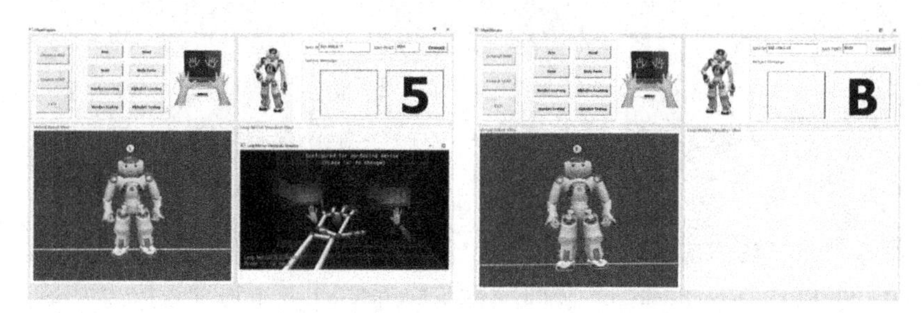

FIGURE 12.11 Number learning (left) and alphabet learning (right).

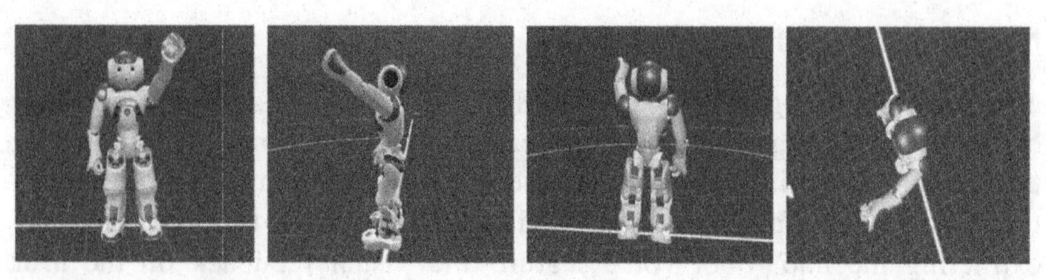

FIGURE 12.12 Left-hand identification by the nao robot (left to right: front view, left side view, rear view, top view, respectively).

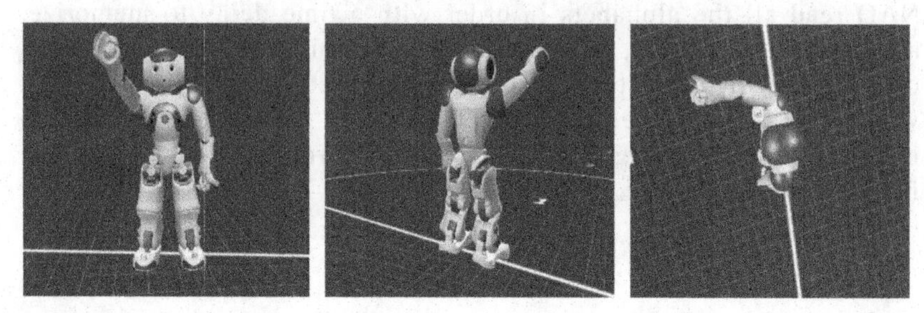

FIGURE 12.13 Right-hand identification by the nao robot (left to right: front view, rear view, top view, respectively).

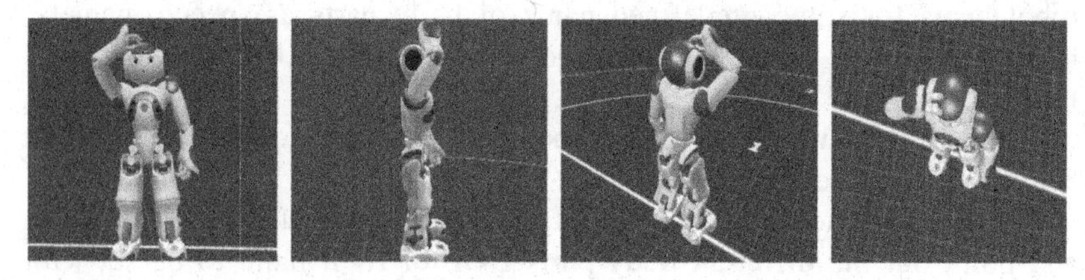

FIGURE 12.14 Identification of head by NAO robot (left to right: front view, right side view, rear view, top view, respectively).

FIGURE 12.15 Identification of chest by NAO robot (left to right: front view, left side view, top view, respectively).

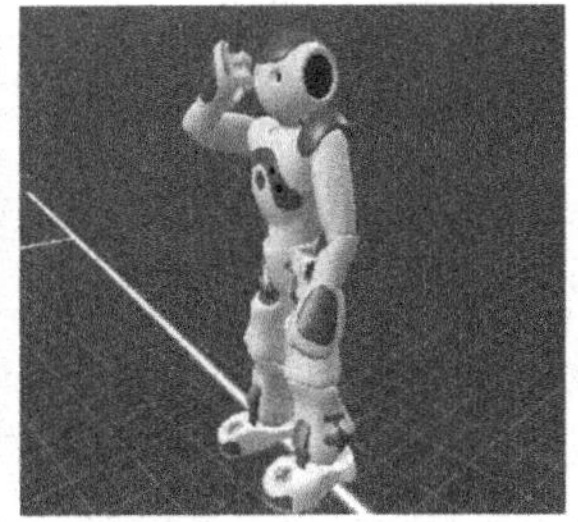

FIGURE 12.16 The NAO robot identification of right eye (left to right: front view, left side view, top view, respectively).

FIGURE 12.17 Identification of nose of the NAO robot (left to right: front view, left side view, top view, respectively).

a different set of values. Table 12.1 shows the different sets of DOFs of the right hand.

Data visualization

The visualizations and comparisons of the different data sets (user hand gesture data and humanoid gesture data along with the timestamp) are visualized below. Nao robot moves his head back and forth with the change of human hand pitch direction, and nao moves his head left and right with the change of human hand yaw direction.

There is a frame delay in mapping user hand gestures to the humanoid gesture with respect to timestamp. The data visualization given below in Fig. 12.18 shows the human hand palm position and Nao hand motion. In this work, palm position y steer nao shoulder pitch, palm position x steer nao shoulder roll, and palm position z navigate nao elbow roll (shown in Fig. 12.19 from left to right, respectively).

TABLE 12.1 Degrees of freedom with values for identifying body parts.

Body part	RElbowRoll	RShoulderRoll	RShoulderPitch	RElbowYaw	RWristYaw
Head	85.7	−19.3	−52.9	28.4	65.2
Nose	86.2	15.0	−4.4	35.3	82.9
Eye	87.1	−8.8	−42.2	4.9	104.3
Chest	84.1	13.0	20.9	16.1	50.4
Right hand	2.4	−10.3	−33.3	10.7	60.0
Left hand	−2.4 (LElbowRoll)	10.3 (LShoulderRoll)	−33.3 (LShoulderPitch)	−10.7 (LElbowYaw)	−60.0 (LWristYaw)

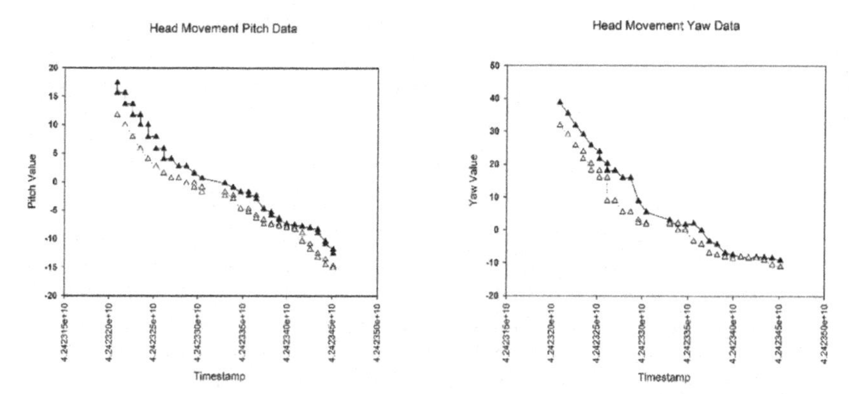

FIGURE 12.18 Mapping data of human hand gestures with Nao head movements.

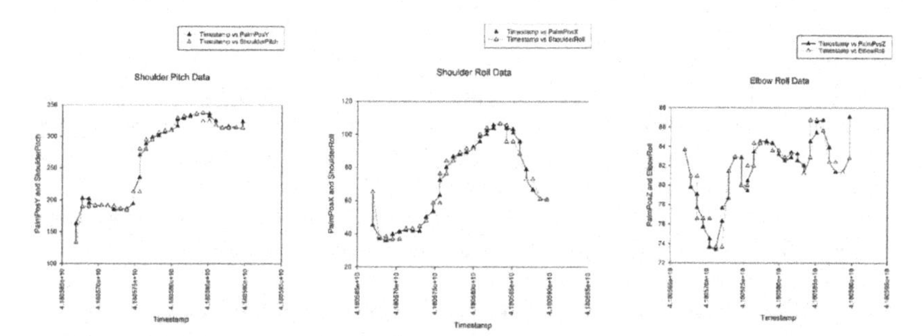

FIGURE 12.19 Mapping data of human hand gestures with NAO hand movements.

Conclusion

The objective of teleportation of human gesture is a functional translation of human body limbs relative positions toward equivalent humanoid robot kinematics representations. Leap motion used here as an interface of immersive user experience is very convenient in elementary brain-body coordination. This interaction scope further offers the extension to realize the pleasurable learning experience logic and mental modeling of eye estimation. In this work, we tried to explore leap motion usage and it significantly translated the robot interaction experience in teaching elementary sensor estimation. This work further progresses toward effective user interaction modeling and profound experiment designs to enhance the strategic logic learning experience in self body parts movements with a humanoid robot limb constructs. This work successfully transformed the human limb's complex skeletal movement to robot movement in a responsive manner with a series of algorithmic Percept action sequence. Extensive usability testing for outcome-based interaction is going on.

The functional faultlessness of the crusade fidelity is already established. The close domain usability testings have confirmed the extensive application possibilities of the proposed system in the areas of assistive HRI and heuristic-based rapid robot limb control.

References

Avola, D., Cinque, L., Levialdi, S., Petracca, A., Placidi, G., & Spezialetti, M. (2014). *Markerless hand gesture interface based on LEAP motion controller. DMS 2014* (pp. 260–266).

Chan, A., Halevi, T., & Memon, N. (2015). Leap motion controller for authentication via hand geometry and gesture. In T. Tryfonas, & I. Askoxylakis (Eds.), *Human aspects of information security, privacy, and trust. HAS 2015. Lecture notes in computer science* (Vol. 9190). Switzerland: Springer. Available from https://doi.org/10.1007/978-3-319-20376-8_2.

Chang, C.-W., Lee, J.-H., Chao, P.-Y., Wang, C.-Y., & Chen, G.-D. (2010). Exploring the possibility of using humanoid robots as instructional tools for teaching a second language in primary school. *Educational Technology & Society, 13*(2), 13–24.

General, A., Silva, B. D., Esteves, D., Halleran, M., & Liut, M. (2016). A comparative analysis between the mouse, trackpad, and the leap motion. In *CHI Conference*.

Gouda, W., & Gomaa, W. (2014). *Complex motion planning for NAO humanoid robot. International conference on informatics in control, automation, and robotics (ICINCO)*. IEEE.

Guo, C., & Sharlin, E. (2008). *Exploring the use of tangible user interfaces for human-robot interaction: A comparative study*. CHI *2008*: Proceedings of the SIGCHI conference on human factors in computing systems, April 2008 (pp. 121–130).

Hussein, M., Torki, M., Gowayyed, M., & Saban, M. E. (2013). Human action recognition using a temporal hierarchy of covariance descriptors on 3D joint locations. In *International joint conference on artificial intelligence*. Beijing, China.

Jamet, F., Masson, O., Jacquet, B., Stilgenbauer, J.-L., & Baratgin, J. (2018). Learning by teaching with humanoid robot: A new powerful experimental tool to improve children's learning ability. *Hindawi Journal of Robotics*, 11. Available from https://doi.org/10.1155/2018/4578762.

Jamet, F., Masson, O. J., Stilgenbauer, J.-L., & Baratgin, J. (2018). Learning by teaching with humanoid robot: A new powerful experimental tool to improve children's learning ability. *Hindawi Journal of Robotics*, 11. Available from https://doi.org/10.1155/2018/4578762.

Mohandes, M., Deriche, M., & Oladimeji, S. (2014). *Arabic sign language recognition using the leap motion controller. 2014 IEEE 23rd international symposium on industrial electronics (ISIE)*. Istanbul: IEEE.

Mubin, O., & Ahmad, M. I. (2016, November 7). Robots are likely to be used in classrooms as learning tools, not teachers—The conversation.

Naidu, C., & Ghotkar, A. (2016). Hand Gesture recognition using leap motion controller. *International Journal of Science and Research, 5*(10), 436–441.

NAO Documentation—Aldebaran 2.1.4.13 documentation. (2019). Retrieved from Doc.aldebaran.com: https://doc.aldebaran.com/2-1/homenao.html2.

Nifal, M., Logine, T., Sopitha, S., & Kiruthika, P. (2017). Space mouse - hand movement and gesture recognition using leap motion controller. *International Journal of Scientific and Research Publications, 7*(12), Retrieved from. Available from http://www.ijsrp.org/research-paper-1217.php?rp = P727039.

Nikolaidis, S., Lasota, P., Ramakrishnan, R., & Julie, S. (2015). Improved human-robot team performance through cross-training, an approach inspired by human. *The International Journal of Robotics Research, 34*(14), 1711–1730.

Nikos, K. (2013). Complete analytical inverse kinematics for NAO, in *2013 International conference on autonomous robot systems*, 1–6. https://doi:10.1109/Robotica.2013.6623524.

Python SDK Documentation—Leap Motion Python SDK v3.2 Beta documentation. (2019). Retrieved from Developer-archive.leapmotion.com: https://developer-archive.leapmotion.com/documentation/python/index.html.

Rodriguez, I., Astigarraga, A., Jauregi, E., Ruiz, T., & Lazkano, E. (2014). *Humanizing NAO robot teleoperation using ROS. 2014 IEEE-RAS international conference on humanoid robots* (pp. 179–186). IEEE. Available from https://doi:10.1109/HUMANOIDS.2014.7041357.

Savur, C., & Sahin, F. (2015). *Real-time American sign language recognition system by using surface EMG signal. 14th International conference on machine learning and applications.* Miami, FL: IEEE.

Shamsuddin, S., Ismail, L. I., Yussof, H., Zahari, N. I., Bahari, S., Hashim, H., & Jaffar, A. (2011). *Humanoid robot NAO: Review of control and motion exploration. 2011 IEEE International Conference on Control System, Computing, and Engineering.*

Shamsuddin, S., Yussof, H., Ismail, L. I., Mohamed, S., Hanapiah, F. A., & Zahari, N. I. (2012). Humanoid robot NAO interacting with autistic children of moderately impaired intelligence to augment communication skills. *Procedia Engineering* (41), 1533–1538. Available from https://doi:10.1016/j.proeng.2012.07.346.

Shang, J., Jong, M., Lee, F., & Lee, J. (2007). A pilot study on virtual interactive student-oriented learning environment, In *2007 First IEEE international workshop on digital game and intelligent toy enhanced learning (DIGITEL'07)*, 65–72. https://doi:10.1109/DIGITEL.2007.6

Sinha, M., & Deb, S. (2017). An interactive elementary tutoring system for oral health education using an augmented approach. In *Proceedings of the 16th IFIP TC 13 international conference* Mumbai, India, Part II, September 25–29.

Staretu, I., & Moldovan, C. (2016). Leap motion device used to control a real anthropomorphic gripper. *International Journal of Advanced Robotic Systems*, *13*(3), 113.

Taheri, A., Alemi, M., Meghdari, A., Pouretemad, H., & Basiri, N. M. (2016). *Social robots as assistants for autism therapy in Iran: Research in progress. International conference on robotics and mechatronics (ICRoM).*

Tan, J., & Biswas, G. (2007). *Simulation-based game learning environments: Building and sustaining a Fish tank. 2007 first IEEE international workshop on digital game and intelligent toy enhanced learning (DIGITEL'07)* (pp. 73–80). IEEE. Available from https://doi:10.1109/DIGITEL.2007.44.

Yu, N., Xu, C., Wang, K., Yang, Z., & Liu, J. (2015). *Gesture-based telemanipulation of a humanoid robot for home service tasks. Proceedings of the IEEE international conference on cyber technology in automation, control, and intelligent systems (CYBER)* (pp. 1923–1927). Shenyang, China: IEEE, June 8–12, 2015.

Chapter 13

Recognition of trivial humanoid group event using clustering and higher order local auto-correlation techniques

K. Seemanthini[1] and S.S. Manjunath[2]
[1]*DSATM, Bangalore, India,* [2]*ATME, Mysore, India*

Introduction

Surveillance system provides protection, security, and helps to keep an eye on anything from a house environment to public areas. Components of surveillance are cameras, cables, power adapter, and monitors, and video recorder. Surveillance system gives better security than traditional security task. It helps to optimize processes, prevent accidents, and reduce costs. Video technology or CCTV helps to protect commercial-related objects systematically depicted in Fig. 13.1.

(a) CCTV Camera (b) Commercial Survillence

FIGURE 13.1 Video surveillance system.

Cognitive Computing for Human-Robot Interaction. DOI: https://doi.org/10.1016/B978-0-323-85769-7.00001-X

FIGURE 13.2 Traffic control in surveillance system.

FIGURE 13.3 Violence detection.

- Protecting trade secrets: determines potential criminals or identifies them by facial recognition technique.
- Securing offices and commercial buildings: For thieves warehouses, salesrooms and offices are the main target.
- Monitoring car parking lots in open areas: helps to avoid events such as illegal parking and vandalism respectively depicted in Fig. 13.2.
- Efficiently tracking logistics and storage: it prevent the disappearance of goods from warehouses and are increasingly being used to monitor processes in packing and production systems.
- The other application using surveillance systems are retail, gastronomy, private households, mobile surveillance, public places, sports, and agriculture depicted by Fig. 13.3.

Human detection in video surveillance system

Development of powerful computers, low cost and high resolution sensors such as CCD or CMOS a great interest in the area of computer vision (CV) has been generated. Intelligent cameras are ubiquitous in automatic vision and surveillance system to track people. Surveillance system locates people position and track people in video, which facilitate understanding of human behavior for better event detection. In real-time application, tracking automatically finds players and paths of players' movement, differentiating players can significantly help sports expert. In CV perspective the major issue arises during vision-based tracking are to detection of objects and tracking across different frames especially when variation exist in physical appearance, movement, occlusion, and pose (Ciptadi, Goodwin, & Rehg, 2014).

Many approaches have come up to deal with this problem, over many decades. The accuracy of tracking multiple people is not satisfied due to three reasons given below.

1. To design generic and robust object detection model is still a problem due to huge background clutter, illumination, and appearance variations and limited training samples depicted by Fig. 13.4. Thus it's challenging to extract a part belonging to human region from background image scene. The problem increases more as the perspective is not yet improved, which gives important future activities information. Therefore, this important information enhances tracking performance, especially when dealing with complex interaction and occlusions.

Human detection in surveillance system does an excellent job because of anomaly event detection. It focuses at making difference between the things in video supervision system. Some applications of automatic vision-based tracking are shown in below Fig. 13.5.

A system should able to detect and extract motions of objects to classify object accurately. Further this object is tracked to analyze at high level. A tricky task in machine vision (MV) view is detecting humans, due to pose changes, light,

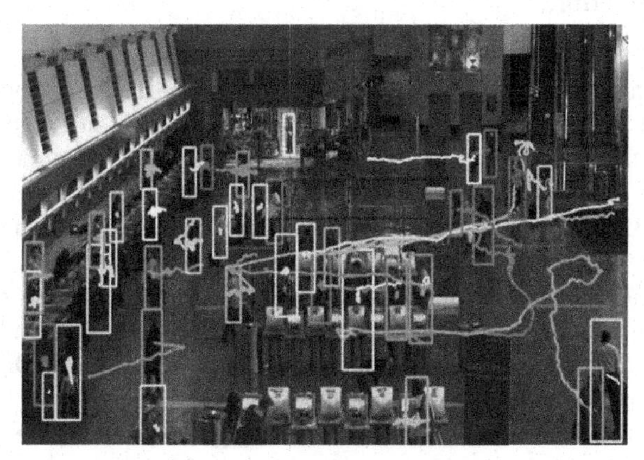

FIGURE 13.4 Multiple people detection and tracking in presence of occlusion and background scene.

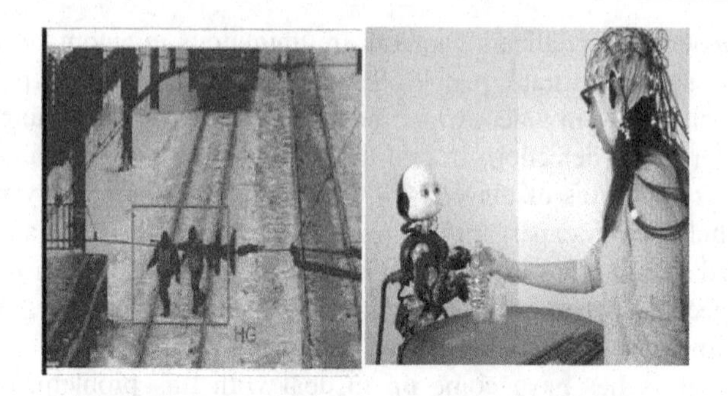

(a) Video Surveillance (b) Human-Robot interaction

(c) Automotive driving assistance system.

FIGURE 13.5 Application of video tracking system.

background and cloth colors, this limitation decreases the detection performance. The human detection processes by two ways object detection and classification.

Object detection

An object is tracked and identified by classifying and segmenting the image. The conventional methods for detecting object are spatio-temporal method, background segmentation and Optical Flow (OF). Background segmentation is the most effective method, it identifies entity by segmenting forefront from background scene. The camera could be fixed, pure translational or mobile in nature. Background methods try to identify moving objects in current and reference image in a pixel-by-pixel form. The reference frame referred as "background image," depicted by Fig. 13.6. Approaches to perform background deletion are adaptive Gaussian mixture, nonparametric background, temporal differencing, warping background and hierarchical models. The OF technique uses moving objects flow vectors to detect regions.

OF is robust to noise, color, nonuniform lighting and requires enormous computational and are complex to gesticulation. Spatial-temporal based gesture recognition considered 3D spatio-temporal information.

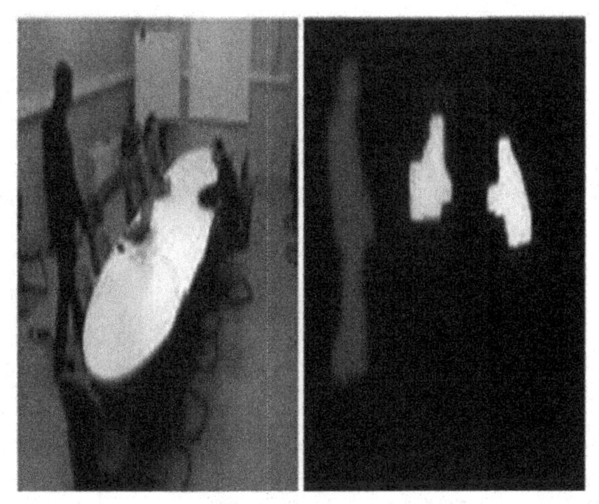

FIGURE 13.6 (A) Original image (B) background segmentation with one color per object (Manfredi, Vezzani, Calderara, & Cucchiara, 2014).

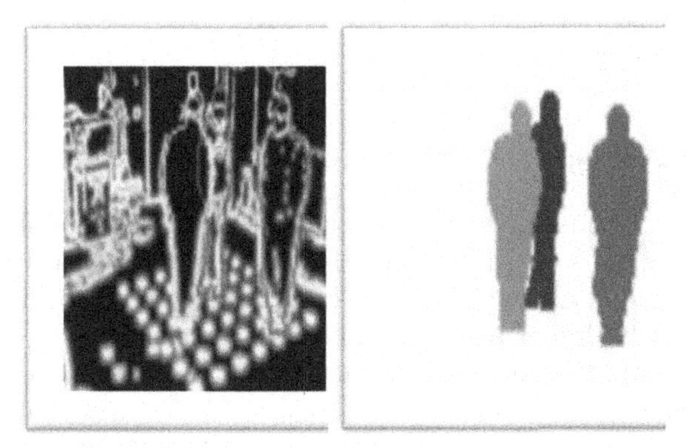

FIGURE 13.7 Shaped based segmentation without background subtraction (Tran, Gala, Kakadiaris, & Shah, 2014).

To analyze the group three steps is required segmentation, feature extraction that helps for tracking and recognizing event. Feature extraction is useful to obtain object information (Fig. 13.7).

Pixel level object detection like edge detection or background subtraction mainly provides low level features by density computation instead of counting. Texture level detection estimates people in numbers instead of identifying individual. The object detection by image patches is required for modeling which can extract high level features. After feature extraction objects is track.

Object tracking minimizes the occlusion like color intensity, illumination, appearance. Tracking multiple human approaches has been introduced to recognize and detect behavior in group, by identifying each person location in

FIGURE 13.8 Human detection and tracking in a group (John, Mita, Liu, & Qi, 2015; Yang, Zhang, Yang, Chen, & Yang, 2015).

each and every video frame sequence. Region, active contour and feature based approaches are more commonly to detect and to track human object in group depicted by Fig. 13.8.

Object classification

Classification of object is categorized into three ways such as shaped based, texture and motion based. Shaped based method gives shape information which is points, boxes and blobs. However due to articulations it is challenging to precisely differentiate between moving human and moving objects. HOG, part-based template matching and SVM are incorporated to deal with this challenge.

Violence detection is now forming a serious task for recognizing action. Automatic human action recognition in videos such as realistic videos is now becoming an important aspect in applications like CBIR, and surveillance system. Recently proposed method introduced action recognition categorized into local, global and frame-based respectively. Spatio-temporal key points identify human activity in local-based method.

Unsupervised learning same as BOW method is introduced to known probability of distribution extract features. However, when feature points are less then unsupervised learning fails to get much information. Global based techniques like optical flow describe motion in each frame with particular instant of time.

Analyzing and tracking of human activity is forming a most trending and active research topics. The activity analysis is necessary in surveillance system to improve machine and human interaction. However various activities are involved in multiple people with their different interactions which cause difficult in task of detection, because number of human variations such as human actions, interactions variation reveals among human groups (Tran et al., 2014).

Latest readings experiments combined behavior of cluster underneath numerous perceptions. Cluster actions are distinguished by distinct actions and their interactions with supplementary individuals (Manfredi et al., 2014). The group activities are differentiated by the localities and individual movements. The numerous cluster activities can be exists like walking, hand shaking, hugging and kicking etc. Nevertheless the factors involving such as background clutter, clatter, obstructions, presence of performers, audiovisual excellence and lighting changes have subsidized to the effort for the recognition and discovery of events (Ciptadi et al., 2014; Milan, Schindler, & Roth, 2013).

The proposed model provides effective way for trivial humanoid cluster and discovery of events. The adaptive FCM approach is employed for segmenting the desired human region. It assessments the required groups and fuses collections to detect humanoid area (Lajevardi and Hussain 2012). By using Completed Local Binary Pattern (CLBP) (Mazzon, Poiesi, & Cavallaro, 2013) and High Order Local Auto-Correlation (HLAC) (Karpathy et al., 2014) algorithms features are collected. To extract statistical features like counting the humanoid domes and calculate the wideness among the individual domes, an innovative method Spatial Gray Level Difference Method (SGLDM) has been employed. RNN is incorporated for classification. To recognize group activity, an effective and compact Local Group Activity (LGA) descriptor has been proposed which encodes movements and human postures within the group. The Bag-Of-Words (BOW) approach creates finite features set which are obtained from LGA descriptors.

Literature survey

The literature survey is carried out on the human detection, event detection and tracking of human in a group.

Ciptadi et al. (2014) has presented action recognition description based on temporal movement action encoding. The presented action consists of histogram pattern of movement which will encode the actions based on global temporal dynamics. The key observations are temporal dynamics actions are very strong to viewpoint changes and variance in appearance, and makes helpful to retrieve action and recognition. This section describes about the Movement Pattern that encodes global feature pattern for action without the requirements of the tracking features over time. It demonstrates the model significance for cross view action using IXMAS dataset.

Manfredi et al. (2014) have discussed the surveillance of the crowded scenes environments. It mainly works with static crowds, where human group interacts and remains in the same position. The SVM classifier employed to spatial localization and to detect static crowds. The OCSVM works on texture, the motion or movement features are identified for spatial region extraction, identification and filtering. Filtering is employed to reduce noise and false alarms because of moving flaws of people. Using inner texture descriptors and one class classifications, the model can able to get training set and can sufficiently obtain crowd model used for scenarios to shares the similar view points. Evaluation using public and real data setups validates the given system.

Tran et al. (2014) has described group behavior analysis frame work. In the crowded environment people can be represented by undirected graph. The graph works on vertices and edges of people. Social signaling cues describes about the interaction degree among the two people. The system proposes the graph-based clustering method to determine groups. Two signaling cues are given to detect group. If people are together it helps to consider it as a group engaged in governing activity effectively eliminates dataset defects. With the help of created interacting groups, the system creates a descriptor that captures interaction movement of people in a group. BOW method is employed for recognizing group activity and SVM for activity detection.

John et al. (2015), Detection of pedestrian is dominant in ADAS and autonomous driving. Computer vision technology since finds several applications in security and surveillance etc. Usually, detection of pedestrian in images is performed in visible light spectrum, which is unsuitable during night time recognition. Thermal imaging and Infrared (IR) are preferred in night time pedestrian detection because of its capacity to absorb emitted energy form pedestrian. The identification procedure firstly extracts the pedestrian form IR captured image. To characterize the object robust feature descriptors are employed. The adaptive fuzzy c-mean clustering method is employed for image segmentation and to retrieve the pedestrian. The CNN networks are employed simultaneously for learnt relevant features and to perform binary classification.

Yang et al. (2015) has suggested an action recognition model in combination with Depth Motion Maps (DMM) and CLBP by introducing Multi-Class Boosting Scheme (MBS). DMM characterizes the action motion energy. A multiple boosting algorithm is employed for the effective decision level classifier. CLBP is the enhancement of LBP, it achieves more significant for classification of rotation invariant texture. Evaluations are conducted on datasets such as MSRAction3D and MSRGesture3D to indicate the system performance.

Vandit Gajjar et al., introduced surveillance system countering against limited human resources. A computer vision technique such as HOG, Deep Neural Networks (DNN) with visual and saliency prediction model is incorporated to detect ROI. The dataset considered for demonstration is osu color-thermal pedestrian from Ohio State University. HOG structures are grouped by means of k means procedure. The detection precision and recall rate obtained is 83.11 and 41.27 percent respectively. Multiplying the saliency model with images results in salience-windowed images; these images reduce time and recall compared to normal images. The classification accuracy obtained is 76.86%. To achieve the 76.86 classification accuracy SVM is adopted.

Yuan Gao et al., introduced violence detection in surveillance system since violence detection problem is greater in action recognition it incorporates oriented violent flows (OViF) feature extraction algorithm. Adaboost with linear SVM adopted for improving the system performance. ViF feature descriptor describes the variation in observed motion magnitudes but still ViF is not significant. ViF loses important information when same pixel vectors flows in two video frames having same magnitude with different directions. ViF assumes no difference exist between two flow vectors, however vectors flows differ a lot. Therefore new feature representation method called OViF vector counts.

Shih-Chung Hsu et al., Human abnormal behavior detection is introduced to monitor the psychiatric patient. A normal human behavior features are characterized easily. However difficulty lies in detecting abnormal human behavior due to its unpredictable and complicated behavior. Abnormal behavior changes motion and appearance. Unsupervised learning such as N-cut algorithm and SVM technique segments the video objects. Condition random field (CRF) is incorporated by adaptive threshold for classifying normal and abnormal behavior events. In behavior abnormal events, occurrence of motion activities is large. Motion energy image (MEI) computed in different directions by Hu moment described it as feature vector. Abnormal behavior detection represents the behavior matching with the normal behavior. For input behavior, behavior template matching is applied if behavior does not match with template then it is abnormal behavior.

Reshma Khemchandani et al., in CV system the active area is human activity detection and recognition. The challenging issue arises while recognizing human in video is the background noise. These problems are addressed by robust least squares twin -SVM (RLS-TWSVM) with large datasets. Experimental demonstration is conducted using Weiz-mann, IXMAS, UMD and UIUC1 human action datasets to find effectiveness in RLS-TWSVM.

Tahmina Khanam et al., in CV automatic surveillance system in video is the active topic to detect crimes in public places and to provide security. The objects required to detect are illegal carriage materials in baggage. Baggage detection concept is described by considering dynamic human motion. To detect human parts background subtraction concept is adopted rather than using sliding window technique to increase significance of given system and HSI model to handle uneven illumination. Rotational Signal Descriptor (RSD) model is extracted from ROI to improve efficiency in HOG. Finally, dynamic method makes system to classify carried baggage.

Domonkos Varga et al., Detecting and finding different human objects in video is a fundamental task in CV domain. Objects detection in surveillance system helps in robotics and automotive safety. A new descriptor based on Multiscale CS-LBP operator is introduced. Experiment is demonstrated on CAVIAR pedestrian dataset which helps for detecting pedestrians on real-time. Pedestrian detection with multiscale CS-LBP operator and experimental results are reported.

Ahmad Jalal et al., Human Activity Recognition (HAR) system is introduced by developing image depth technologies without using optical markers/motion sensors in human body. A multifused feature descriptor in online is used to recognize activity of human from depth map sequence which is continuous in nature. HAR system fragments human being depth silhouettes by sequential human being activity information and spatial information to obtain the skeleton joint features. Spatial-temporal features are extracted and concatenate with four skeleton joint features. Skeleton joint vectors consist of torso based distance feature (DT), key joint based features (DK), spatio-temporal feature such as magnitude, direction and angle. HOG-DDS Shape feature are the outcomes of the depth differential silhouettes (DDS) among two successive images.

Harihara Santosh Dadi et al., Human recognition in surveillance system using camera became a hot topic in CV area. Gaussian mixture technique is used for tracking human. Gaussian mixture Mask divided into four regions to track human simultaneously. HOG extract feature vectors of face region which are later fed as input to SVM. 3 different approaches of experiment are conducted while training images. Every 3rd, 5th and 10th frame of the initial 100 frames are considered. The current remaining frames are kept for validation. Three datasets such as AITAM1, AITAM2 and AITAM3 is considered for demonstration work. The experimental results shows, as complexities of dataset increases the performance metrics are getting decreased.

K. Seemanthini and S. S. Manjunath et al., Human detection in surveillance system forms hot research area for recognizing abnormal action, identification of person, recognizing the human activity and finding actions such as walking and talking etc. This entire work is motivating towards human object detection and tracking. Cluster based segmentation approach is illustrated where video gets divided into frames followed by clustering and feature extraction by HOG. Classification by using SVM; each object activity is detected and classified by SVM. The model accuracy is calculated for each object detection is up to 89.59%.

Sultan Daud Khan et al., Automatic social pedestrian group detection is presented in crowd. Instead of finding pedestrian trajectories similarity, trajectories are clustered into groups; here pedestrians are clustered into groups by locations of pedestrian's trajectories. This approach evaluates by using different datasets. Experimental results achieves significant accuracy under both di-chotomous and tri-chotomous systems. Investigational grades elaborates that this method is fewer computationally, affluent than the existing state-of-the- art approaches.

Minaeian, Liu, and Son (2018) Fast and accurate moving target detection is now forming a challenging task when computational resources are in limited condition. Here object identify and fragment numerous moving forefront objects since a videocassette series captured by using a monocular stirring camera (unmanned aerial vehicle (UAV)). Here camera motions are analyzed by tracking key points with pyramidal Lucas–Kanade 2 foreground segmentation integrated with a confined gesture times gone by occupation and spatio- temporal differencing to detect multiple moving objects. For effectiveness of image registration, perspective homograph is employed using thumb rule technique intervals are adjusted. This model tested on various scenarios by using UAV camera. Accuracy of detecting targets in real-time scenario has been successfully demonstrated by comparing with existing schemes.

Andrea Pennisi, Bloisi, and Iocchi (2016), online and real-time crowd detection in surveillance system is described using FSCB algorithm. Strong or stable key point features are tracked in video sequence. Temporal features are extracted and moving blobs are segmented for detecting abnormal event activity using two measures such as temporal variation and entropy. Experiment measures are conducted by having online available datasets consisting of different multitude circumstances and diverse event categories which determine efficiency such as UMN, PETS and AGORASET. For PETS and AGORASET, ground truth values are accessible at FSCB website.

Wang, Oneata, Verbeek, and Schmid (2016) motion estimation and trajectory features are improved. Features points are matched between consecutive frames by SURF algorithm and dense OF technique. Further matched features estimates homography with RANSAC. To enhance homography estimation, outlier matches are removed using human detector as camera does not constrain camera. Due to motion by camera, trajectories consistent are removed and also homography removes from OF. Fisher vector another feature extraction approach is explored to BOW histogram. Experimental results are evaluated by considering classification of actions using six datasets, localization actions in lengthy movies and complex events recognition.

Jalal, Mahmood, and Hasan (2019), Human action analysis and tracking now becomes an interesting task for computer vision field as it finds much application in industrial and commercial sectors. However efforts are made to make machines for understanding human behaviors in outdoor environment by a new recognizing human interactions scheme. Humanoid action recognition (HAR) system distinguish eight multifaceted different humanoid actions by using BIT interface dataset which involves boxing, hand shaking, kick, high-five and push respectively. Various features algorithms are designed with CNN for measuring the system performance.

Nikouei et al. (2018), Human detection, recognition of its behavior and estimate in insolent video surveillance comes under a category, where huge video streaming data transition takes time and creates pressure on networks. Generally video processing includes object detection which demands calculating and is exclusive to handle using limited resource. Inspired by convolution and Single Shot Multi-Box Detector (SSD), CNN is introduced for pedestrian detection in affordable computation workload. CNN remained accomplished by means of portions of ImageNet and VOC07 datasets that gives the location of objects of interest. Lightweight CNN obtains 2.03 and 1.79 max and average FPS respectively, which is very fast Haar Cascaded algorithm having 6.6% false positive rate.

Deng, Wang, Jiang, Pan, and Sun (2018), Multiobject tracking concept in surveillance system is introduced by means of Long Short- Term Memory (LSTM) and deep learning algorithm. Object detector called as YOLO V2 detects the multiple objects. The problem behind tracking of single object is Markov Decision (MD) because MD process setting enables sequence decisions. To track single object, tracker combined with network such as CNN followed by an LSTM. Each tracker is trained with CNN network followed by data association using LSTM. The system results elaborate that given tracker provides better performance comparing with the other existing methods.

Irfan, Tokarchuk, Marcenaro, and Regazzoni (2018), Different patterns and people movements are detected in crowd and those detected pattern features are classified as normal and abnormal behaviors. By sliding window, Statistical features are collected from knowledge base. A model used for classification and training the movement patterns of people is named as Random Forest model. Two datasets are tested by system model which is obtained from phones during social events gathering. Evaluation results depict that cell telephone data detects anomalies in crowd's substitute to video supervision sensors by giving important achievements. To segment abnormal activities in swarm by using non-viable statistics is straightforward to instruct and organize.

Suriani, Hussain, and Zulkifley (2013), Event recognizing systems provides safety, security, convenience and good human lifestyle. Accurately detecting events and giving alerts to sudden changes during physical threat, fire and bomb alert etc. is necessary. Performance of recognizing the events depends exclusively on low level processing accuracy such as detecting objects, tracking, recognition and ML techniques. Event occurring significance over abnormal event and frameworks used to recognize sudden event is briefly explained. Several decision constructing techniques for recognizing unexpected event is elaborated. The rewards and limitation of employing 3D imageries using numerous cameras in real-time solicitation are briefly discussed.

Yimengzhang, Chang, and Liu (2012), Recognition of events in videos is now forming a challenging task. However attempts are made for addressing this issue by introducing interaction concept between group people. Some events which are recognized are fighting, chasing and flanking respectively. To recognize complex events a new method is introduced to study the group scenario from individual trajectories. These features are arranged using the BOW scheme. Finally, SVM incorporated to categorized video segment into six different categories. In real-time system, model is implemented by tracking multitarget and to recognize events in groups involving 20 people.

Tran, Yuan, and Forsyth (2013), proposed spatial-temporal path to detect and localize video event in group and cluttered scenes. This model handles variation in events, shape, scale and invariant to motions. Spatial-temporal paths gives video event trajectories against event detection by spatial-temporal sliding windows, thus Spatial-temporal paths is better to handle events generated by objects and achieving optimal solution with low complexity. Experimental results are demonstrated using real video dataset having anomaly such as running, walking etc. A Max-Path technique improves the event smoothness, missed or weak detections and reduces false positives occur due to occlusions and low quality of image.

Wang and Snoussi (2015), OF descriptor and histogram representation are introduced for abnormal event detection in video. SVM combined with kernel PCA methods to recognize unusual events in current video frame after learning and characterizing normal behaviors. Histogram of the Optical Flow Orientation (HOFO) features are collected and analyzed from foreground image. Future work minimizes the false alarms and considers online sample training. To collect more efficient feature vectors using OF or by interchanging the OF with other methods which can represent the event information. Due to large normal event samples it is difficult to train on one batch.

Yafengyin and Man (2013), abnormal human activity recognition in videos, cognitive semantics event detection is proposed. For a given video human detection and tracking is achieved by describing cognitive linguistic which involves paths, actions, place and human. Then the semantic distance is computed. Similarly the entire human is combined to diminish semantic entropy. Once group of individual human is identified, then their activity is tracked and classified into group activities by using spatio-temporal semantics features. Similarly group activities are analyzed and a probabilistic context is achieved from activities using Minimum Description Length (MDL) criterion. The syntax rules are then rummage-sale to characterize test videos by cognitive linguistic. Video event representation describes and recognizes multifarious humanoid actions, including both distinct and cluster actions.

Kylestephens and Bors (2016), unlike recognition of single human being action in crowd, the confined moving activity of human is inclined by other individuals. By taking into the account, relationships among activity flows and moving location a new scheme is introduced to illustrate the discriminative in group activity. The input given to system model is both in time and time- movement space. Further this space are modified by Kernel Density Estimation (KDE) and then given ML algorithms as input to classify. KDE model the period location and gesture spaces for describing such connections. The system efficiency is relying on without manual annotation of tracks.

Stone and Skubic (2014) proposed the two stage system and Microsoft Kinect system for older adults fall detection. Initial stage involves detecting of person's in vertical state for each individual frames. A utility of vertical time series helps to segment or detect person from their ground events by tracking persons over time. In second stage decision tree is ensemble to get confidence of a fall that proceeded as ground event. Experimental demonstration conducted by considering older adults. Given model consists of 13 apartments to conduct the experiment. The database contains 454, 445 falls from trained stunts, 9 naturally occurred during resident falls. Results are achieved by cross validation of sitting, standing and lying down positions. Significantly outputs are received after scrutinized with five fall detection algorithms.

Vemulapalli, Arrate, and Chellappa (2014) proposed human action recognition scheme using costly depth sensor and skeleton based estimation algorithm. Recently many researchers employed joint locations or angles to analyze human skeleton. This model exploits skeletal algorithm which basically works on 3D geometric translations and rotations between different body parts. Skeletal algorithm representation is the representation of action which is nothing but curved manifold. Human actions are modeled as curves using the skeletal representation. Since it is difficult for skeletal model to map action curves to its algebraic form called vector space. By incorporating the Fourier temporal pyramid, by combining SVM and dynamic time wrapping it then performs classification. Three action datasets are made to use which shows that given system perform better than existing algorithms.

Cheng, Qin, Huang, Jiang, and Tian (2010) introduced broadcasting video investigation, human action analysis and mobility. Recently there were limited algorithms introduced for recognizing human activities and cluster communication. However in video supervision system numerous human being intercommunication give a challenging task. This representation detects interactions in group of people which contains few persons. For identification of dense crowd the proposed scheme illustrates a layered model which provides uniform statistical representation. By using the layer model, proposed scheme represent group actions flexibly with different features at different action scales. The given scheme uses a Gaussian processes for motion trajectories, to deal variety of movements inside the group. The proposed schemes use visual information and depict participants, visual style features and group shapes characteristics. Hence proposed scheme recognize crowd activities and enhances accurateness of identification.

Naifanzhuang, Yusufu, Ye, and Hua (2017) describes group activity tracking. Activity detection plays an important role in technological pertinences resembling civic safety and video supervision system. This scheme exhibits end-to-end solution to detect human using Differential Recurrent CNN (DRCNN) and stacked differential LSTM (DLSTM) networks. The input is video data without individual object contrast to previous methods. Thus it is possible to generate more effective algorithm. DLSTM, DRCNN understand composite behaviors in series and cluster dynamics. This system provides automatic activity recognition and accuracy is quite more contrast to subsequent form of recognized models.

Sharif, Saha, Arefin, and Sharif (2011) introduced both restricted and unrestricted space of video supervision system and occasion detection. The proposed model focuses to detect event by incorporating the OF techniques. OF performs stack inferior instructions similar to point of attention. By employing k-means algorithm, key points are trailed and clustered.

To compute PC of every crowd together, statistical location, oversight, key point movements of every group are estimated. Further they are considered as the standard distinguishing mechanisms of apiece group relatively than distinct feature facts. The computed observations clusters are defined either high or low activity clusters. Then displacement and direction thresholding are calculated as final result.

Shao, Ji, Liu, and Zhang (2012) introduced internal atmosphere human being action investigation and segmentation. This scheme recognizes diverse activities and types of behavior concurrently. Two segmentation techniques are used which are activity based and color severity approaches. Both methods are capable to segment movements in temporal cycles efficiently. Finally descriptors extract features which contain both local shape and spatial layout information; hence this system is efficient for action modeling and well suitable for detection or recognizing variety of actions. Milan, Leal-Taixé, Reid, Roth, and Schindler (2016) proposed human tracking and detection for multiple objects in non-restricted scenarios. A set of sparse detection algorithm are given as input to the large intensity tracker. The tracker is to accurately correlate these "dots" over point in time. This model exhibits inadequacy such as images sequences are ignored by thresholding, applying non-maximum suppression and weak detection response. The multitarget tracker gives image low information and connects all pixels to particular objective or differentiates it at the same time as background, video segmentation in adding together to the conventional bounding box illustration in unrestrained real world videos. To achieve excellent result of many standard algorithms and extensively outplays state-of-the art human group tracking and recognition methods in dense scenes.

Ahmad and Kamal (2014) proposed automatic human group detection problem, tracking of 3D human poses. Human poses from monocular depth camera are related to ML applications. By making use of ridge data depth maps this model presents a real-time tracking method for corpse parts to pose recognition. The features depths are extracted first and then processed with ridge data of binary silhouettes as human body skeleton shape. Pose estimation initializes each and every body parts with joint paints information by defining pose. According to simplified torso-center, features are extracted to track body parts. This helps to present the 3D body joint angles based on forward kinematic analysis. The investigational model provides efficiency for the complex human pose and skeleton for even dynamic scenes.

Ge, Collins, and Ruback (2012) discussed detection of individual groups based on their interaction. Hierarchical clustering is introduced based on bottom-up approach to detect individual group. Hierarchical clustering uses hausdorff distance method in support of speed and to form a pair of propinquity. This system elaborates the small pedestrian group recognition by using trajectories. Human perception small pedestrian human group prediction is explained in superior way for agglomerative clustering. Results are scrutinized by robotic tracing which in particular helpful to endow with correlated features of heavy crowds. Identified course particulars are sufficient amount to envision appendage gesticulation and gawk information.

Chen, Zhang, Su, Li, and Wang (2016) proposed a CLBP for classifying scene in remote sensing land. To extract and accurately categorize the primary surface characteristics of numerous resolutions and enhanced comprehensive CLBP description is endorsed. Kernel-based training mechanism is used to execute Ms-CLBP and is discriminated in terms of correctness, categorization and computational complication. This model is tried and true on 19 categories of asteroid images and 21 categories of terrain datasets. This representation illustrates the heightened work.

Alahi, Ramanathan, and Fei-Fei (2014) proposed human identification in dense environment alike railway terminals and capital ventures commonly. This system is designed and executed by taking into consideration of benchmark dataset including 42 million directions of railway terminals. The modules described here represent perambulator terminus and crew transportability in outsized range of exertion. Problems correlated to particular annotations of sporadic video cameras and perambulator abnormality representation beyond various video cameras has been consigned. With the view of bridging disintegrated and unrecognized movements, a newest description named as Social Affinity Maps (SAM) is recommended. The provisional outcome presents the betterment in the execution by SAM features and beginning by the side of target earlier.

Sun, Wang, and Sheu (2017) interpreted the deepness of the image information rejuvenation from high-fidelity vision system. Deepness of the image information is fetched using RGB-D video camera or by double video camera. Many times detection of objects is possible by retrieving the deepness of the image information or by using the chromatic knowledge of the image. The pros and cons of deepness of the image information and chromatic knowledge of the image information are aggregated to accomplish enhanced result. The model presents moving object detection based on color information and depth image. This method detects and identifies the motion objects with less noise in the background. The segmentation accuracy obtained by interpreting the above model is 84.4%.

Manfredi et al. (2014) have delineated the methods to inspect the dense cramped sequences. Manfredi spotlight on human crowd intercommunication in similar places at the same time and also on unchanging crowd. To categorize characteristics of dense crowd SVM is endorsed. Swarm with contiguous scenes are filtrated in account to lessen clatter along with forged counting suitable to action flows of human being group. One internal surface description and SVM is obtained from the distinct instruction set, which intern is used for dissimilar representations to inlays the swarm that analyze the crowd expression. The obtained results are evaluated by applying universal datasets.

Tran et al. (2014) have conferred activity analysis in group. Humans in a group can be efficiently analyzed with indiscriminate graphic representation in which y-axis is indicating people and edge between 2 humans is measured to calculate their interaction. To compute interaction between people societal-signalize clues is illuminated in detail. A graph-based clusters finds interaction region in swarming environment. Binary social-signals are used and evaluated for the group discovery. A BOW method represents group of actions and SVM classifier for action detection. The experimental result shows better result in contrast to other existing method.

Feng, Wang, Dlay, Naqvi, and Chambers (2017) has inspected the confrontations like numeral variants, occluded objects and sounds intricate in human object detection. To conquer and to represents previously mentioned obstacles, Probability hypothesis density (PHD) filtering method is endorsed to design and develop Markov Chain Monte Carlo (MCMC). Social Force Model (SFM) a latest technique is further used to examine the interfacing amongst the objects. SFM legitimizes the possibilities within the MCMC samples in support of PHD filter. One class SVM (OCSVM) used for moderating the noise. The measured classification is from the TUD, CAVIAR and PETS2009 datasets. Outcomes are compared and verified with the existing algorithms. The chosen selected algorithm accomplishes better results.

Gauzere, Ritrovato, Saggese, and Vento (2015) has developed a novel based knowledge and deep-down method to identify and recognize human clusters. Initially the developed model investigates the problems occurred in identifying and tracking, subsequently it describes deep-down approach instead of bottom-up approach to determine the information. The deep-down approach concedes us to maintain uniformity in trajectory eradication. The representation depreciates the low down substance of the fewer detected aspects and it considers one image per sec slightly than 7 from dataset. The introductory demonstrations are operated by means of standard PETS 2009 dataset. The sculpt represents the mixture of facts and little information are adequate to track human trajectories.

Ciptadi et al. (2014) described action recognition using descriptor based on temporal features like movement action encoding. The considered actions involve movement with its histogram pattern which encode the actions based on global temporal dynamics. The temporal actions are very strong with changes and variance in appearance. This temporal information is helpful to retrieve action and recognition. Movement Pattern Histogram (MPH) extracts global pattern for action without having the requirements to track features over time. IXMAS dataset is made in use to demonstrate the significance of system for cross view action retrieval.

Manfredi et al. (2014) proposed the surveillance system in crowded scenarios. This system mainly focuses on group detection having interaction between people in crowd and remained in same place for a while. The SVM classifier classifies spatial features detect human in static crowds. The OCSVM uses texture features that extracted at pitch level. The movement features are identified to get spatial region, identification and filtering. Motion information is filtered to decrease noise and to prevent false alarms because of moving flaws of people. The texture descriptors along with classifier make model capable to get training set and can sufficiently detect crowd. Evaluation is done by publicly available and real data setups to validates the given system performance.

John et al. (2015) introduced detection of pedestrian for Advanced Driver Assistance Systems (ADAS). The key technology in CV system is detecting of pedestrian since it finds several applications in security and surveillance etc. Generally, pedestrian detection is performed in visible light spectrum and thus not possible to detect during night time recognition. Thermal imaging and Infrared (IR) light are preferred in night time detection as this light can able to absorb reflected energy from pedestrian. The procedure is to first extract the pedestrian form IR captured image. Feature descriptor helps to characterize the pedestrian. The Adaptive FCM (AFCM) segments pedestrian image and helps to extract pedestrian attributes. The CNN networks are employed simultaneously to learn relevant features and to perform binary classification.

Yang et al. (2015) introduced action recognition model using combination of Depth Motion Maps (DMM) and CLBP with Multi-Class Boosting Scheme (MBS). DMM represents action movement's energy whose features, extracted by CLBP. A multiple-class boosting (MB) is a decision level classifier to effectively classify the relevant features. CLBP is the enhance modification of LBP to achieve significant classification accuracy. Evaluations are done by using datasets such as MSRAction3D and MSR Gesture to indicate the system performance.

Priya and Nawaz (2016) introduced a new system for distinguishing image which discriminate image object features using SGLDM and LBP method to exhibit the effectiveness. Spatial distribution, gray level dependency is extracted using SGLDM. Run difference method is quite similar with SGLDM for extracting spatial features which describes prominent texture size. LBP give robustness against variations of illuminations. The feature obtained gives better interpretation based on sharpness, correlation, contrast, energy and homogeneity. KNN solves problems such as classification and regression.

Lajevardi and Hussain (2012) proposed feature extraction, selection and classification algorithms for detecting facial expression. Modules like face detection, feature collection, optimal feature selection and finally classification to recognize face. Generally, AdaBoost algorithm detects face followed by feature extraction. Input frames are processed for feature characteristics by various attributes extraction algorithm namely log Gabor filter, Gabor filter, LBP and HLAC. Informative features are obtained using filter feature and wrapper selection methods. Experiments are conducted using several facial expression databases shows comparison of different algorithms.

Riccardo Mazzon et al. proposed detection and tracking scheme which is having a hot search in CV system. To detect and track human interactions Social Force Model (SFM) is adopted. SFM methods detect human behaviors to recognize human interactions in crowd by reducing the error among the measurements and predictions. The recognized groups are tracked by interactions centers and time by employing buffered graph-based tracker. The model shows the system frame work performs well over publically available datasets with improvement in group detection.

M.S. Priya et al. to categorize the image, the system is investigated to discriminate features using SGLDM and LBP method to exhibit the effectiveness. Spatial distribution and dependence between the gray levels can be extracted using SGLDM. Run difference method collect spatial features that describe prominence and texture size. LBP features offer robustness against variations of illuminations. The extracted feature gives better interpretation based on correlation, contrast, energy, sharpness and homogeneity. KNN classifier is used to solve classification and regression. The system classifies the image by employing KNN.

Seyed Mehdi Lajevardi et al. has investigated related to feature extraction, selection and classification methods to recognize the facial expression. This system consists of modules such as face detection, extraction of features, selecting feature and finally classification. LGF, Gabor filter, LBP and HLAC are used to extract face features detected by AdaBoost classifier. Informative features are obtained by using filter feature and wrapper methods. Experiments are conducted using various databases showing comparison of different algorithms (Figs. 13.9 and 13.10).

Riccardo Mazzon et al. has given a method to detect and also to track interactions of using Social Force Model (SFM). The methods analyze human behaviors to predict people interactions in crowd by reducing the error among

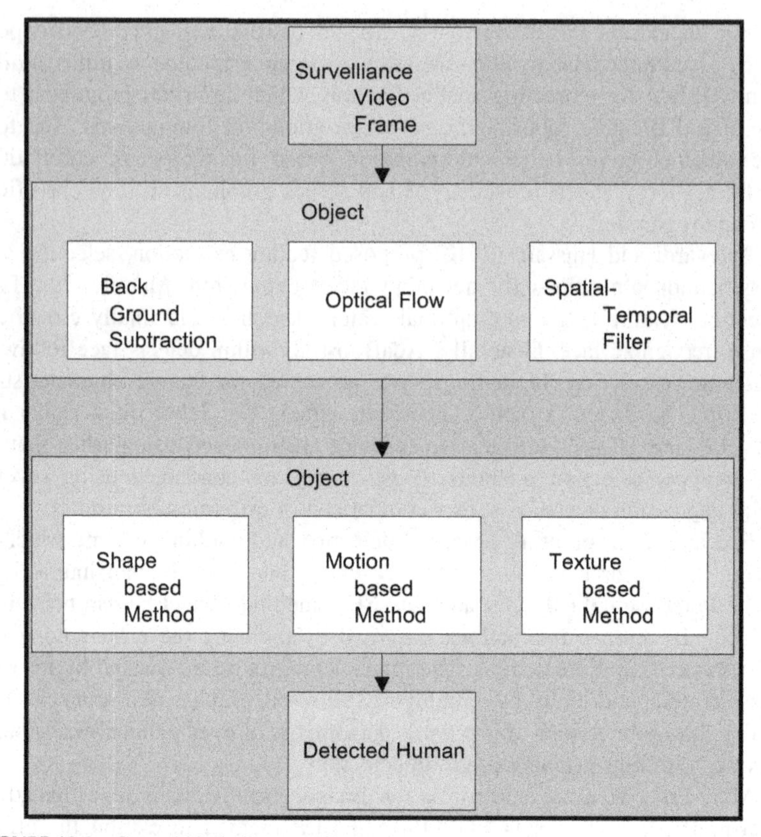

FIGURE 13.9 General block diagram of human detection.

the measurements and predictions. The recognized groups are tracked by considering the interactions centers and time by employing buffered graph-based tracker. The model shows the system frame work performs well over datasets that are publically available with improvement in group detection.

Methodology

The recommended method is represented in Fig. 13.11. The given model gives description about the event in the collaboration of trivial humanoid cluster. Originally the humanoid region is projected by frame by frame grouping step. To recognize and detect events in trivial humanoid cluster FCM followed by extraction of appropriate structures, CLBP and HLAC algorithms is required. The further process is validation of number of humans present in the small human group by their relevant statistical features which includes detected human heads and width between the discovered individual heads. SGLDM has been used to extract statistical feature and

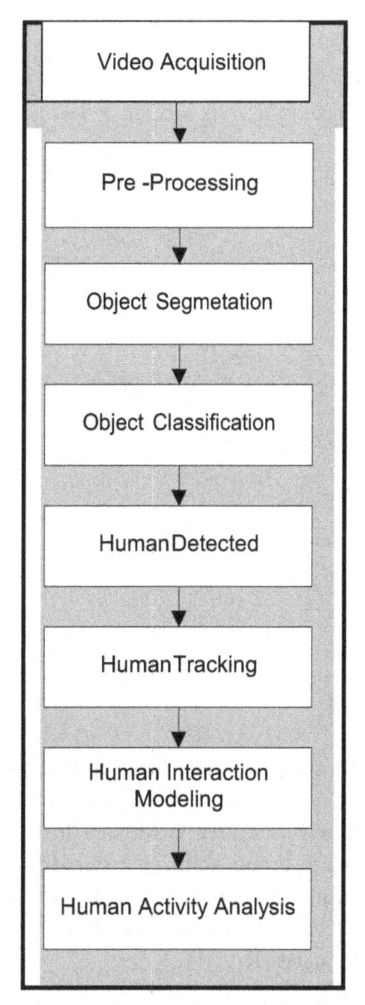

FIGURE 13.10 General flow of human detection and activity analysis.

RNN to classify human and nonhuman regions. Final step is to detect group activity. LGA descriptor has been proposed to encode movements and pose of people within a group. The BOW approach employed to create vocabulary features set which are derived from LGA descriptors. The CNN classifier classifies people activities. Each step in detail is described in below sections.

The recommended method is divided in two stages, that is, testing and training phase which is further described in 3 stages i.e. Rule 1, Rule 2 and Rule 3. Rule 1 gives description about the detection of human region. Rule 2 explains about the human validation and Rule 3 provides detail about event detection. The below sections describes Rule 1, Rule 2 and Rule 3 in detail.

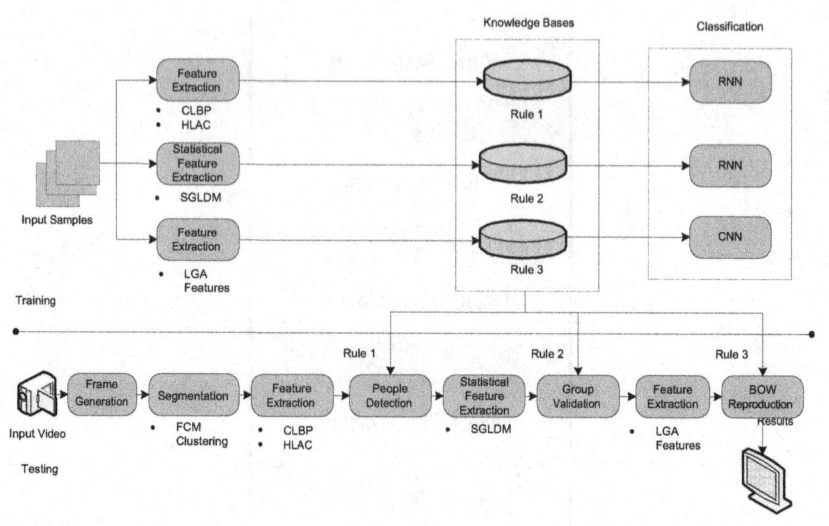

FIGURE 13.11 Architecture of event detection.

Results

Rule 1

Frames are generated from input video. The input video frames are enhanced by applying sharpening filter. The resultant enhanced frames are passed to segmentation block known as Fuzzy C-Mean Clustering (FCM), which are best suits for overlapped group data. Cluster features are extracted by employing novel techniques CLBP and HLAC. These features and knowledge base stored features are compared, based on which human region will be separated using RNN classifier.

Segmentation

A brief description about the segmentation approach is discussed here. An adaptive FCM extract human region from input frame is employed here. The input image $I(a,b)$ gives pixel intensity located at (a,b). FCM is a soft iterative process for partitioning image into numbers of clusters by reducing the objective function J which is described in Eq. (13.1).

$$J(U,v) = \sum_{c=1}^{C}\sum_{n=1}^{N}(u_n^c)^q d^2(x_n, v_c) \tag{13.1}$$

Where $X = \{x_n\}_{n=1}^{N}$ gives intensity in input frame. C indicates clusters. $U = \{u_n^c\}$ Defines fuzzy partition matrix, where u_n^c is degree membership of

n^{th} intensity. q Depicts weighting exponent while $V = \{v_c\}_{c=1}^{C}$ represent weighted cluster centroid. d^2 measures the distance between intensity and cluster centroid. The estimation of cluster number is done using image intensity information. More specifically, the cluster number C is obtained from μ_I using the following measure $C = \frac{\mu_I}{\eta}$ here η is an empirical constant. An optimized U generates the cth cluster using Eq. (13.2).

$$P^{c}_{(a,b)} = u^{c}_{x(a,b)} \tag{13.2}$$

Where $u^{c}_{x(a,b)}\, \epsilon U$ corresponding to partition membership degree. Find $K \leq C$ cluster centroids to detect human region, $\{v_k\}_{k=1}^{k}$ and fuse cluster maps P^k to get region map $\tilde{P}$ using Eq. (13.3).

$$\tilde{P} = \sum_{k=1}^{k} e^{-(k-1)} P^k \tag{13.3}$$

By binarizing $\tilde{P}$ and connecting components, human regions and nonhuman regions are estimated. Hence inferior segmentation results can be observed.

Feature extraction

The segmentation result given as input to feature extraction block for further process. Two novel algorithms termed as CLBP and HLAC are employed to extract features which are described below.

Completed local binary pattern

LBP descriptor is used in image processing and computer vision applications to provide effective texture description. Consider a centered pixel t_c, its neighboring pixels are uniformly spaced having a radius $r(r > 0)$ with the center at t_c, if the coordinates of t_c are $(0, 0)$ and m neighbors $\{t_i\}_{i=0}^{m-1}$ are considered, the coordinates of t_i are $\left(-rsin\left(\frac{2\pi i}{m}\right), rcos\left(\frac{2\pi i}{m}\right)\right)$. By thresholding the neighbors LBP is computed $\{t_i\}_{i=0}^{m-1}$ with center pixel t_c and m-bit binary number. LBP is expresses in the Eq. (13.4).

$$LBP_{(m,r)}(t_c) = \sum_{i=0}^{m-1} s(t_i - t_c)2^i = \sum_{i=0}^{m-1} s(d_i)2^i \tag{13.4}$$

Where $d_i(t_i - t_c)$ shows distance between center pixel and each neighbor, $s(d_i) = 1$, if $(d_i) \geq 0$ and $s(d_i) = 0$, if $(d_i) \geq 1$. The LBP considers only signed description of d_i while neglecting magnitude value. However the sign and magnitude are complementary to reconstruct the difference d_i. In CLBP, LBP difference divided into sign and magnitude (absolute values of $d_i\ i.e.|d_i|$). Fig. 13.12 depicts the sign CLBP_S and magnitude component CLBP_M.

(a)

26	42	16
20	24	26
40	12	18

(b)

2	18	-8
-4		2
16	-12	-6

(c)

1	1	-1
-1		1
1	-1	-1

(d)

2	18	8
4	24	2
16	12	6

FIGURE 13.12 (A) 3 × 3 sub image, (B) local difference, (C) sign CLBP, and (D) magnitude CLBP.

The CLBP_M operator is defined by the Eq. (13.5) as follows,

$$CLBP\_M_{m,r} = \sum_{i=0}^{m-1} p(|d_i|, c)2^i, p(u,c) = \begin{cases} 1, u \geq c \\ 0, u < c \end{cases} \tag{13.5}$$

where c represents threshold set to $|d_i|$ from the whole image. The CLBP center region codes the center pixel which has discriminate information. It is coded as in the Eq. (13.6).

$$CLBP\_C_{m,r} = p(t_c, c_1) \tag{13.6}$$

Each CLBP operator component is applied and histogram of all the blocks is combined to obtain a feature vectors.

Higher order local auto-correlation

The HLAC is the extension of auto-correlation (AF) higher order statistics (HOS) and it can extract the features. The N-th order AF is defined in Eq. (13.7).

$$x(a_1, a_2, \ldots, a_n) = \int f(r)f(r + a_1)\ldots f(r + a_n)d_r \tag{13.7}$$

Here $f(r)$ denotes intensity of pixel r and $(a_1, a_2, \ldots, a_n)$ as N displacements. HLAC features are represented by number of mask window with 0, 1 and 2 displacements. Each pattern is scanned and for every position white marked pixels is computed.

The products mask obtain is then summed to give one features. As compared to Fourier transform, window size resemble to frequency components while displacements distribution corresponds to directional components. HLAC uses the two-dimensional information distributions and directions to analyze the image more closely.

Algorithm: HLAC feature extraction
Input: Input segmented image either color or gray. Output: HLAC Features $x_1, x_2, \ldots x_n$
Step. 1: Traverse the input segmented image pixel and compute average value.
Step. 2: Calculate the absolute difference between every pixel and average value; further each and every pixel value is updated.
Step. 3: Determine the newer pixel values.

Step. 4: For i = 1:n Using template p_i, match local auto-correlation.
Step. 5: Calculate x_i. value and update End
Step. 6: Maximum value is determined $x_1, x_2, \ldots x_n$.
Step. 7: For i = 1:n.
Step. 8: $x_i = \frac{x_i}{max}$
End algorithm

Recurrent neural networks

RNNs are machine learning networks useful in speech recognition, natural language and speech processing. RNN can able to handle data dependencies. Where the output at time t-1 together with next input is fed as input to neuron at time t. Typical RNN neuron is depicted in Fig. 13.13. The input of an RNN unit is a sequence of variable $x = \{x_1, x_2 \ldots \ldots \ldots x_t\}$ where x_t is a feature vector at time step t. σ and tanh representing sigmoid and hyperbolic tangent function, respectively. Given the previous hidden layer state $h_t - 1$, the current hidden layer state h_t and the output layer state y_t can be calculated by

$$h_t = \sigma_h(w_h x_t + U_h h_{t-1} + b_n \tag{13.8}$$

$$y_t = \sigma_y\left(w_y h_t + b_y\right) \tag{13.9}$$

w_h, w_y input to hidden and hidden to output weight matrices, respectively. U_h matrix between hidden layer and itself. b_n and b_y are biases. σ_h and σ_y denote the activation functions. RNN failed to store memory for a long time due to the problem of vanishing and exploding gradient. Thus different models of RNN are LSTM and gated recurrent unit (GRU).

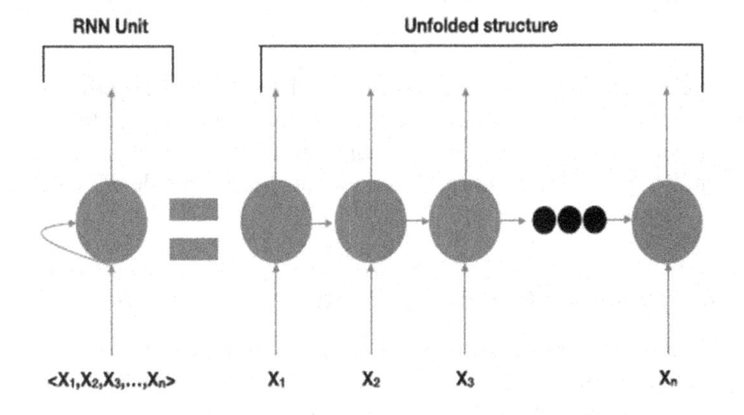

FIGURE 13.13 Recurrent Neural Network neuron.

LSTM deals with the vanishing and exploding gradient problems of RNN. The LSTM uses memory state Ct and the hidden state h_t with three gates: input i_t, forget f_t and output o_t gate. The three gates control the flow of information and combine the current input x_t with hidden state h_{t-1} coming from the previous timestamp. The σ returns values between 0 and 1. The temporary cell y_t rescales the current input and made with tanh that returns values between -1 and 1. i_t Control the current information required to maintain. f_t Indicates how much of the previous memory needs to be retained at the current step. o_t Impacts the new hidden state h_t deciding how much information of the current memory will be outputted to the next step.

$$i_t = \sigma(W_{ix}x_t + W_{ih}h_{t-1} + b_i) \tag{13.10}$$

$$f_t = \sigma\left(W_{fx}x_t + W_{fh}h_{t-1} + b_f\right) \tag{13.11}$$

$$y_t = tanh\left(W_{yx}x_t + W_{yh}h_{t-1} + b_y\right) \tag{13.12}$$

$$c_t = i_t \odot y_t + f_t \odot c_{t-1} \tag{13.13}$$

$$o_t = \sigma(W_{ox}x_t + W_{oh}\, h_{t-1} + b_o) \tag{13.14}$$

$$h_t = o_t \odot \tanh(c_t) \tag{13.15}$$

Rule 1 concludes that the clustering results and extracted features from both CLBP and HLAC are compared with stored trained features in training phase based on which RNN classifier differentiate human and nonhuman objects.

Rule 2

The output human region obtained from rule 1 will be further processed for human validation. The validation is done by using novel algorithm termed SGLDM. The statistical features like human heads count, features extraction from head pose and width between the individual heads are determined using SGLDM method.

Spatial gray level difference method

SGLDM construct cooccurrence matrices of gray level in region of interest. The algorithm works on second order conditional probability density $g(i,j,d,\theta)$ where element at location (i,j) gives chances of having two different cells in given angle θ and distance d. The direction of texture is evaluated using angle and distance values provide size of the texture. Hence by

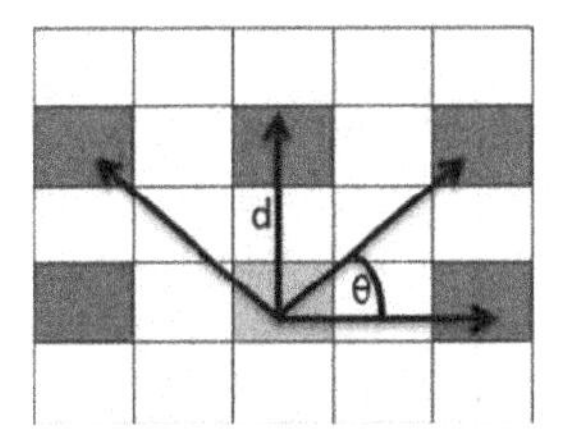

FIGURE 13.14 GLCM of yellow pixel with d = 3 and θ.

using various θ and d values, various SGLD Matrices can be generated. The restricted values of angle θ are 0, 45, 90, and 135°. Fig. 13.14 depicts cooccurrence for yellow pixel having d as 3 and angle θ.

Different parameters of texture reflect diverse properties of image.

- Contrast: contrast gives measurement of local variations in gray values. An image with high intensity level differences has high contrast. The parameter also illustrates the matrix dispersion from its diagonal. Contrast is defined by the Eq. (13.16).

$$Contrast = \sum_{i,j} |i-j|^2 p(i,j) \tag{13.16}$$

Where $p(i,j)$ denotes the GLCM elements, Homogeneity: is the inverse difference moment that measures the image local homogeneity shown in Eq. (13.17).

$$Homogeneity = \sum_{i,j} \frac{p(i,j)}{1+|i-j|} \tag{13.17}$$

- Energy: Energy will reflect pixel-value repetitions. Homogeneous type of images consists of few dominant gray tone transitions, those results into higher energy. It is defined as in the Eq. (13.18).

$$Energy = \sum_{i,j} p|i-j|^2 \tag{13.18}$$

- Correlation: it measures the correlation of neighboring pixel.

$$Correlation = \sum_{i,j} \frac{(i-\mu i)(j-\mu j)p(i,j)}{\sigma_i \sigma_j} \tag{13.19}$$

Finally, the statistical features includes counting human heads, feature extraction of head pose and calculating the width between the detected heads are compared and classified by RNN classifier with the previously stored features in knowledge base based on which human validation is done.

Rule 3

The obtained validation output will be further processed for detection of group activity. Group activity gives number of interaction of people, which is determined using location and individual's movements. The proposed system employs activity descriptor, that is, LGA descriptor which will encode behaviors and interactions between people [03].

Local group activity (LGA) descriptor

The interactions between people describes spatial information regarding group and also allows locating the unique interactions in the scene. To extract behaviors inside the group, LGA encodes movements and poses of people in a group. Let g represents group at time t and n denotes number of people interacting in a group g. LGA descriptor extracts the behavior of group. The movements of persons in group g are very important to determine specific activity. Consider the activities walking, kicking, hugging and hand shaking, etc. Motion information is considered which includes pose characterization to build a proper descriptor to recognize group activity. The set of head poses are denoted by The set of head poses are denoted by $P = \{1, \ldots, p\}$, movements is defined by $V^{\rightarrow} = \{\vec{V}_1, \ldots, \vec{V}_n\}$. The LGA is a 2D symmetric histogram of size $p \times p$ and value of individual bin $(x, y) \in p \times p$ is determined in Eq. (13.20).

$$LGA(x, y) = \sum_{i,j \in g, p_i = x, p_j = y} w(i, j) |\vec{V}_i| |\vec{V}_j| \tag{13.20}$$

Where $\vec{V}_i$ and $\vec{V}_j$ are the motion magnitudes. To analyze the group activity efficiently, the proposed model employed Bag-of-Words (BoW) approach. The BOW symbolizes the histogram code word belongs to vocabulary set. To learn the vocabulary of code-words, trained video data extracted from LGA descriptor is used. LGA methods are assigned to distinct codeword to denote group behavior sequence as one Dimensional histogram code-words. As group action is denoted by BoW, CNN is used to learn and classify the group activities.

Convolution neural network (CNN)

CNN consists of more than one convolution layers and then tracks by more than one flatteringly associated layers. The architecture of CNN is developed to yield benefit of two- dimensional input image. Another advantage of CNN is that it requires easy training.

CNN is a FFANN used to analyze visual imagery. To determine match features of input patch image, multiply feature pixel value with corresponding pixel value in the image. Then add the answer and divide by total pixel of the image. In both way matching pixel value results 1. Similarly if pixels are mismatched the results value is -1. If all features match, then add features and divide by total number of pixels gives a 1. Similarly if features are not matching the image patch features, then the results is -1. Repeat the process by lining feature with image patch. Determine the outcome of each convolution task to make a newer 2D array from it. This matching map shows filtered version of original image. Values nearer to 1 exhibit strong matches while near to-1 gives strong match feature. Value close to 0 tells no match for the feature. The Rule 3 concludes the detected event in the small human group using Bag-of-words (BoW) approach and LGA descriptor.

Artificial Intelligence (AI) facing a low growth to bridge gap between human and machine abilities. Many researchers are working on various aspects to make intelligence in systems. One of such research areas is CV. AI enable machines to see humans world, analyze and perceive it in away as human do and even able to get knowledge of multiple tasks such as image, video analysis and recognition, Recommendation Systems and NLP (Natural Language Processing) respectively. The advancement involved in CV due to concept and construction of Deep Learning made more precise and perfection with time over one particular algorithm (John et al., 2015).

CNN is a Deep Learning network, it's takes input image and assigns weights, biases to different objects in image for differentiating one from the other. The CNN architecture is depicted by Fig. 13.15. Image is nothing but a pixel matrix fed to input layer. A convolution layer can capable to capture the spatial-temporal features of image by convolving relevant filter/Kernels with input matrix elements (Fig. 13.16).

The matrix multiplication using strides helps to down sample feature. The stride length shows how many steps required for move or slide window filter across an image. A standard CNN consists of stride length as 1. Pooling layer works similar to as convolution layer. It reduces convolved feature size to reduce the power required for dimensionality reduction. Furthermore, pooling layer extracts interesting features which are not varying with rotation. Fully-Connected layer (FCC) usually learns non-linear high-level features of convolution and pooling layer.

Experimental results

The evaluated result for the overhead deliberated procedure is described here. The existing viciousness detection dataset is considered for demonstration. Algorithm has taken the video frames as the input generated from the video input. This input video frame of size 256×256 compromises three to four persons from the dataset. Finally the performance of human detection and action recognition is summarized. The observations of the proposed model are shown in Figs. 13.6 and 13.7. Fig. 13.17A–C shows the unique frame, gray transformation, and two-fold conversion frame correspondingly.

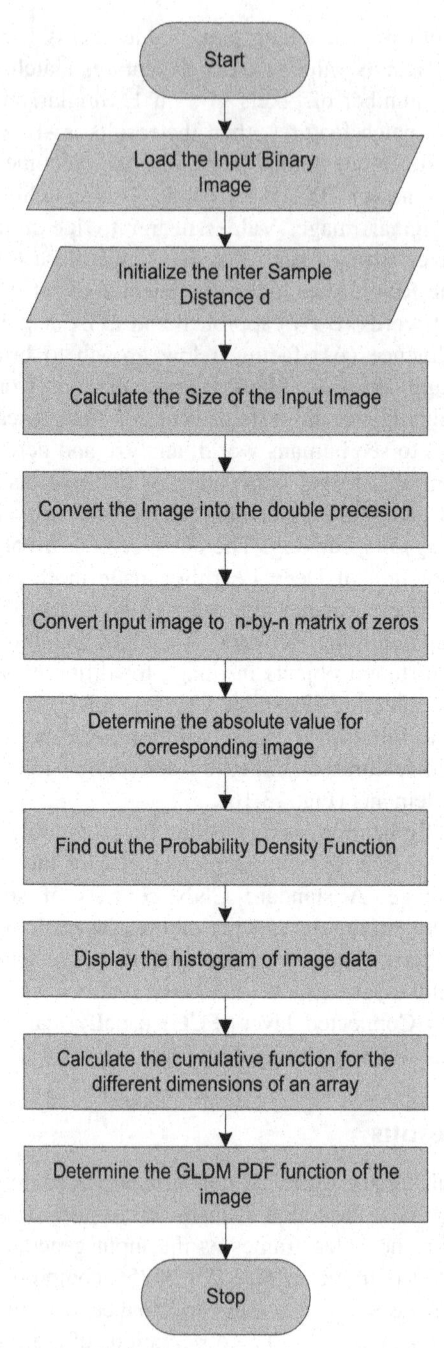

FIGURE 13.15 Flowchart for SGLDM.

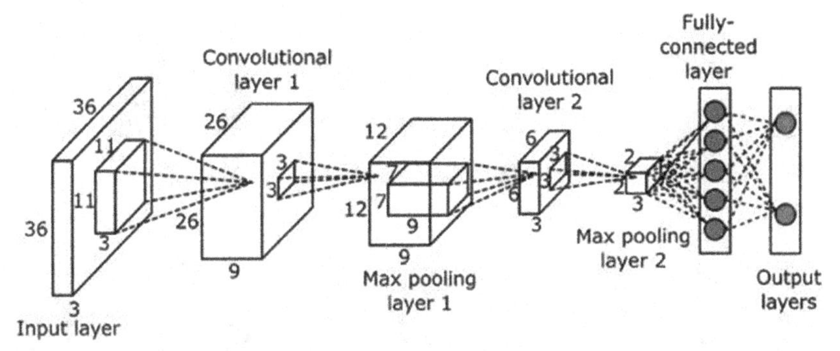

FIGURE 13.16 CNN model.

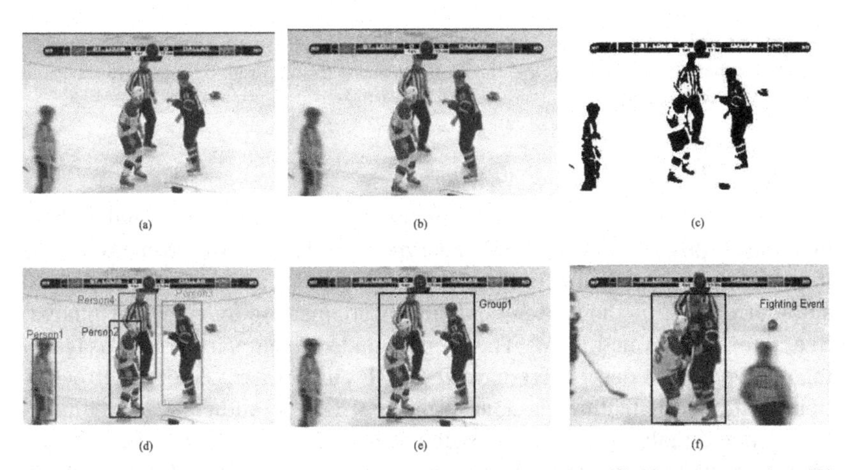

FIGURE 13.17 (A) Input frames of video 1; (B) gray conversion; (C) binary conversion; (D) human validation results of respective frames; (E) group detection result of respective frames of video 1; (F) event detection.

To validate humans, FCM segmentation and RNN classifier are employed as shown in Fig. 13.17D. To determine human group, SGLDM is employed as shown in Fig. 13.6E. Once group is detected and the corresponding features are extracted using LGA, BOW method followed by CNN classifier to detect the event activity in a group as shown in Fig. 13.17F. Likewise the similar process is measured.

Additional video frames shown in Fig. 13.18A–F respectively. The event activity which is considered in proposed system is violence detection activity example fighting is considered in given dataset. The recommended archetypal illustrates the investigational outcomes in terms of accurateness which is moderately worthy when matched to further prevailing models. Our recommended system shows that the accuracy is 93% (Fig. 13.18).

FIGURE 13.18 (A) Input video 1 frames; (B) gray conversion; (C) binary conversion; (D) human validation results of respective frames (E) group detection result of respective frames of video 1; (F) event detection.

Conclusion

A unique methodology for trivial humanoid cluster recognition and authentication is proposed which will encrypt the specific occurrences in the videotape structure using fuzzy c-mean clustering and higher order local auto-correlation methods. Segmentation is attained by by means of innovative algorithm termed FCM. The recommended exemplary marks usage of innovative article descriptors termed CLBP and HLAC. The further detection of statistical features is done using SGLDM to authenticate the quantity of individuals present in the video sequence. The achieved features are then coordinated with the features deposited in the knowledge base using RNN classifier. This technique is wide-open to offer precise small human groups detection and validation. The Bag-of-Words (BoW) approach based on local group activity (LGA) descriptor has been employed to encode the movements and to discover the interaction. The CNN classifies the group activity. The results show the good performance of interacting group and group activity recognition.

References

Ahmad, J., & Kamal, S. (2014). Real-time life logging via a depth silhouette-based human activity recognition system for smart home services, In *2014 11th IEEE international conference on advanced video and signal based surveillance (AVSS)* (pp. 74–80). IEEE.

Alahi, A., Ramanathan, V., & Fei-Fei, L. (2014). Socially-aware large-scale crowd forecasting. In *2014 IEEE conference on computer vision and pattern recognition* (pp. 2203–2210). IEEE.

Andrea Pennisi, D. D., Bloisi., & Iocchi, L. (2016). *Online real-time crowd behavior detection in video sequences, . Computer vision and image understanding* (Vol. 144, pp. 166–176). Elsevier.

Atrevi, D. F., Vivet, D., Duculty, F., & Emile, B. (2017). A very simple framework for 3D human poses estimation using a single 2D image: Comparison of geometric moments descriptors. *Pattern Recognition*, *71*, 389–401.

Chen, C., Jafari, R., & Kehtarnavaz, N. (2016). A real-time human action recognition system using depth and inertial sensor fusion. *IEEE Sensors Journal*, *16*(3), 773–781.

Chen, C., Zhang, B., Su, H., Li, W., & Wang, L. (2016). Land-use scene classification using multi-scale completed local binary patterns. *Signal, Image and Video Processing*, *10*(4), 745–752.

Cheng Z., Qin, L., Huang, Q., Jiang, S., & Tian, Q. (2010). Group activity recognition by Gaussian Processes Estimation. In *2010 20th international conference on pattern recognition* (pp. 3228–3231). IEEE.

Ciptadi A., Goodwin M.S., & Rehg J.M. (2014). Movement pattern histogram for action recognition and retrieval. In *Conference on computer vision* (pp. 695–710). Springer.

Colque, R. V. H. M., Caetano, C., de Andrade, M. T. L., & Schwartz, W. R. (2017). Histograms of optical flow orientation and magnitude and entropy to detect anomalous events in videos. *IEEE, Transactions on Circuits and Systems for Video Technology*, *27*(3), 673–682.

Deng, C., Wang, L.-F., Jiang, M.-X., Pan, Z.-G., & Sun, X. (2018). *Multiobject tracking in videos based on LSTM and deep reinforcement learning*, *Abstract and applied analysis* (2018, pp. 1–12). Hindawi.

Duta, I. C., Uijlings, J. R., Ionescu, B., Aizawa, K., Hauptmann, A. G., & Sebe, N. (2017). Efficient human action recognition using histograms of motion gradients and VLAD with descriptor shape information. *Multimedia Tools and Applications*, *76*, 1–28.

Feng, P., Wang, W., Dlay, S., Naqvi, S. M., & Chambers, J. (2017). Social force model-based MCMC-OCSVM particle PHD filter for multiple human tracking. *IEEE Transactions on Multimedia*, *19*(4), 725–739.

Gauzere, B., Ritrovato, P., Saggese, A., & Vento, M. (2015). Human tracking using a top-down and knowledge based approach. In *Image analysis and processing—ICIAP* (pp. 257–267). Springer.

Ge, W., Collins, R. T., & Ruback, R. B. (2012). Vision-based analysis of small groups in pedestrian crowds. *IEEE Transactions on Pattern Analysis and Machine Intelligence*, *34*(5), 1003–1016.

Irfan, M., Tokarchuk, L., Marcenaro, L., & Regazzoni, C. (2018). Anomaly detection in crowds using multi sensory information. In *2018 15th IEEE international conference on advanced video and signal based surveillance (AVSS)* (pp. 1–6). IEEE.

Jalal, A., Mahmood, M., & Hasan, A. S. (2019). Multi-features descriptors for human activity tracking and recognition in indoor-outdoor environments. In *2019 16th International Bhurban conference on applied sciences and technology (IBCAST)* (pp. 371–376). IEEE.

John, V., Mita, S., Liu, & Qi, B. (2015). Pedestrian detection in thermal images using adaptive fuzzy C-means clustering and convolutional neural networks. In *IEEE, IAPR international conference* (pp. 246–249). IEEE.

Karpathy, A., Toderici, G., Shetty, S., Leung, T., Sukthankar, R. & Fei-Fei, L. (2014). Large-scale video classification with convolutional neural networks. In *IEEE conference on computer vision and pattern recognition* (pp. 1725–1732), IEEE.

Kylestephens, & Bors, A.G. (2016). Human group activity recognition based on modelling moving regions interdependencies. In *2016 23rd international conference on pattern recognition (ICPR)* (pp. 2115–2120). IEEE.

Lajevardi, S. M., & Hussain, Z. M. (2012). Automatic facial expression recognition: Feature extraction and selection. *Image and Video Processing*, *6*(1), 159–169.

Liu, L., Shao, L., Li, X., & Lu, K. (2016). Learning spatio-temporal representations for action recognition: A genetic programming approach. *IEEE Transactions on Cybernetics*, *46*(1), 158–170.

Manfredi, M., Vezzani, R., Calderara, S., & Cucchiara, R. (2014). Detection of static groups and crowds gathered in open spaces by texture classification. *Pattern Recognition Letters*, *44*, 39–48.

Mazzon, R., Poiesi, F., & Cavallaro, A. (2013). Detection and tracking of groups in crowd. In *IEEE international conference* (pp. 202–207). IEEE.

Milan, A., Leal-Taixé, L., Reid, I., Roth, S., & Schindler, K. (2016). MOT16: A benchmark for multi-object tracking. *arXiv Preprint arXiv:1603.00831*.

Milan, A., Schindler, K., & Roth, S. (2013). Detection- and trajectory-level exclusion in multiple object tracking. In *IEEE conference on computer vision and pattern recognition* (pp. 3682–3689). IEEE.

Minaeian, S., Liu, J., & Son, Y.-J. (2018). Effective and efficient detection of moving targets from a UAV's camera. *IEEE Transactions on Intelligent Transportation Systems, 19*(2), 497–506.

Naifanzhuang, T., Yusufu, J., Ye, & Hua, K.A. (2017). Group activity recognition with differential recurrent convolutional neural networks. In *2017 12th IEEE international conference on automatic face and gesture recognition (FG 2017)* (pp. 526–531). IEEE.

Nikouei, S. Y., Chen, Y., Song, S., Xu, R., Choi, B.-Y., & Faughnan, T. R. (2018). Real-time human detection as an edge service enabled by a lightweight CNN. In *2018 IEEE international conference on edge computing (EDGE)* (pp. 125–129). IEEE.

Priya M.M. and Nawaz D.G.K., MATLAB based feature extraction and clustering images using K-nearest neighbor algorithm, 2016.

Shao, L., Ji, L., Liu, Y., & Zhang, J. (2012). Human action segmentation and recognition via motion and shape analysis. *Pattern Recognition Letters, 33*(4), 438–445.

Sharif, M. H. U., Saha, A. K., Arefin, K. S., & Sharif, M. H. (2011). Event detection from video streams. *International Journal of Computer and Information Technology, 1*(1), 108–114.

Stone, E. E., & Skubic, M. (2014). Fall detection in homes of older adults using the microsoft kinect. *IEEE Journal of Biomedical and Health Informatics, 19*(1), 290–301.

Sun, C. C., Wang, Y., & Sheu, M. H. (2017). Fast motion object detection algorithm using complementary depth image on a RGB-D camera. *IEEE Sensors Journal, 17*(17), 5728–5734.

Suriani, N. S., Hussain, A., & Zulkifley, M. A. (2013). Sudden event recognition: A survey. *Journal* (13), 9966–9998.

Tran, D., Yuan, J., & Forsyth, D. (2013). Video event detection: From subvolume localization to spatiotemporal path search. *IEEE Transactions on Pattern Analysis and Machine Intelligence, 36*(2), 404–416.

Tran, K. N., Gala, A., Kakadiaris, I. A., & Shah, S. K. (2014). Activity analysis in crowded environments using social cues for group discovery and human interaction modelling. *Pattern Recognition Letters, 44*, 49–57.

Vemulapalli, R., Arrate, F., & Chellappa, R. (2014). Human action recognition by representing 3D skeletons as points in a Lie group. In *Proceedings of the IEEE conference on computer vision and pattern recognition* (pp. 588–595). IEEE.

Wang, H., Oneata, D., Verbeek, J., & Schmid, C. (2016). A robust and efficient video representation for action recognition. *International Journal of Computer Vision, 119*(3), 219–238.

Wang, T., & Snoussi, H. (2015). Detection of abnormal events via optical flow feature analysis. *Sensors, 15*(4), 7156–7171.

Yafengyin, G. Y., & Man, H. (2013). Small human group detection and event representation based on cognitive semantics. In *2013 IEEE seventh international conference on semantic computing* (pp. 64–69). IEEE.

Yang, Y., Zhang, B., Yang, L., Chen, C., & Yang, W. (2015). Action recognition using completed local binary patterns and multiple-class boosting classifier. In *Third IAPR Asian conference on pattern recognition (ACPR2015)* (pp. 336–340). IEEE.

Yimengzhang, W. G., Chang, M.-C., & Liu, X. (2012). Group context learning for event recognition. In *2012 IEEE workshop on the applications of computer vision (WACV)* (pp. 249–255). IEEE.

Chapter 14

Understanding the hand gesture command to visual attention model for mobile robot navigation: service robots in domestic environment

E.D.G. Sanjeewa, K.K.L. Herath, B.G.D.A. Madhusanka, H.D.N.S. Priyankara and H.M.K.K.M.B. Herath

Department of Mechanical Engineering, Faculty of Engineering Technology, The Open University of Sri Lanka, Nugegoda, Sri Lanka

Introduction

The elderly population, identified as those aged 65 and over, has risen significantly over the past 45 years (Maduwage, 2019) due to advancements and improved access to healthcare services (Maduwage, 2019). Fourteen percent of the world's population is now older than 65 (Herath & de Mel, 2019) for the first time in history. In the European Union, 11.5% of the population was old in 1970. But the figure had risen to 20.5% by 2016 (Ismail & Tahir, 2017). All signs are that this pattern will continue. The implications are that an ever-growing number of people face disabilities such as reduced mobility or diminished cognitive ability, all sorts of chronic diseases, and very often loneliness and social isolation. Coping with these clinical and social concerns is often associated with high and dynamic costs and realistic difficulties (Manresa, Varona, Mas, & Perales, 2005).

Robots were primarily built for routine, rough, or other dangerous activities in factories and hazardous field conditions until recently. In these implementations, robots did not occupy the same workspace as humans, and all communications were facilitated by machine interfaces, control panels, joysticks, etc. However, the development of service robots, i.e. robots conducting functional human or equipment roles outside of industrial automation, has been rendered feasible by technical improvements (Dutta & Chaudhuri, 2009). Because

Cognitive Computing for Human-Robot Interaction. DOI: https://doi.org/10.1016/B978-0-323-85769-7.00003-3

service robots are expected to operate in individuals' vicinity, there is also a need for social interaction skills (Luo & Wu, 2012). There are many difficulties in designing physical systems that can work virtually near us and have motivated creating a dedicated discipline known as human-robot interaction (HRI). Robots are being introduced now to render much of the tasks of our world more manageable. These robots are expected to have a general sense of the consequences of their behavior in interactive human–robot environments (Anjaly, Devanand, & Jisha, 2014). Many robots joining social environments are specialists in either one or a few activities. Cleaning robots (Sun, Meng, & Ang, 2017), rescue robots (Muñoz-Salinas, Aguirre, & García-Silvente, 2007), shopping assistants (Wang et al., 2018), and healthcare robots (Sünderhauf et al., 2016) are some instances of task-specific robots needing lower overall emotional intelligence. However, robotic structures have been a show of social and emotional interaction between humans and machines in the deployment of robots in social settings (Manap, Sahak, Zabidi, Yassin, & Tahir, 2015). In the coming decades, however, artificial agents' hospitality and emotional intelligence are projected to increase with their wide variety of social applications (Moladande & Madhusanka, 2019). Most importantly, robots must match their emotional behavior with a real situation (Mantoro & Ayu, 2018).

A robot may take decisions on the built interaction pattern under some conditions. These conditions need understanding. This kind of robotic machine is now used in school, childcare and other areas of society. These networks do need to be developed to respond to unforeseeable conditions where citizens travel. It's a challenging but sought-after function integrated into a robot's personage (Salvini et al., 2012), so that diverse social interaction are imitated and learned. In comparison, humanoid robots may emulate traditional human characteristics in physical presence, movements, sentences, facial expressions, etc. In Maimone & Fuchs (2011), an example structure in the adaptive behavior of robot reactions is shown.

The system proposed involves the social status of hand movements and relevant social contextual knowledge for domestic scenarios. By ignoring socially practical obstacles and approaches to the object, the comfortable human structure, center on the domestic situation (Appuhamy & Madhusanka, 2018), will approximate a relative target position in the world. The domestic situation involves the robot's motion planning system, consisting of the mobile robot motion control algorithm for Simultaneous Location and Mapping (SLAM). Through testing under the newly created comfortable human indexes, we test the utility of the proposed system. This paper presents the visual attention model for service robot navigation with a hand gesture command. Experiments were performed in an artificially created domestic setting. The proposed definition will also enhance the ambiguous capability of robots' knowledge assessment and can effectively transmit entity references through nonverbal behavior to a human partner. This model could attract a user's attention, concentrating on computer vision and human–robot interaction.

To enter HRI, this chapter focuses on the management by gesture awareness of maintenance robots' activity. For visual processing, a Kinect V2.0 camera is used. According to Kinect V2.0, a robot maintenance control device must be set up. Instructions for obtaining human skeleton details must be submitted to the robot through a serial port, and bone-recognition movements must monitor the robotic movement. This finishes the contact with the robot so that engineers can experience the problem easier. To search for predefined artefacts, the Kinect V2.0 camera rotates on the robot platform depicted with a hand motion number. The computer then calculates, using the Kinect V2.0 sensor, the desired object's depth and navigates to the object. To identify hazards, three IR sensors are mounted on the platform. The platform rotates and navigates the right side until an entity is found, so it retracts. The spinning angles of the platform were determined with a gyro sensor.

As follows, the chapter is structured. In Human–robot interaction section, human–robot interaction is discussed. Hand gesture recognition section discusses the hand gesture recognition system. A proposed service robot platform design is provided in Proposed design section. Results section discusses the hand gesture algorithm, object recognition, mobile robot navigation, and mapping. Finally, Conclusion section concludes the investigation.

Human–robot interaction

Healthcare service robots can support users in numerous areas, including physical, mental, psychological, and cognitive. The user interface, various sensors, such as cameras and laser scanners, a mobile base, and often a pair of weapons are used in a standard service robot. Healthy examples of intelligent support robots are Care-O-bot (Starner & Pentland, 1997), Homemade (Li, Zhang, & Liu, 2010) and PR2, Toyota HSR (Lewis, 2013), and TIAGo (Ben-Ari & Mondada, 2018). For contact with their cameras, they may select and hold the items of individuals who navigate indoors individually, perceive, and approach a human. Since the arms should be stable and willing to bear payloads, the technological requirements are high. The weapons render the robot multifunctional, but still challenging and pricey, so it is not accessible for end-users today. There are many more specialized robots with fewer functionalities, in comparison to such multifunctional robots. One example is Pepper (Xu, 2017), which has weapons only to perform movements to reinforce the user-interaction phrase. Some robots have no weapons but can also conduct physical activities, such as the Robotic Support Assistant (Wang, Zhang, & Leung, 2015) or SMOOTH demonstrator (Liu & Wang, 2018).

Currently, regardless of the dynamic socio-economic behavior of culture, people lack time to handle their domestic workload. It is also not straightforward to find trustworthy and professional servants or caregivers to take charge of everyday tasks (Sharma & Verma, 2015). Therefore, the domestic populace prefers to use service robots to help their everyday routine (Piana, Stagliano,

Odone, Verri, & Camurri, 2014). Due to the problems associated with an increasing elderly population, the usage of service robotics has emerged (Regnier et al., 2004). Domestic service tasks also include direct communications with these robots and their consumers, who are non-robotic experts. Among these service robots, human-friendly interactive characteristics are, therefore favored (Shibata, 2004). Human-like behavior and reasoning ability can be integrated into the architecture of robotics to reflect human-friendly interactive characteristics (Madhusanka & Jayasekara, 2016). A primary need for achieving human-like actions in service robots is the capacity to recognize user activity accurately and adequately to react to customers.

When engaging with each other in various situations/contexts, human beings tend to establish different distances from their peers (Pathak, Jalal, Agrawal, & Bhatnagar, 2015). During such encounters, a service robot can maintain a reasonable distance from its customer, strengthening the relationship between the robot and the consumer. It should also be possible for the robot to reach the user so that the motion of the robot would not impede or distract the user's current activity (Stolzenwald & Bremner, 2017). Proxemics between two (Herath, Sanjeewa, Madhusanka, & Priyankara, 2020) persons depends on both people's present actions and the contact context (Pieskä, Luimula, Jauhiainen, & Spiz, 2012). Service robots should also be able to interpret their customers' actions and settle on proxemics relevant to the current situation. Several experiments have been performed to establish methods for preserving real proxemic properties between robots and human users (Islam, Siddiqua, & Afnan, 2017). However, in various engagement modes/contexts, specific methods have been introduced to preserve the user's necessary distance, and the way to address the user is not seen. Also, the robot's navigation route is of little concern to these approaches. The approaches often consider that during encounters, only the consumer of concern is involved in preserving proxemics and is overlooked by other persons in the area. Therefore, the approaches presented will not effectively address a person who is discussing with another person (Sanjeewa, Herath, Madhusanka, & Priyankara, 2020).

A service robot was proposed for shopping centers, which can meet a visitor for services (Ong & Bowden, 2004). However, the cited work's main contribution was to establish an efficient technology to attract the Customer's attention to initialize the touch and enhance the human-friendly method of the robot (Madhusanka & Ramadass, 2021). The technique is also only appropriate for one person. The computer's method would not be more fun for the other person who speaks to the client. To maneuver the robots with their users (Fritsch, Lang, Kleinehagenbrock, Fink, & Sagerer, 2002) recommended methods for constructing human-friendly trajectories. However, these methods cannot be pursued to assess a service robot's method with two individuals having a conversation as the inputs to the job are confined to the increase of robotics and humans' side-by-side navigation.

A lot of tests with the path direction of a machine were performed to assess consumer preferences. A robot has been sent to a pair of users in eight different methods in the study described in Chen, Georganas and Petriu (2007). The users were then questioned in each case to determine their degree of comfort. Similarly, the research addressed in Vasudevan, Gächter, Nguyen and Siegwart (2007) explored consumers' happiness with the approach of the robot in single customer and group circumstances. According to these findings, consumers strongly supported the path from the front, and the least favorable route is to approach them. It examined the impact on the comfortable stop distances when a robot approaches to communicate with him (Szafir, Mutlu, & Fong, 2014). Any of the features and actions of robotics affect individuals' willingness in at least a short meeting to interact with the system. The study discussed in Matarić (1997) defines social rules for comfortable and humanly suitable robot behavior. Therefore, the computing room of the HRI studies expects a robot companion to do the right things and satisfy his task requirements acceptably and conveniently for humans. Therefore, its consumers would be more accepted over a long time by the robot's real-time performance that fulfills human social standards and expectations (Bera, Randhavane, Prinja, & Manocha, 2017; Singh et al., 2017).

A machine with the ability to talk to the humanoid face is depicted in Nygaard et al. (2019). Display the emphasis, then switch the robot to the human. The authors have indicated that these characteristics are the minimum requirements for meaningful social interaction between humans and robots. Nevertheless, this approach has failed to boost the robot's ability to understand the state of the customer until utilizing the two features described above in a meeting. The robot often practices a morally prudent behavior in which people are interested during this technique. It relies regardless upon the people on the behavior of the machine. According to the model proposed in Herath, Jochum and Vlachos (2017), witness and judgment are also fundamental attributes in man-robot interaction. A computer, who is also an intelligent weight loss coach, was introduced in Kidd and Breazeal (2007). This is an unusual situation in which the robot embodies the information of the incident. And the robot was used to minimize human obesity. In this scenario, the robot is approved by its customers for a long-term touch. But this instrument cannot grasp market conditions and behaviors in social environments that are not important for physical well-being. Higher relational information and enhanced customer understanding can be ensured by integration into a robotic architecture with elements of emotional and intuitive actions (Daglarli, Temeltas, & Yesiloglu, 2009).

Hand gesture recognition

Interaction between persons and computers is currently mostly carried out via the mouse, the keyboard, remote control, touch screens, and other means of close contact. Conversely, noncontact, more automatic, and organic touch are achieved by sound and physical movements. By standard and intuitive noncontact methods, contact is commonly called flexible and efficient. Various experiments also tried to use noncontact techniques such as tone, facial expressions (Hashimoto, Hitramatsu, Tsuji, & Kobayashi, 2006), physical movements (Beetz et al., 2015), and gestures (Nehaniv et al., 2005) to perceive in the machine the intentions and information of others. The motion is among them, the most significant aspect of the human language and the meaning and richness of HRI is influenced by its development (Rahman & Ikeura, 2018).

In the last few decades, data glove usage (Rekimoto, 2001) to obtain the angle and the position of each joint in the motion has been widely described and measured. Due to the costs and inconvenience of shipping the sensor, it is challenging to use widely. However, the noncontact visual evaluation methods provide the low cost and comfort for the body, which are the most typical methods for evaluating motion.

In the presence of different illumination conditions, Chakraborty, Bhuyan, and Kumar (2017) and Song and Sacan (2012) implemented color patterns which use the pixel distribution in each color space to improve the accuracy of recognition significantly. However, because of the light sensitivity in the imaging process (Verma, Roy, Pandey, & Mittal, 2019), it wasn't easy to achieve the necessary results utilizing model-based approaches. Visual inspection methods focused on algorithms, for example, the hidden Markow model (Beal, Ghahramani, & Rasmussen, 2001), the partition filter, (Carpenter, Clifford, & Fearnhead, 1999) and the Ada-Boost learning algorithm, were commonly used to perform gesture detection (Babu, Achanta, Murty, & Swapna, 2012). Today the complex equations make real-time impossible to do. While only poor 2D pictures have been used, the results above are not effective in real-time to obtain motions. Thus, it is likely that 3D is supplemented with profound expertise in identification of 2D image expressions. Usually, 3D data are obtained utilizing binocular cameras (Karunachandra & Herath, 2020), Kinect sensors (Zhang, 2012), Leap Motion (Lu, Tong, & Chu, 2016) and other equipment. These devices may typically be used to collect depth knowledge, which can easily be collected by various directions in spatial relations (Kunze, Doreswamy, & Hawes, 2014) or infrared reflection (Benet, Blanes, Simó, & Pérez, 2002), to define and distinguish noncontact images instead of using complicated devices.

It is necessary to remove the action from a complicated picture, and more data usually tends to render the movements more precise and possible. To achieve more depth of information, a Kinect sensor is commonly favored, and a critical light sensitivity performance in the identification of gesture has been seen. However, questions regarding locating manuals remain unanswered, and it is necessary to further address segmentation in depth (Singh, Gahlot, & Mittal, 2019). Recently, the study has centered on recognizing the hand gesture identification problem rather than the problem of gestural segmentation.

The gesture segmentation techniques were employed by setting the direct gap (Ma & Peng, 2018) or by the hand (Ma & Peng, 2018) as the key object. The difference between the hand and the sensor is, therefore constrained. It differentiates the hand's motions by only moving the hand to a specific position and keeping distance during the process. More basic techniques are simple and efficient. Researchers have usually done their best to seek normal relations between human beings, such as communication between individuals. In the latter and present literature, however, the disadvantage of a lack of distance is unnatural. Approaches for manual identification, such as pattern matching (Nguyen-Dinh, Roggen, Calatroni, & Tröster, 2012) or finite-state machines (Hong, Turk, & Huang, 2000), often have incredible detection speeds. However, only gestures can grasp the above methods. The finger hull distinguishes the gestures by the convex hull recognition algorithm (Ganapathyraju, 2013) and allows the individual fingertips to be separated from each position. It will have more awareness of movements and has a future advantage. The algorithm for convex hull identification was used to identify the fingertip gesture, and thus any fingertip of the human hand was detected (Dash, Acharya, Mittal, Abraham, & Kelemen, 2020).

Proposed design

In this chapter, an enhanced threshold segmentation approach centered on the Kinect sensor with short distance depth recognition information is suggested. In a wide range of circumstances and complex environments, the recommended solution benefits from light exposure. It can reliably detect gestures, where the chosen objects fully cover up the hand signals. The first is the median filtering preprocesses the RGB image data and the Kinect sensor's depth image data. Secondly, an improved contextual stratification method is proposed in tandem with the threshold of information depth and skin color; therefore, expressions may be observed in broad contexts in dynamic contexts. Finally, the neighboring treatment is performed with the human side's ROI section. The Haar Cascade Classifier technique is also used to detect the fingertips and consider the magnitude of motions to check the planned treatment's viability. The experimental findings demonstrate that the proposed solution will achieve adequate performance and strong robustness. And the signal of the consumer is taken as a computer entry. The robot rotating in any direction decides the signal course. For the experiment findings, three specimens were selected and called "1," 2' or "3," as can be shown in Fig. 14.1, displaying the cup, glass and first-aid boxes.

After receiving the signal, the Kinect sensor is used to measure the required objects' lengths and angles. It sends that recorded data to the system. A Haar Cascade Classifier is used after receiving the object number to identify the objects according to the input signal (Vithanawasam & Madhusanka, 2018). After the input and classified items are compared to the

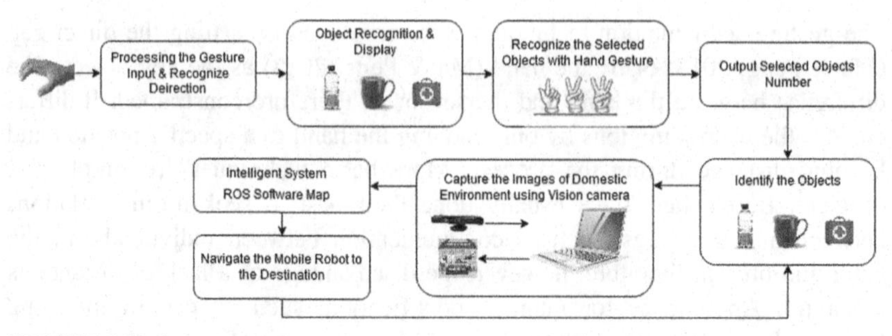

FIGURE 14.1 System overview.

object, the robot starts navigating with previously recorded measurements. This platform is equipped with three sharp IR sensors to prevent hindrances on the way. The robot moves for 50 cm on the right and re-starts the measurement process when an obstacle is identified. Feedback on the robotic angle of rotation is provided through a gyro sensor. After reaching the object's location, the robot rotates back 180°. Then it searches the human again using the cascade classifier. The distance and the angle to the user are measured using the Kinect camera, and the previous process is repeated up to the user's robot platform.

Fig. 14.2 shows the mobile robot platform with actuating component with four 120 rpm and 12 V DC motors. A dual H-bridge driver controls the four motors. Obstacles are avoided if the mobile robot platform moves along the route calculated. As the microcontroller of this ATmega 1280, Arduino mega boards are used. There is a 16 MHz crystal oscillator, and it can supply 14 PWM output pins which can be used to control motor movement.

Results

A hand gesture recognition system was developed using image processing algorithms from the OpenCV Library to detect numbers according to the displayed gesture. The identification of segmentation and edge includes hand detection. An edge traversal algorithm is applied to the hand's segmented contour to eliminate unwanted background noise once the hand gestures are segmented.

Hand gesture and direction recognition

The algorithm is classified layer by layer by space stratification after capturing the depth image and the RGB image. The gesture ROI is received through two points. The first step is to classify the hand signal's approximate location using GetJoints in the Kinect V2.0 API to discern the correct or incorrect hand from other exposed areas of the body. The second step is to

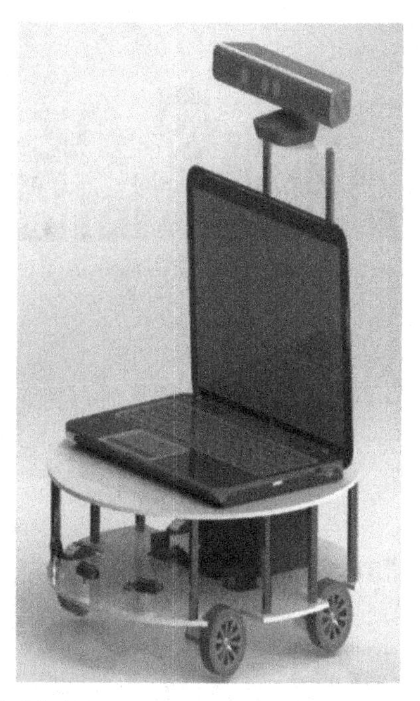

FIGURE 14.2 Mechanical design.

FIGURE 14.3 Object selecting window using hand gestures.

detect a fingertip in the estimated picture frame, as seen in Fig. 14.3. This measures the same ROI of the motions. Identification of items is carried out along with the object's deep awareness close the sensor. Therefore if the RGB knowledge criteria are fulfilled, the target may be conveniently found. The improved threshold segmentation method is used for gesture extraction according to the design sense in Fig. 14.1, and the gesture algorithm can be seen in Fig. 14.5. The results of gesture extraction in real-time are seen in Fig. 14.4, according to the gesture algorithm. Fig. 14.4 demonstrates that the improved threshold segmentation framework can automatically and effectively identify the hand gesture's position in real-time. Moreover, correct

FIGURE 14.4 Hand gesture for the fingertip count recognition.

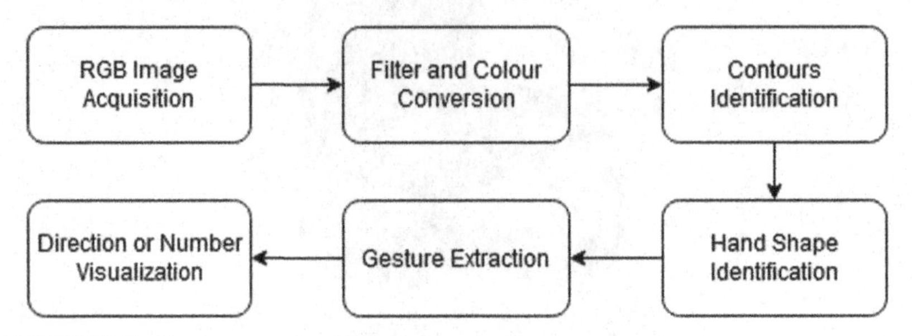

FIGURE 14.5 Hand gesture algorithm.

hand signals in the ROI region, lead to greater accuracy of fingertip detection, can be obtained via the improved procedure.

The probability of noise disturbance or other uncertainties can be eliminated when the amount of points of interest matches the requirement provided in this chapter by a certain number of points. These points are then calculated to be the color of the skin in the color transfer. In this way, the place where the hand is placed will be calculated. Following the contours of the hand and the form of the hand is the next stage. After the method of extraction of activity detected by hand type, the features are removed. The findings show the number of fingers and position attributable to the consumer's behavior (Fig. 14.5).

After the manual motion's binary image is retrieved from the complex sense, the contour profile is withdrawn to position the hand in the ROI field. The palm print is then discovered, and the fingertips can finally be set to the obtained contours. The FindContours algorithm (Kondakor, Törcsvari, Nagy, & Vajk, 2018) is used to remove hand gesture contours from the ROI region in this section. The extraction of the contour points is usually achieved by calculating the size of the nearby pixels. The FindContours algorithm's fundamental idea is to define the outline by defining the border of the binary picture between black and white areas. A binary representation of the gesture was produced in the previous segment; the algorithm of FindContours, as seen in Fig. 14.6, is then used to draw from the ROI region for gestures.

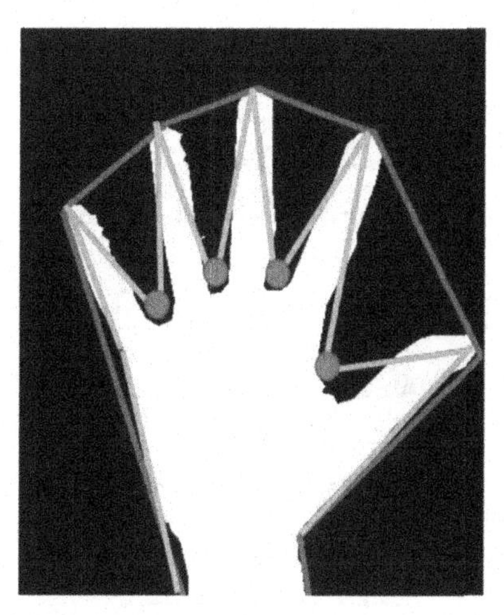

FIGURE 14.6 ROI area of hand gesture.

Hand gesture algorithm

For the vision-based hand gesture recognition method, which we described via an algorithm, there were distinct stages in the system we proposed and created—the gesture recognition flowchart of the operation as shown in Fig. 14.7.

The suggested algorithm of hand gesture recognition has the following steps.

1. From the recorded video stream, extract a frame, that is, a hand image.
2. The extracted frame is converted from the color space of RGB to the color space model of YCbCr. Then using skin color-based detection techniques, the hand is detected in the image.
3. The machine transformed the picture into black and white after hand recognition (i.e., the skin pixels were identified as white and nonskin pixels as black). Some preprocessing strategies, such as picture filling, morphological erosion utilizing 15×15 structuring elements, etc., were implemented to enhance image clarity and eliminate noise.
4. The equivalent diameter, field, perimeter, and orientation of detected objects are found in the frame for the function extraction centroid. All the features were used before we had the nonconflicting production.
5. The gesture is recognized by counting the number of white items in the picture and its direction. Lastly, an instruction is transmitted to the device's programs, referring to the known motion.

The Kinect V2 camera can provide sequential static 2D images in a series. Static 2D images, an intuitive method for gesture recognition is based on convexity defects or curvature. The convexity defects detected

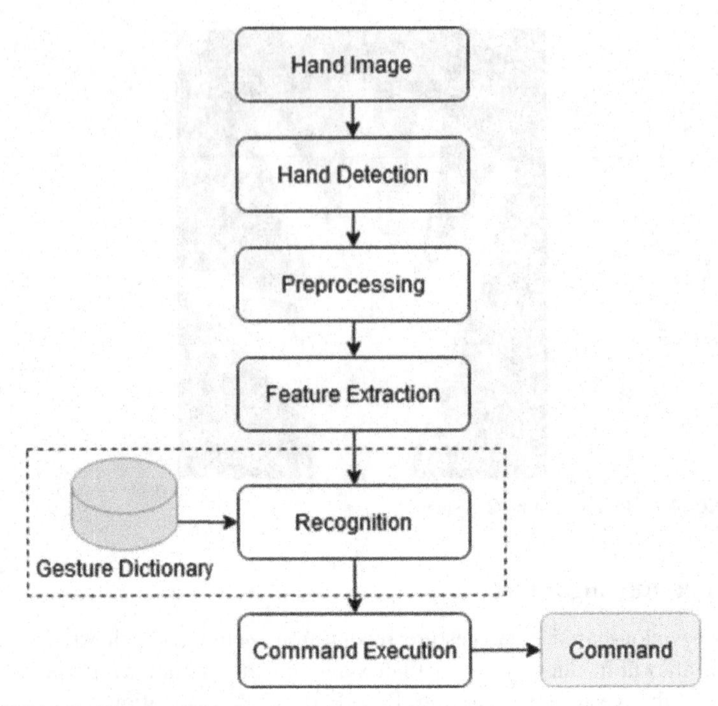

FIGURE 14.7 Flowchart of the hand gesture recognition process.

from a binary image of a palm using the algorithm is shown in Fig. 14.8. Based on convexity defects, the curvature-based method is quite like that. Detecting fingers is the nature of these methods. A model is needed to filter invalid convexity defects and to ensure that each defect represents a finger to create this method effect. The length and intersection angle of each defect are two standard parameters used to build the model. Features such as imperfection, elongations, or other features that can be directly extracted from the hand image's contour information are considered. A more complex model can be constructed. However, a general model for directly describing hand gestures based on these contour characteristics is still lacking. Even in cases where we only count fingers, the empirical model does not always work satisfactorily during our testing.

Experiments are measured both left and right directions. Actual results in real situations to test the accuracy of estimating the pointing direction in an indoor environment are shown in Fig. 14.9. Before that, the user was asked to stand before the mobile robot platform and align his body with it.

Object recognition

Identification of artefacts using Haar feature-based cascade classifiers is a required method of artefact identification proposed by Viola and Jones (2001).

FIGURE 14.8 The detected fingertips after filtering invalid convexity defects.

FIGURE 14.9 Direction of gesture detection.

Haar cascade classifiers based on features are a machine-based learning method in which both positive and negative photos learn a cascade function. It's then used to classify items in other pictures. Objects are drawn from both positive and negative photos in this method, as seen in Fig. 14.10. The positive picture is an image having just the right thing. The gloomy picture is an illustration that doesn't have the necessary individual. In cascades' production, a favorable image is taken, and negative images are mounted on it. In this method, positive images are resized and distorted at different angles into negative images. One thousand positive images and 2000 negative images from each object type were used for this process.

The Object Identification Process is shown in Fig. 14.11. Three objects were trained in this study to be identified using the method of the Haar cascade. In typical indoor environments, the lightning state was used. The process of human identification is indicated in Fig. 14.12. For this task, the Haar cascade method is used. Both standing and sitting samples of humans at different angles were used to train the model with high accuracy when implementing this technique on humans.

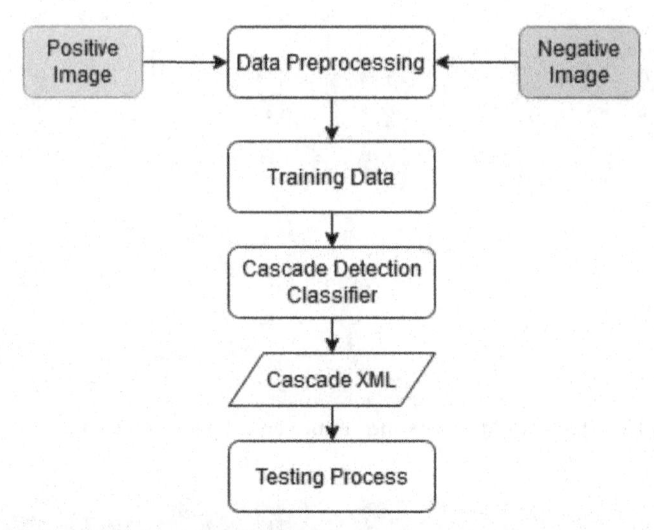

FIGURE 14.10 Object recognition algorithm.

FIGURE 14.11 Object identification.

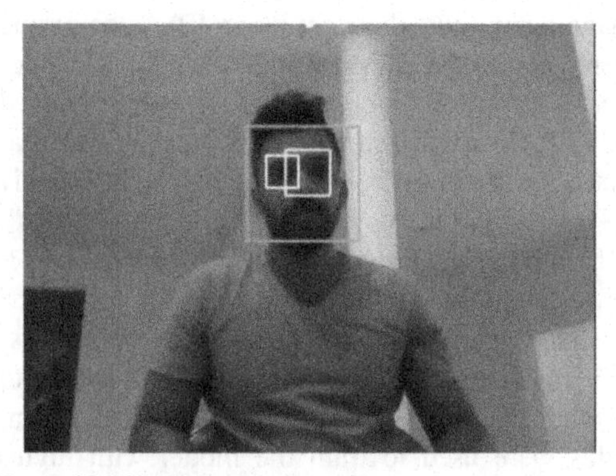

FIGURE 14.12 Human identification.

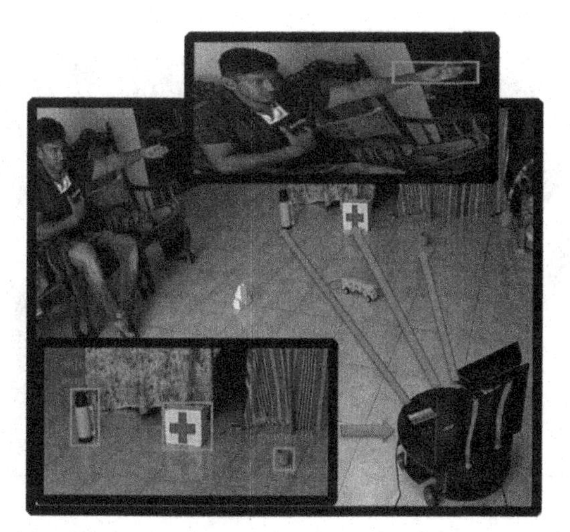

FIGURE 14.13 Input direction from the user.

FIGURE 14.14 Identifying the direction and all objects.

Fig. 14.13 shows the user's input direction input process and the objects' direction identification process. To be identified as object numbers 1, 2, and 3, three objects are fed to the system. First, the robot always looks at the user, identifying him as a human being. The system identifies it as left in the right direction after the person shows the objects' direction, as shown in Fig. 14.14. The robot then rotates in the desired direction and identifies available objects. Also using the Kinect camera, the distance values for each object are calculated. Then it is turned back again to the user.

Fig. 14.15 shows the system identification process of the hand gestured number detected, and the method of detection of convexity defects is used to

FIGURE 14.15 Hand gestured object number identification.

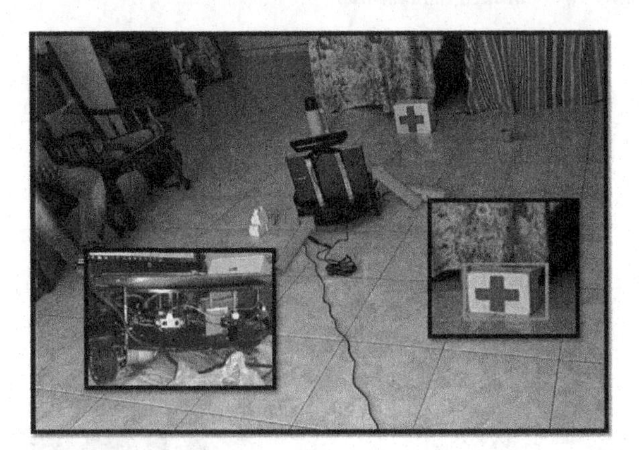

FIGURE 14.16 Navigating by obstacle avoiding.

recognize the hand gesture. Fig. 14.16 shows the navigation of the object by preventing obstacles. One object was placed directly in front of the robot and the other two objects at 45° degree angles. By avoiding obstacles, the robot then navigates to the desired object. Sharp IR sensors are used to detect obstacles. Within 15 cm, it detects obstacles and rotates and moves 50 cm to the right side, then starts the distance measuring process again. A gyro sensor is used to give feedback on the angle of rotation of the robot.

The target touch is shown in Fig. 14.17, and the frame breaks at 15 cm from the target. At this point, it was called because the framework reached and selected the appropriate target. The robot spins backwards at 180 degrees. It uses the cascade classifier once again to search for the individual. The user's distance and angle are then measured using the Kinect camera, and

FIGURE 14.17 Reaching the object.

FIGURE 14.18 Returning to the user.

FIGURE 14.19 Reaching the user.

the previous process is repeated until the robot platform user is returned and reached, as shown in Figs. 14.18 and 14.19 as a result.

Mobile robot navigation

There has lately been much interest in service robotics, but simultaneous positioning and mapping (SLAM) pose an impediment in robotic mobile service. When dealing with the mapping and positioning of robots, which involve constructing a map, locating oneself, recognizing routs, and preventing obstacles (Durrant-Whyte & Bailey, 2006), four significant problems are involved. These issues are neither autonomous nor interdependent. The productivity of the robot can only be improved by reducing errors in each relation. Address this SLAM topic, and many studies have been undertaken. In Yuan, Li, and Su (2016) suggested a way to construct rectangular maps for indoor mobile robotics in environmental mapping. As parallel and perpendicular to standard feature-based maps, the rectangular maps should be maintained and the lightweight maps resulting are useful for human contain, typical home environments used by people who live mainly in this application. This needs more requirements and greater mapping flexibility.

Equally complimentary are self-position and map formation. It can trigger the wrong mapping if a self-localization error occurs. In a huger error, the wrong map will give rise to self-location. In the indoor self-location segment, where conventional positioning methods like GPS cannot be used, robot auto-location depends on recognizing known scenes through autonomous activity. However, most current approaches are matched to the scene or recognized through landmarking.

The typical basic specifications for robot simulators are an accurate simulation of physics, high-quality object features, alignment with the Robot Operating System (ROS) architecture and multiplatform support. This presents excellent prospects for modeling robots and their sensors, creating robot control algorithms, the realization of mobile robot simulation, animation, and navigation in practical 3D settings. ROS provides databases and resources to support software engineers to build robotic applications. ROS is often abstracted from hardware, system drivers, databases, viewers, message delivery, package control, and more. ROS is available under license as an open-source. Fig. 14.20 displays the ROS-based visual SLAM techniques, while the Kinect V2 sensor and scales are used for mapping the area.

FIGURE 14.20 Robot Operating System interface with a service robot platform.

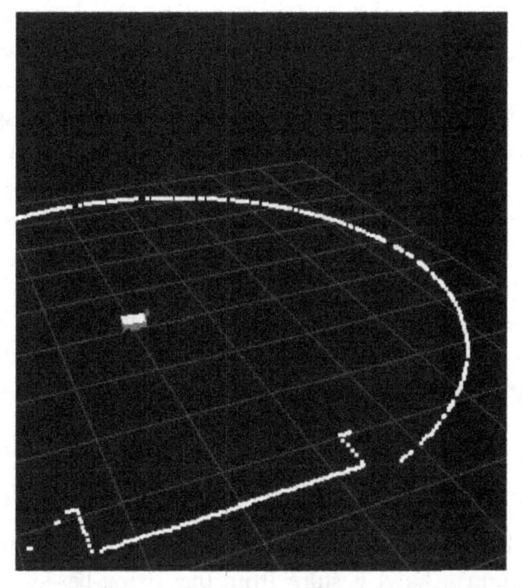

FIGURE 14.21 Location of the mobile robot inferred from a map and the location of the map.

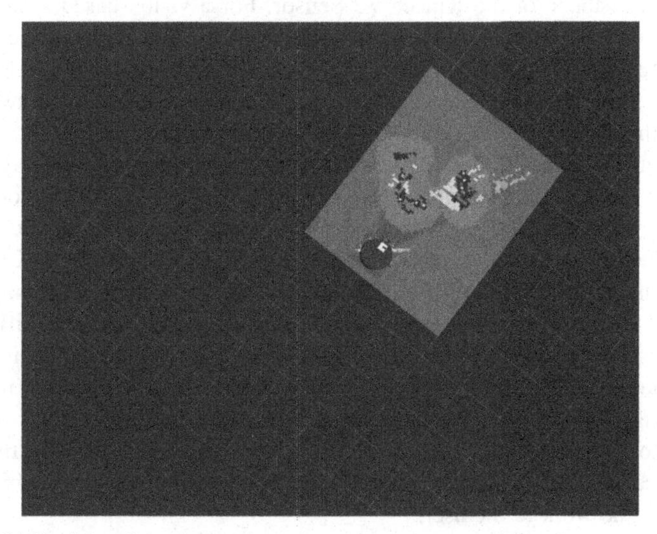

FIGURE 14.22 Map of a real dynamic environment.

Fig. 14.21 shows the mobile robot's location inferred from a map and the location of the map. There were simulating through ROS. Initially, the mobile robot needs to map the environment to get information around the considered domestic environment (e.g., room). The obstacles and user-required objects were located relative to the location of the mobile robot. Then the robot gets distance from the objects and the obstacles using ultrasonic sensors. The mobile robot scans the domestic environment and gets information about the position of required objects and where the person is where the obstacle is relative to the mobile robot's position.

Mapping is the process of using its sensors to create a spatial model of the robot's environment. For localization and navigation, the map is then used. A mobile robot platform with a simulated ultrasonic construction has been set up to map a SLAM algorithm's environment. The workspace environment was assumed to have been created. The concept of packages, nodes, topics, messages, and services are used in this framework. A node is an executable program that takes data and passes it to other nodes from the robot's sensors. To create the map, all these nodes allow the robotic system to move. A map developed during mapping is shown in Fig. 14.22.

Conclusion

This study aimed to develop an interactive service robot platform capable of identifying the human hand gesture from the vision-based system. The system's efficiency was found to be approximately 90% within the optimal 4–5-foot distance of the Kinect V2 Sensor. For a vision-based system with high efficiency, the Kinect V2 sensor might work. According to the results, with OpenCV, gesture recognition, direction recognition and object recognition were performed effectively. It can also be observed that the reflection caused the shiny and curved target to be misidentified.

We also investigated the SLAM method, providing all the trajectories obtained during the experiment with service robot motion in an indoor environment by processing the Kinect depth sensor. We compared such visual SLAM techniques based on ROS as Kinect V2 depth sensor data. Kinect V2 sensors have given good results and created absolute values for a map and a location. In terms of resource consumption, there is no vital difference. Various variables such as surface difference, object widening at its lower components, sensor orientation, or loops are also considered during mapping and navigation. Specific forms of artefacts will be tested at various distance ranges for further study. From the findings, it can be concluded that the robot has succeeded in navigating through user input and reaching the object and returning the work to the user.

References

Anjaly, P., Devanand, A., & Jisha, V. R. (2014). Moving target tracking by mobile robot. In *2014 International conference on power signals control and computations (EPSCICON)* (pp. 1–6). IEEE.

Appuhamy, E. J. G. S., & Madhusanka, B. G. D. A. (2018). *Development of a GPU-Based Human Emotion Recognition Robot Eye for Service Robot by Using Convolutional Neural Network* (pp. 433–438). IEEE.

Babu, M. P., Achanta, V., Murty, N. R., & Swapna, K. (2012). Development of maize expert system using ada-boost algorithm and Navie Bayesian classifier. *International Journal of Computer Applications Technology and Research, 1*(3), 89–93.

Beal, M., Ghahramani, Z., & Rasmussen, C. (2001). The infinite hidden Markov model. *Advances in Neural Information Processing Systems, 14*, 577–584.

Beetz, M., Bartels, G., Albu-Schäffer, A., Bálint-Benczédi, F., Belder, R., Beßler, D., & Weitschat, R. (2015). Robotic agents capable of natural and safe physical interaction with human co-workers. In *2015 IEEE/RSJ International conference on intelligent robots and systems (IROS)* (pp. 6528–6535). IEEE.

Ben-Ari, M., & Mondada, F. (2018). *Robots and their applications. Elements of robotics* (pp. 1–20). Cham: Springer.

Benet, G., Blanes, F., Simó, J. E., & Pérez, P. (2002). Using infrared sensors for distance measurement in mobile robots. *Robotics and Autonomous Systems, 40*(4), 255–266.

Bera, A., Randhavane, T., Prinja, R., & Manocha, D. (2017). Sociosense: Robot navigation amongst pedestrians with social and psychological constraints. In *2017 IEEE/RSJ International conference on intelligent robots and systems (IROS)* (pp. 7018–7025). IEEE.

Carpenter, J., Clifford, P., & Fearnhead, P. (1999). Improved particle filter for nonlinear problems. *IEE Proceedings-Radar, Sonar and Navigation, 146*(1), 2–7.

Chakraborty, B. K., Bhuyan, M. K., & Kumar, S. (2017). Combining image and global pixel distribution model for skin colour segmentation. *Pattern Recognition Letters, 88*, 33–40.

Chen, Q., Georganas, N. D., & Petriu, E. M. (2007). Real-time vision-based hand gesture recognition using haar-like features. In *2007 IEEE instrumentation & measurement technology conference IMTC 2007* (pp. 1–6). IEEE.

Daglarli, E., Temeltas, H., & Yesiloglu, M. (2009). Behavioral task processing for cognitive robots using artificial emotions. *Neurocomputing, 72*(13–15), 2835–2844.

Dash, S., Acharya, B. R., Mittal, M., Abraham, A., & Kelemen, A. G. (Eds.), (2020). Deep learning techniques for biomedical and health informatics. Cham: Springer.

Durrant-Whyte, H., & Bailey, T. (2006). Simultaneous localization and mapping: Part I. *IEEE Robotics & Automation Magazine, 13*(2), 99–110.

Dutta, S., & Chaudhuri, B. B. (2009). A color edge detection algorithm in RGB color space. In *2009 International conference on advances in recent technologies in communication and computing* (pp. 337–340). IEEE.

Fritsch, J., Lang, S., Kleinehagenbrock, A., Fink, G. A., & Sagerer, G. (2002). Improving adaptive skin color segmentation by incorporating results from face detection. In *Proceedings of the 11th IEEE international workshop on robot and human interactive communication* (pp. 337–343). IEEE.

Ganapathyraju, S. (2013). Hand gesture recognition using convexity hull defects to control an industrial robot. In *2013 Third international conference on instrumentation control and automation (ICA)* (pp. 63–67). IEEE.

Hashimoto, T., Hitramatsu, S., Tsuji, T., & Kobayashi, H. (2006). Development of the face robot SAYA for rich facial expressions. In *2006 SICE-ICASE international joint conference* (pp. 5423–5428). IEEE.

Herath, D. C., Jochum, E., & Vlachos, E. (2017). An experimental study of embodied interaction and human perception of social presence for interactive robots in public settings. *IEEE Transactions on Cognitive and Developmental Systems, 10*(4), 1096–1105.

Herath, H. M. K. K. M. B. & de Mel, W. R. (2019). Electroencephalogram (EEG) controlled anatomical robot hand. In *Faculty of engineering technology student academic conference 2018*, p. 73.

Herath, K. K. L., Sanjeewa, E. D. G., Madhusanka, B. G. D. A., & Priyankara, H. D. N. S. (2020). HAND GESTURE COMMAND TO UNDERSTANDING OF HUMAN-ROBOT INTERACTION. *GSJ*, *8*(7).

Hong, P., Turk, M., & Huang, T. S. (2000). Gesture modeling and recognition using finite state machines. In *Proceedings of the fourth IEEE international conference on automatic face and gesture recognition (Cat. No. PR00580)* (pp. 410–415). IEEE.

Ismail, A. P., & Tahir, N. M. (2017). Human gait silhouettes extraction using Haar cascade classifier on OpenCV. In *2017 UKSim-AMSS 19th International conference on computer modeling & simulation (UKSim)* (pp. 105–110). IEEE.

Islam, M. M., Siddiqua, S., & Afnan, J. (2017). Real time hand gesture recognition using different algorithms based on American sign language. In *2017 IEEE international conference on imaging, vision & pattern recognition (icIVPR)* (pp. 1–6). IEEE.

Karunachandra, R. T. H. S. K., & Herath, H. M. K. K. M. B. (2020). Binocular vision-based intelligent 3-D perception for robotics application. *International Journal of Scientific and Research Publications (IJSRP)*, *10*(9), 689–696.

Kidd, C. D., & Breazeal, C. (2007). A robotic weight loss coach. In *Proceedings of the national conference on artificial intelligence* (Vol. 22, No. 2, p. 1985). Menlo Park, CA; Cambridge, MA; London; AAAI Press; MIT Press; 1999.

Kondakor, A., Törcsvari, Z., Nagy, A., & Vajk, I. (2018). A line tracking algorithm based on image processing. In *2018 IEEE 12th international symposium on applied computational intelligence and informatics (SACI)* (pp. 000039–000044). IEEE.

Kunze, L., Doreswamy, K. K., & Hawes, N. (2014). Using qualitative spatial relations for indirect object search. In *2014 IEEE international conference on robotics and automation (ICRA)* (pp. 163–168). IEEE.

Lewis, M. (2013). Human interaction with multiple remote robots. *Reviews of Human Factors and Ergonomics*, *9*(1), 131–174.

Li, W., Zhang, Z., & Liu, Z. (2010). Action recognition based on a bag of 3D points. In *2010 IEEE computer society conference on computer vision and pattern recognition-workshops* (pp. 9–14). IEEE.

Liu, H., & Wang, L. (2018). Gesture recognition for human-robot collaboration: A review. *International Journal of Industrial Ergonomics*, *68*, 355–367.

Lu, W., Tong, Z., & Chu, J. (2016). Dynamic hand gesture recognition with leap motion controller. *IEEE Signal Processing Letters*, *23*(9), 1188–1192.

Luo, R. C., & Wu, Y. C. (2012). Hand gesture recognition for human-robot interaction for service robot. In *2012 IEEE international conference on multisensor fusion and integration for intelligent systems (MFI)* (pp. 318–323). IEEE.

Ma, X., & Peng, J. (2018). Kinect sensor-based long-distance hand gesture recognition and fingertip detection with depth information. *Journal of Sensors*, *2018*, 5809769.

Madhusanka, B. G. D. A., & Jayasekara, A. G. B. P. (2016). Design and development of adaptive vision attentive robot eye for service robot in domestic environment. In *2016 EEE International Conference on Information and Automation for Sustainability (ICIAfS)*, (pp. 1–6). IEEE.

Madhusanka, B. G. D. A., & Ramadass, S. (2021). Implicit Intention Communication for Activities of Daily Living of Elder/Disabled People to Improve Well-Being. *IoT in Healthcare and Ambient Assisted Living* (pp. 325–342). Singapore: Springer.

Maduwage, S. (2019). Sri Lankan 'silver-aged'population. *Journal of the College of Community Physicians of Sri Lanka*, *25*(1).

Maimone, A., & Fuchs, H. (2011). Encumbrance-free telepresence system with real-time 3D capture and display using commodity depth cameras. In *2011 10th IEEE international symposium on mixed and augmented reality* (pp. 137–146). IEEE.

Manap, M. S. A., Sahak, R., Zabidi, A., Yassin, I., & Tahir, N. M. (2015). Object detection using depth information from Kinect sensor. In *2015 IEEE 11th international colloquium on signal processing & its applications (CSPA)* (pp. 160–163). IEEE.

Manresa, C., Varona, J., Mas, R., & Perales, F. J. (2005). Hand tracking and gesture recognition for human-computer interaction. *ELCVIA Electronic Letters on Computer Vision and Image Analysis*, *5*(3), 96–104.

Mantoro, T., & Ayu, M. A. (2018). Multi-faces recognition process using Haar cascades and eigenface methods. In *2018 Sixth international conference on multimedia computing and systems (ICMCS)* (pp. 1–5). IEEE.

Matarić, M. J. (1997). Learning social behavior. *Robotics and Autonomous Systems*, *20*(2–4), 191–204.

Moladande, M. W. C. N., & Madhusanka, B. G. D. A. (2019). Implicit Intention and Activity Recognition of a Human Using Neural Networks for a Service Robot Eye. In *2019 International Research Conference on Smart Computing and Systems Engineering (SCSE)*, (pp. 38–43). IEEE.

Muñoz-Salinas, R., Aguirre, E., & García-Silvente, M. (2007). People detection and tracking using stereo vision and color. *Image and Vision Computing*, *25*(6), 995–1007.

Nehaniv, C. L., Dautenhahn, K., Kubacki, J., Haegele, M., Parlitz, C., & Alami, R. (2005). A methodological approach relating the classification of gesture to identification of human intent in the context of human-robot interaction. In *ROMAN 2005. IEEE international workshop on robot and human interactive communication*, 2005. (pp. 371–377). IEEE.

Nguyen-Dinh, L. V., Roggen, D., Calatroni, A., & Tröster, G. (2012). Improving online gesture recognition with template matching methods in accelerometer data. In *2012 12th International conference on intelligent systems design and applications (ISDA)* (pp. 831–836). IEEE.

Nygaard, T. F., Nordmoen, J., Ellefsen, K. O., Martin, C. P., Tørresen, J., & Glette, K. (2019). *Experiences from real-world evolution with dyret: Dynamic robot for embodied testing. Symposium of the Norwegian AI society* (pp. 58–68). Cham: Springer.

Ong, E. J., & Bowden, R. (2004). A boosted classifier tree for hand shape detection. In *Sixth IEEE International conference on automatic face and gesture recognition, 2004.* Proceedings. (pp. 889–894). IEEE.

Pathak, B., Jalal, A. S., Agrawal, S. C., & Bhatnagar, C. (2015). A framework for dynamic hand gesture recognition using key frames extraction. In *2015 Fifth national conference on computer vision, pattern recognition, image processing and graphics (NCVPRIPG)* (pp. 1–4). IEEE.

Piana, S., Stagliano, A., Odone, F., Verri, A., & Camurri, A. (2014). Real-time automatic emotion recognition from body gestures. *arXiv preprint arXiv*, *1402*, 5047.

Pieskä, S., Luimula, M., Jauhiainen, J., & Spiz, V. (2012). Social service robots in public and private environments. *Recent Researches in Circuits, Systems, Multimedia and Automatic Control*, 190–196.

Rahman, S. M., & Ikeura, R. (2018). Cognition-based variable admittance control for active compliance in flexible manipulation of heavy objects with a power-assist robotic system. *Robotics and Biomimetics*, *5*(1), 7.

Regnier, G., Makineni, S., Illikkal, I., Iyer, R., Minturn, D., Huggahalli, R., & Foong, A. (2004). TCP onloading for data center servers. *Computer*, *37*(11), 48–58.

Rekimoto, J. (2001). Gesturewrist and gesturepad: Unobtrusive wearable interaction devices. In *Proceedings of the fifth international symposium on wearable computers* (pp. 21–27). IEEE.

Salvini, P., Cecchi, F., Macrì, G., Orofino, S., Coppedè, S., Sacchini, S., & Dario, P. (2012). *Teaching with minirobots: The local educational laboratory on robotics. Advances in autonomous mini robots* (pp. 27–36). Berlin, Heidelberg: Springer.

Sanjeewa, E. D. G., Herath, K. K. L., Madhusanka, B. G. D. A., & Priyankara, H. D. N. S. (2020). VISUAL ATTENTION MODEL FOR MOBILE ROBOT NAVIGATION IN DOMESTIC ENVIRONMENT. *GSJ*, *8*(7).

Sharma, R. P., & Verma, G. K. (2015). Human computer interaction using hand gesture. *Procedia Computer Science*, *54*, 721–727.

Shibata, T. (2004). An overview of human interactive robots for psychological enrichment. *Proceedings of the IEEE*, *92*(11), 1749–1758.

Singh, R., Gahlot, A., & Mittal, M. (2019). IoT based intelligent robot for various disasters monitoring and prevention with visual data manipulating. *Int. J. Tomogr. Simul*, *32*(1), 90–99.

Singh, R., Gehlot, A., Mittal, M., Samkaria, R., Singh, D., & Chandra, P. (2017). *Design and development of a cloud assisted robot. International conference on next generation computing technologies* (pp. 419–429). Singapore: Springer.

Song, B., & Sacan, A. (2012). Automated wound identification system based on image segmentation and artificial neural networks. In *2012 IEEE international conference on bioinformatics and biomedicine* (pp. 1–4). IEEE.

Starner, T., & Pentland, A. (1997). *Real-time American sign language recognition from video using hidden markov models. Motion-based recognition* (pp. 227–243). Dordrecht: Springer.

Stolzenwald, J., & Bremner, P. (2017). Gesture mimicry in social human-robot interaction. In *2017 26th IEEE international symposium on robot and human interactive communication (RO-MAN)* (pp. 430–436). IEEE.

Sun, H., Meng, Z., & Ang, M. H. (2017). Semantic mapping and semantics-boosted navigation with path creation on a mobile robot. In *2017 IEEE international conference on cybernetics and intelligent systems (CIS) and IEEE conference on robotics, automation and mechatronics (RAM)* (pp. 207–212). IEEE.

Sünderhauf, N., Dayoub, F., McMahon, S., Talbot, B., Schulz, R., Corke, P., & Milford, M. (2016). Place categorization and semantic mapping on a mobile robot. In *2016 IEEE international conference on robotics and automation (ICRA)* (pp. 5729–5736). IEEE.

Szafir, D., Mutlu, B., & Fong, T. (2014). Communication of intent in assistive free flyers. In *Proceedings of the 2014 ACM/IEEE international conference on human-robot interaction* (pp. 358–365).

Vasudevan, S., Gächter, S., Nguyen, V., & Siegwart, R. (2007). Cognitive maps for mobile robots-an object based approach. *Robotics and Autonomous Systems*, *55*(5), 359–371.

Verma, O. P., Roy, S., Pandey, S. C., & Mittal, M. (Eds.), (2019). *Advancement of machine intelligence in interactive medical image analysis*. Singapore: Springer Nature.

Viola, P., & Jones, M. (2001). Rapid object detection using a boosted cascade of simple features. In *Proceedings of the 2001 IEEE computer society conference on computer vision and pattern recognition. CVPR 2001* (Vol. 1, pp. I-I). IEEE.

Vithanawasam, T. M. W., & Madhusanka, B. G. D. A. (2018). Dynamic face and upper-body emotion recognition for service robots. In *2018 IEEE/ACIS 17th International Conference on Computer and Information Science (ICIS)*, (pp. 428–432). IEEE.

Wang, L., Zhao, L., Huo, G., Li, R., Hou, Z., Luo, P., & Yang, C. (2018). Visual semantic navigation based on deep learning for indoor mobile robots. *Complexity*, *2018*, 1627185.

Wang, R., Zhang, H., & Leung, C. (2015). Follow me: A personal robotic companion system for the elderly. *International Journal of Information Technology (IJIT)*, *21*(1).

Xu, P. (2017). A real-time hand gesture recognition and human-computer interaction system. *arXiv preprint arXiv*, *1704*, 07296.

Yuan, W., Li, Z., & Su, C. Y. (2016). RGB-D sensor-based visual SLAM for localization and navigation of indoor mobile robot. In *2016 International conference on advanced robotics and mechatronics (ICARM)* (pp. 82–87). IEEE.

Zhang, Z. (2012). Microsoft kinect sensor and its effect. *IEEE Multimedia*, *19*(2), 4–10.

Chapter 15

Mobile robot for air quality monitoring of landfilling sites using Internet of Things

Rajesh Singh[1], Anita Gehlot[1], Shaik Vaseem Akram[1], Prabin Kumar Das[1] and Sushabhan Choudhury[2]
[1]School of Electronics & Electrical Engineering, Lovely Professional University, Phagwara, India, [2]University of Petroleum and Energy Studies, Dehradun, India

Introduction

According to the statistics (Joshi & Ahmed, 2016) the waste generation will be steadily increasing by 5% per year, that is, by 2047, waste generation will be around 260 million tons per annum and to dump this bulk of the waste, the prerequisite of the area will be roughly around 1400 sq. kms which is recommended for disposing of this amount of waste. At present, global warming is rising due to the emission of greenhouse gases from landfilling sites (Kumar et al., 2017), and unanticipated emission of the dangerous gas from the landfilling site pollutes the air of surrounding with different gases that severely affects the people living and passing near the landfilling sites. This is where the concept of air quality monitoring comes in. The primary goal of air quality monitoring is to differentiate areas in which emissions exceed environmental requirements of air quality from areas in which they are not. As the requirements of environmental quality are set at contaminant levels which have negative effects on human health, a public air quality organization is mandated to reduce the appropriate pollutant to assess levels exceeding an environmental norm in a region. It is difficult for humans to visit the landfill site for monitoring the emission of gases from the landfilling site. Technologies and strategies are required to be applied for reducing the pollutants in the air and also for achieving the ambient air quality. Recent advancements in networking, sensors, and embedded devices have enabled people in their homes to track and provide assistance. An IoT-based approach is a cost-effective and alternative to conventional air quality monitoring systems. The Internet of Things (IoT) has gained a significant part of

Cognitive Computing for Human-Robot Interaction. DOI: https://doi.org/10.1016/B978-0-323-85769-7.00004-5

human civilization recently, such as home automation, supply-chain, agriculture, health, and structural monitoring (Thu, Htun, Aung, Shwe, & Tun, 2019). However, it is a challenge for establishing wireless communication between the sensor node and the cloud server in the landfilling site due to the unavailability of internet connectivity. The integration of Low Power Wide Area Network (LPWAN)-based LoRa and GPRS communication-based architecture into the mobile robot for delivering reliable and secure transmission from the remote location. LoRa (Long Range) communication is a free licensed spectrum for communicating the low data rate over long-range with reliability and security. Generally, the sensor nodes are energy constraint devices, to meet the sensor node requirement LoRa (Long Range) is an optimal solution (LoRA and IoT Networks for Applications in Industry 4.0, n.d.). LoRa network transmission range is 5 km (urban) 10 km (rural). Due to the presence of a pure ALOHA mechanism, the LoRa consumes low power in the sensor node (Mekki, Bajic, Chaxel, & Meyer, 2019). The low data rate, low power consumption, and long-range data transmission are the requirement of establishing an energy-efficient IoT network (Nikoukar, Raza, Poole, Gunes, & Dezfouli, 2018).

In this study, we are proposing IoT (Internet of Things) and cloud server assisted mobile platform (robot) for continuous monitoring of the air quality in the landfilling sites. Every sensor node senses the air quality of the landfilling and communicates the sensory data to the mobile robot via GPRS and LoRa modem to the mobile robot. GPS (Global Positioning System) in the sensor node provides an accurate location of the sensor node. Here we are integrating two different long-range wireless communication protocols (LoRa radio modem & GPRS) for establishing an energy-efficient and long-range communication network for sensing the air quality parameter in the landfilling site through a mobile robot. LoRa radio modem in the mobile robot act as a transceiver unit for receiving the sensory data from the sensor node. GPRS (Global Packet for Radio Service) communication enables the mobile robot to transmit the data over internet protocol to the cloud server. The cloud server stores the data and displays the data in visualization format. The data of every sensor node and landfilling site are categorized separately for monitoring every landfilling site efficiently. The sensory data available in the cloud server can be used for predicting the patterns of air quality index of a different location. IoT and cloud server-based system enables to implement the real-time and robot for monitoring of air quality from a remote location.

Prior art

Internet of Things is the interconnection of distinct physical devices that communicates with each other through internet protocol (IP) (Singh, Gehlot, Khilrani, & Mittal, 2020). In indoor living conditions, people invest approximately 90% of our lives. It is therefore important to provide indoor air quality control for improved living environments. Internet of Things based real-time air monitoring system is introduced in the city of Dhaka for monitoring the gas leakage and alerting the citizen during the threat. iAir is an air quality monitoring system that acts on the framework of the internet of things (IoT), where MICS-6814 sensor as sensing unit and ESP 8266 module provides internet connectivity (Marques & Pitarma, 2019). Node MCU-based low-cost air quality monitoring system is proposed for gathering data using temperature, moisture, carbon monoxide, and nitrogen dioxide measurements, and then performing data analysis to infer the association between pollutant gases and air quality weather parameters (Aamer, Mumtaz, Anwar, & Poslad, 2018). A monitoring system is proposed for monitoring the air quality and water quality due to the emission of a lot of chemicals from the industries in the air and water (Agarwal, Shukla, Singh, Gehlot, & Garg, 2018). Cloud server enabled air quality monitoring system is implementing with the integration of WSN, Zigbee communication protocol, Nuttyfi, and the system is capable of real-time monitoring from any remote location via internet connectivity (Cookbook For Mobile Robotic Platform Control: With Internet of Things And Ti, 2019; Handbook of Research on the Internet of Things Applications in Robotics and Automation, 2019; Singh, Gehlot, Jain, & Malik, 2020). Monitoring of the landfilling site without the human is possible by deploying the robots for monitoring the air quality (Hu et al., 2019; Singh, Anita, et al., 2019). Robotic systems are widely implemented by scientists as data acquisition tools for thoroughly analyzing the changes in the environment without human intervention and researchers are implementing robotic for monitoring pollution (Salman, Rahman, Tarek, & Wang, 2019), volcanoes (Parcheta et al., 2016), variations in climate (Popa, Carutasu, Cotet, Carutasu, & Dobrescu, 2017). Even the robot implemented Off-Road Quadruped (ORQ) robot is specially designed for monitoring the air quality parameters in uneven terrain (Kumar, Verma, Singh, & Patel, 2016).

Integration of Airborne robotic systems with wireless sensor network is utilizing for monitoring the environmental parameters and communicating over wireless communication protocol (Dunbabin & Marques, 2012). Generally, in order of monitoring indoor air quality (IAQ), there is a requirement of deploying a large sensor system Deployment of a large number of sensor systems increases the infrastructure cost (Hu, Cong, Song, Bian, & Song, 2020; Singh, Gehlot, Thakur, & Mohan, 2019; Singh, Gehlot, Thakur, & Prudhvi, 2019). The airborne robotic system is also replacing the aquatic devices for exploring the deep ocean environment (Bogue, 2011). Wireless Sensor and Robot Networks (WSRNs) based air pollution monitoring system is designed for monitoring the air quality of the urban area (Fu, Chen, & Lin, 2012). Wireless sensor network and internet of robotic things is implemented for monitoring the environmental parameters in the multi-terrain area (Singh, Gehlot, Samkaria, & Choudhury, 2020). To monitor the indoor environment quality, a mobile sensing robot is deployed and the experimental result of ASHRAE 129 air change effectiveness, the mobile sensing robot collate the indoor environment data better than sensor networks (Jin, Liu, Schiavon, & Spanos, 2018). iAirBot is an Internet of Things-enabled robot system that monitors the indoor air quality and automatically triggers in case of exceeding the threshold limit of air quality. iAirBot is facilitated by communicating the alerts information and air quality data with the assistance of social networks (Marques, Pires, Miranda, & Pitarma, 2019).

A remote-controlled mobile robot is implementing for monitoring the emission of gases like SO_2, NO_2, CO_2, O_3, CH_4, and CO during natural and human-caused disasters (Monroy & Gonzalez-Jimenez, 2019; Singh, Samkaria, Gehlot, & Choudhary, 2018). Cloud server and IoT-based robot is implemented for establishing intelligent and wireless monitoring system. The cloud server will enable the user to monitor the environmental parameters and control the robot in a remote location without any human intervention (Saini, Dutta, & Marques, 2020; Samkaria et al., 2018; Singh, Gehlot, Mittal, Samkaria, & Choudhury, 2017). An IoT-based surveillance device is proposed for monitoring the garbage wirelessly through a raspberry pi controller (Rathour, Gehlot, & Singh, 2019).

An IoT-based technology has been proposed to reduce air pollution, which is of specific interest in a heavily populated country, and raspberry bi-based embedded device connects several sensors that gather sensor data of the environment (Kiruthika & Umamakeswari, 2018). IoT-Mobair Mobile App and Wi-Fi-based Arduino IDE Air Quality Monitoring Device are proposed to predict the air quality of the entire route where the person traveling to another location (Dhingra, Madda, Gandomi, Patan, & Daneshmand, 2019). Internet of Things and Cloud server-based air quality monitoring is proposed by integrating Low Power Wide Area Network (LPWAN) and the cloud server enables to visualize the patterns of variation in the air quality (Zheng, Zhao, Yang, Xiong, & Xiang, 2016). LoRa (Long Range) is an LPWAN that enables to transmits sensory data over a long distance. LoRa based sensor node is embedded for sensing the parameters like VOC (volatile organic compound), CO (carbon monoxide), and NO_2 (Nitrogen dioxide). Smart-air is an IoT-based air quality monitoring device that monitors the indoor environment parameters like VOC, CO, CO_2, temperature & humidity, and LTE (Long Term Evolution) technology act as wireless communication for transmitting the data to the cloud server (Jo, Jo, Kim, Kim, & Han, 2020). LoRa (Long Range)-based UAV (Unmanned Air Vehicle) is proposed for monitoring air quality and transmitting over long-range distance and data is logged into the server for further surveillance (Chen, Huang, Wu, Tsai, & Chang, 2018). Pollution is a cloud server & IoT device for monitoring air quality through environmental sensing sensors (Fioccola, Sommese, Tufano, Canonico, & Ventre, 2016). LoRaWAN based low cost and long-range communication-based real-time air quality monitoring system for alerting the quality of air (Addabbo et al., 2019). The cloud server data is employed for predicting the air quality with different datasets. Time-variant analysis model is applied on the dataset for analyzing the air quality index, ARIMA of time-variant analysis model showed better results than another model (Sethi & Mittal, 2020). Supervised learning techniques are applied to the air quality data for estimating the ambient air quality index (Sethi & Mittal, 2019a). A hybrid machine learning algorithm is implemented on the air quality dataset for predicting the air quality index (Sethi & Mittal, 2021). Causality Based Linear method is chosen for selecting accurate parameter that affects the pollution in the environment (Sethi & Mittal, 2019b).

Proposed architecture

A mobile-based robot system is proposing real-time monitoring of the air quality in landfilling sites continuously with the assistance of GPRS (Global Packet for Radio Service), LoRa (Long Range) modem, and Internet connectivity. Generally, in previous studies, the communication protocol which has been implemented is limited to communicate in the short-range. However, implementing GPRS (Global Packet for Radio Service) modem in every sensor node leads to a rise in the infrastructure cost and high energy consumption. We are applying LPWAN (Low Power-based Wide Area Network) based LoRa (Long Range) radio modem in the sensor node for meeting the requirement of IoT in terms of low energy consumption and long-range data transmission. Here we are proposing an LPWAN and GPRS communication-based mobile robot for monitoring the air quality in landfilling sites and it is shown in the Fig. 15.1. "n" number of sensor nodes are deployed at the different positions in landfilling sites for sensing the air quality with different sensors. The mobile robots will be deployed at a specific position for receiving the sensory data from the sensor node.

LoRa modem is embedded in the sensor node for creating a wide area network (WAN) in between the sensor node and the mobile platform. LoRa (Long Range) modem also enables to establish reliable and long-range data transmission. The sensor node (Fig. 15.2) that comprises of LPG sensor, NO sensor, hydrogen sensor, methane sensor, CO_2 sensor, light sensor, DHT sensor, smoke sensor, pressure sensor, and dust particle measurement sensors is for sensing the different environmental parameters in the landfilling site. The location of each sensor node is tracked through GPS (Global Positioning System). DHT sensor in the sensor node senses the temperature and humidity of the landfilling site. LPG sensor detects the concentration of the gas in the air. The smoke sensor detects the smoke generating from the garbage. A dust particle measurement sensor is for sensing the number of dust particles in the air. The computing unit receives the sensory data and its active long-range modem to communicate the sensory data to the mobile robot.

The mobile robot is an integration of the long-range modem and GPRS communication protocol. The computing unit in the mobile robot checks the

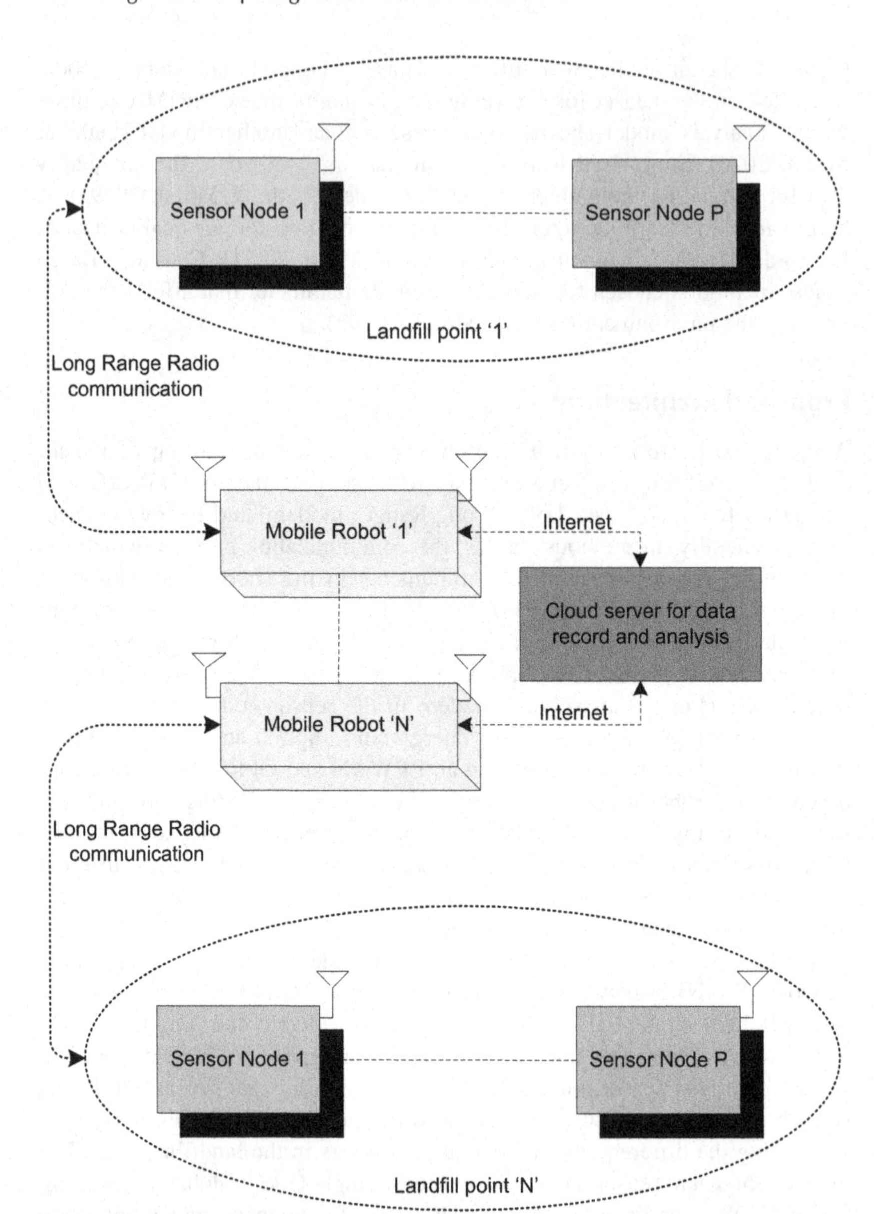

FIGURE 15.1 Architecture for the mobile robot platform.

data received from the sensor node and if yes it activates the Long Range (LoRa) modem to receive the sensory data from the sensor node. The computing unit validates the data received and activates the GPRS modem to transmit the sensory data to the cloud server. The cloud server enables the analyze the data for estimating and predicting the air quality index (Fig. 15.3).

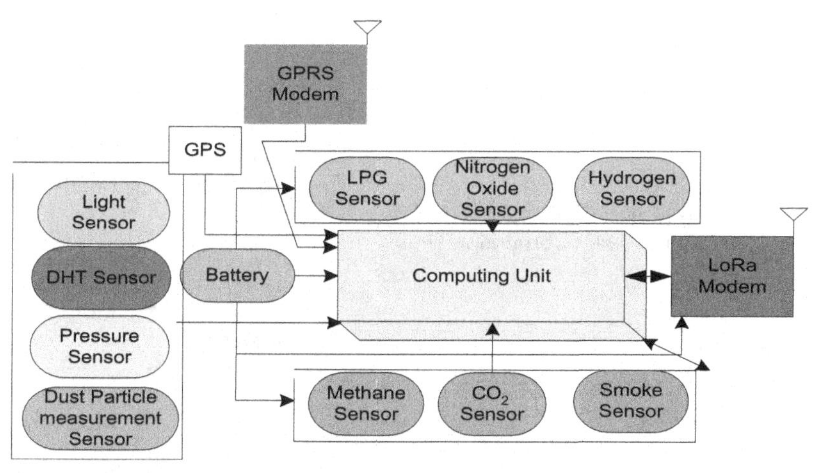

FIGURE 15.2 Sensor node.

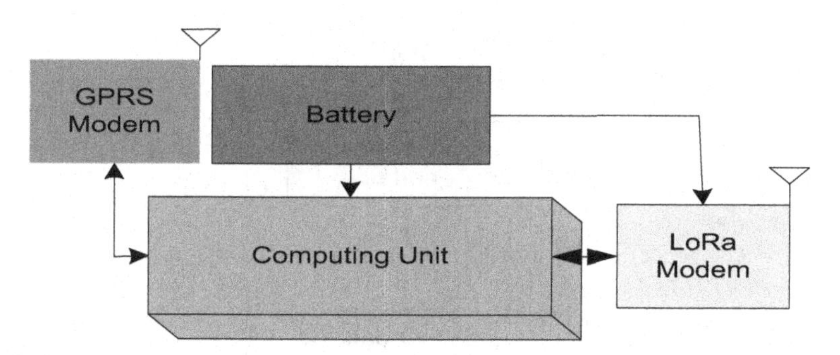

FIGURE 15.3 Sensor node of a mobile robot.

The control unit of the mobile platform that enables the movement of the mobile robot to move from one position to another position is shown in Fig. 15.4. Motors and motor drivers are used for the movement of the wheels in the mobile robot. Charging and booster circuit is for supplying the power to the mobile robot.

In Fig. 15.5, the computing unit in the sensor node checks the sensory data and initiates the LoRa radio modem for transmitting the data to the mobile platform. In Fig. 15.6, the computing unit in the sensor node checks whether serial data received from the sensor node or not. If serial data is received, then it initiates the GPRS for transmission of data to the cloud server via IP (Internet Protocol).

3D Design of mobile robot

Computer-aided design (CAD) refers to devices being used to support the design. With CAD tech, it's feasible to design a whole model in an

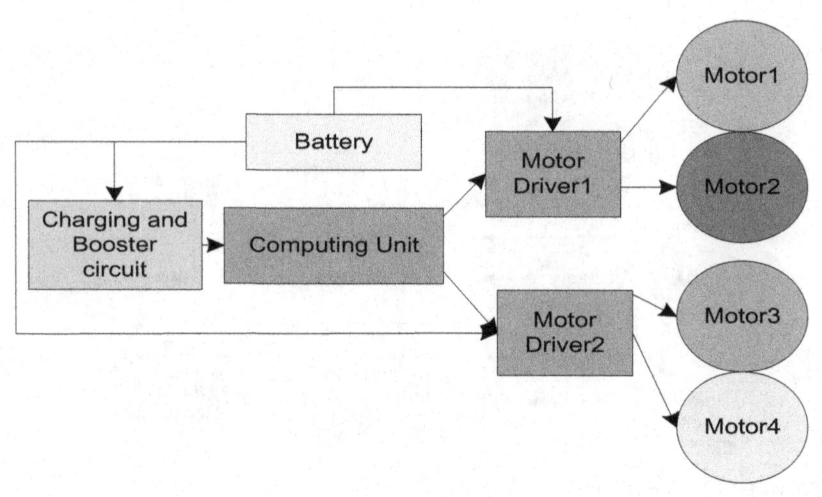

FIGURE 15.4 Control of mobile robot.

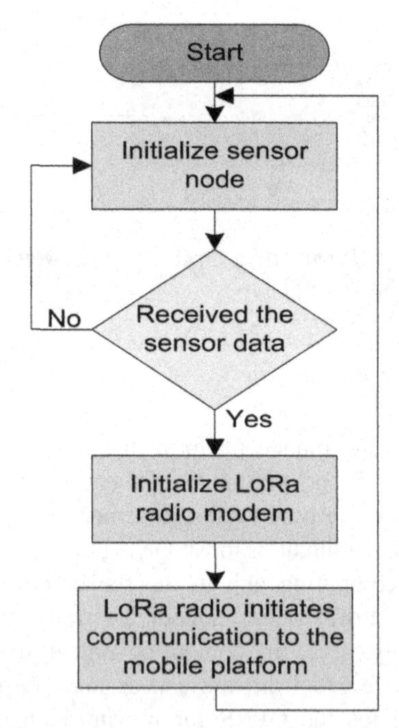

FIGURE 15.5 Sensor node.

imaginary space, allowing imagine properties like height, weight, distance, content, or color until the model is used for a specific use. In this, we have designed the 3D design of the mobile robot on the CAD (computer-aided design) and it is shown in Fig. 15.7.

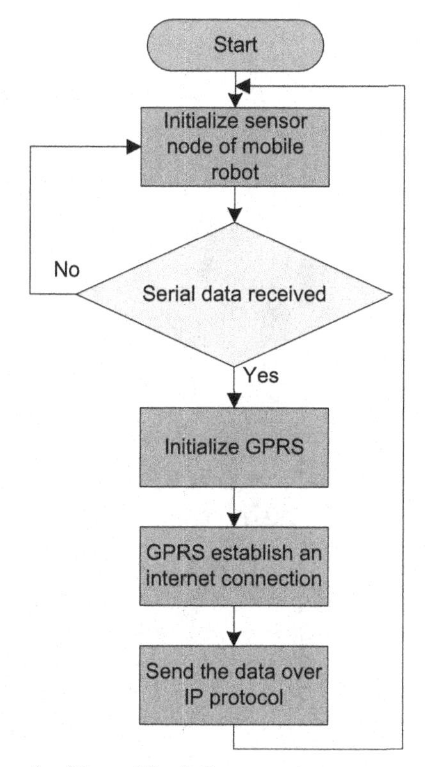

FIGURE 15.6 Sensor node of the mobile platform.

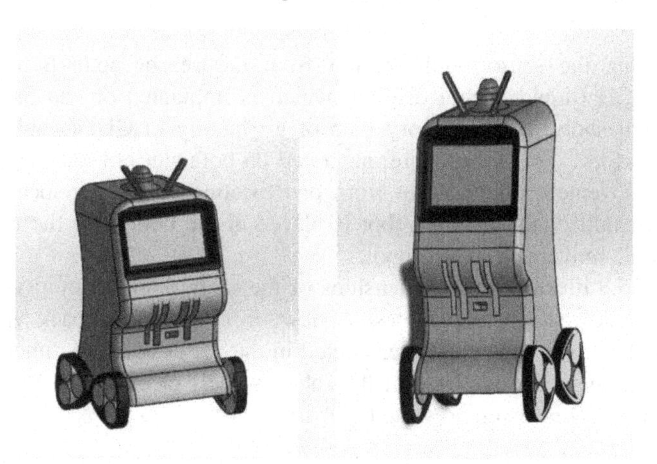

FIGURE 15.7 CAD Model of the system.

The 3D design of a mobile robot comprises of sensor node is embedded in the robot for receiving the sensory information. Two antennas of LoRa and GPRS are embedded on the top of the mobile robot for receiving and

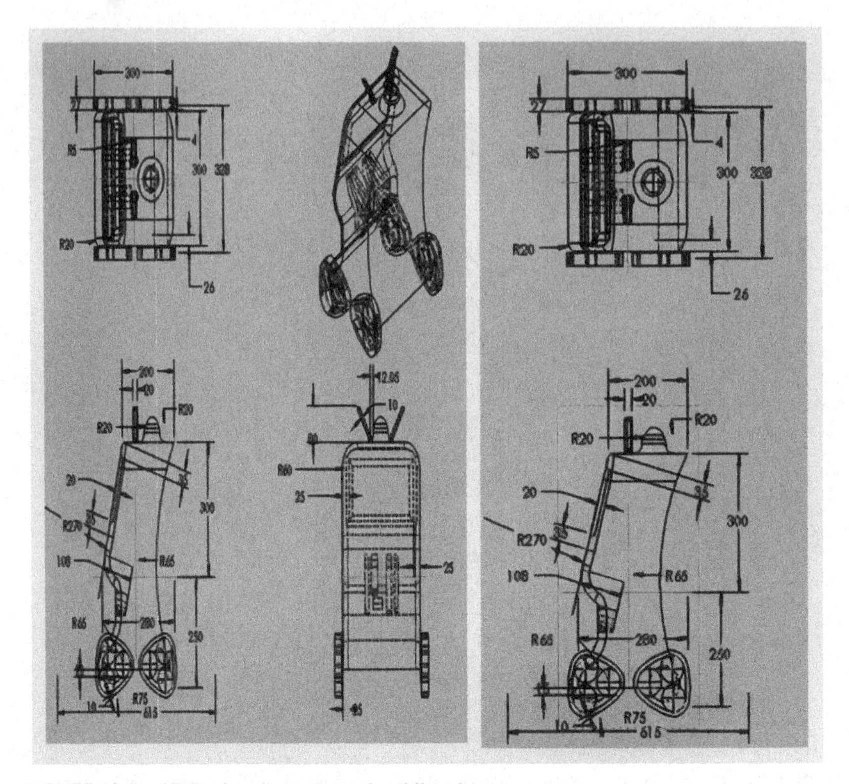

FIGURE 15.8 3D Design dimensions of mobile robot.

transmitting the sensory information from the sensor node to the cloud server. LCD (liquid crystal display) which is implanted on the face of the robot, will show us the sensory data of air quality in GUI (graphical user interface) form. Four wheels are integrated on both sides of the mobile robot for the movement of the robot from one location to another location. The controller unit of the mobile robot is placed at the bottom of the robot that enables the movement of the robot.

Fig. 15.8 illustrates the dimensions of the mobile robot in top view, bottom view, left side view, right side view, front view, and back view. The dimensions of each view are represented in the 3D design and dimensions of each component like the base of the robot, wheels of the robot, the body of the robot, LCD is also represented.

Results

The values obtained from the relevant sensors on the robot were correlated with Table 15.1. of the Air Quality Index. If the sampled data is within the optimal range, then results represented that the air quality is optimal. The

TABLE 15.1 Ideal range of the parameters.

S. No	Parameter	Ideal Range
1	SO_2	(0–850 ug/m^3)
2	NO_2	(0–1000 ug/m^3)
3	CO	(0–16000 ug/m^3)
4	O_3	(0–240 ug/m^3)
5	VOC	(0–250 ug/m^3)
6	Particle	(0–150 ug/m^3)

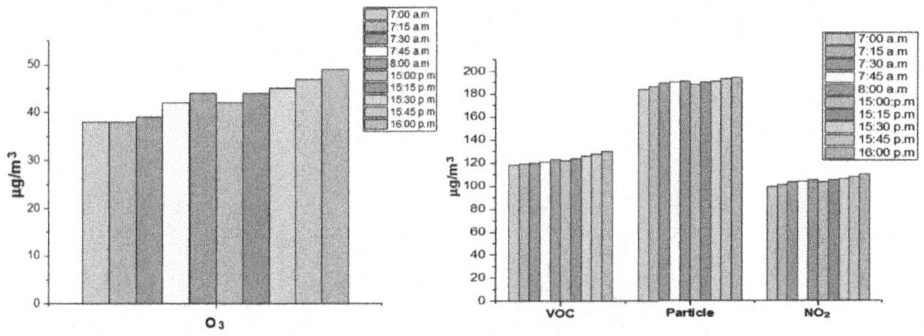

FIGURE 15.9 Sensor data of O_3, VOC, particle and NO_2.

sensor data of the sensor nodes are recorded in the cloud server at different time intervals. We have presented the sensory data at different day timings, 7:00 a.m. and 15:00. In this study we have predetermined the time intervals of 15 minutes, for every 15 minutes of two different timings sensory data is recorded in the cloud server.

Fig. 15.9 illustrates the graphical representation of sensory data of O_3 (Ozone) gas, VOC (Volatile organic compounds), particle, and NO_2 in the environment.

The measurement of the gases is measured in Micrograms per cubic meter (ug/m^3) and it is represented on the y-axis of the graph. The sensory data of the four different gases is represented on the x-axis of the graph. The sensory data of O_3 gas illustrates the amount of O_3 in the air is optimal.

The sensory data of VOC in the above figure represents that the VOC is also in the optimal range at distinct timings. When it comes to the particle, the sensory data illustrates that the particle crossed the optimal range presented in Table 15.1. The sensory data of NO_2 (nitrogen dioxide) reveals that the amount of this gas is in the optimal range.

The graphical representation of CO (carbon monoxide) sensory data is shown in the Fig. 15.10. The sensory data reveals that the CO is also in the optimal range at different time intervals of distinct timings.

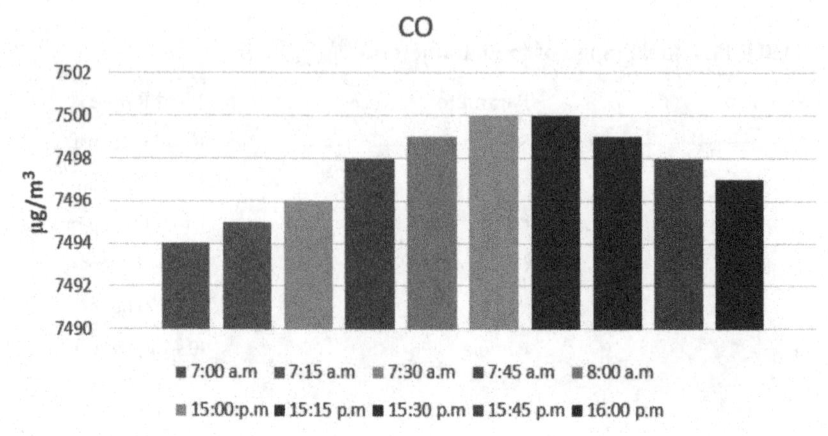

FIGURE 15.10 Sensor data of CO.

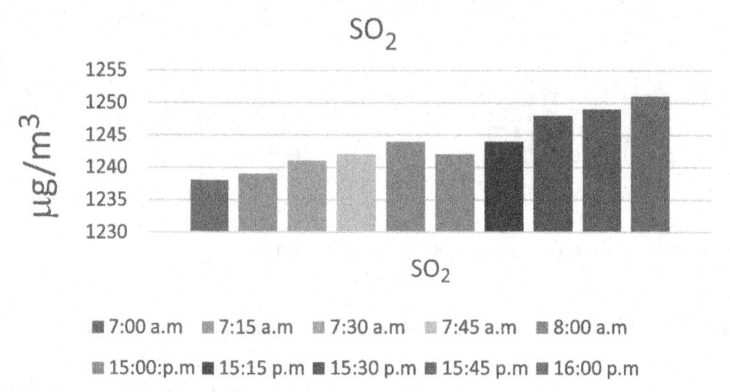

FIGURE 15.11 Sensor data of SO_2.

The graphical representation of SO_2 (sulfur dioxide) sensory data in the Fig. 15.11 The sensory data reveals that the SO_2 exceeds the optimal range (Table 15.1) at different time intervals of distinct timings.

Conclusion

Air quality monitoring at the landfilling sites is an essential task for avoiding the emission of critical gases into the environment. In this study, we have proposed an IoT and cloud server-based mobile robot for monitoring the air quality of the landfill sites in real-time. Low Power Wide Area Network (LPWAN)-based Long Range (LoRa) radio modem and GPRS (Global Packet for Radio Service) is integrated into this study for establishing a reliable and efficient communication system for a mobile robot. The sensory data of the different sensor nodes are recorded in the cloud server at a different time interval.

The proposed system enables real-time monitoring of the landfill sites without human intervention and it assists the municipal authorities to execute necessary measurements for maintaining the air quality in optimal range within a short period. The data recorded in the cloud server can be utilized for further analysis for predicting the causes of the emission of gases.

References

Aamer, H., Mumtaz, R., Anwar, H., & Poslad, S. (2018). A very low cost, open, wireless, internet of things (IoT) air quality monitoring platform. In *2018 15th International conference on smart cities: Improving quality of life using ICT & IoT (HONET-ICT 2018)* (pp. 102–106). IEEE. Available from https://doi.org/10.1109/HONET.2018.8551340.

Addabbo T., Fort, A., Mecocci, A., Mugnaini, M., Parrino, S., Pozzebon, A., & Vignoli, V. (2019). A LoRa-based IoT sensor node for waste management based on a customized ultrasonic transceiver. In *SAS 2019–2019 IEEE sensors applications symposium conference proceedings*. IEEE. Available from https://doi.org/10.1109/SAS.2019.8705980.

Agarwal, A., Shukla, V., Singh, R., Gehlot, A., & Garg, V. (2018). Design and development of air and water pollution quality monitoring using IoT and quadcopter. In *Intelligent Communication, control and devices* (pp. 485–492). Springer.

Bogue, R. (2011). Robots for monitoring the environment. *Industrial Robot*, *38*(6), 560–566.

Chen L. Y., Huang, H. S., Wu, C. J., Tsai, Y. T., & Chang, Y. S. (2018). A LoRa-based air quality monitor on unmanned aerial vehicle for smart city. In *ICSSE 2018: IEEE international conference on system science and engineering 2018*. IEEE. Available from https://doi.org/10.1109/ICSSE.2018.8519967.

Cookbook For Mobile Robotic Platform Control: With Internet of Things And Ti ... - Dr. Anita Gehlot, Dr. Rajesh Singh, Dr. Lovi Raj Gupta, Bhupendra Singh - Google Books. (2019). <https://books.google.co.in/books/about/Cookbook_For_Mobile_Robotic_Platform_Con.html?id = 9pCwDwAAQBAJ&redir_esc = y>. (Accessed November 22, 2020).

Dhingra, S., Madda, R. B., Gandomi, A. H., Patan, R., & Daneshmand, M. (2019). Internet of things mobile-air pollution monitoring system (IoT-Mobair). *IEEE Internet of Things Journal*, *6*, 5577–5584. Available from https://doi.org/10.1109/JIOT.2019.2903821.

Dunbabin, M., & Marques, L. (2012). Robots for environmental monitoring: Significant advancements and applications. *IEEE Robotics & Automation Magazine*, *19*, 24–39.

Fioccola, G. B., Sommese, R., Tufano, I., Canonico, R., & Ventre, G. (2016). Polluino: An efficient cloud-based management of IoT devices for air quality monitoring. In *2016 IEEE secondnd international forum on research and technologies for society and industry leveraging a better tomorrow (RTSI)*. IEEE. Available from https://doi.org/10.1109/RTSI.2016.7740617.

Fu, H.-L., Chen, H.-C., & Lin, P. (2012). APS: Distributed air pollution sensing system on wireless sensor and robot networks. *Computer Communications*, *35*, 1141–1150.

Handbook of Research on the Internet of Things Applications in Robotics and Automation: 9781522595748: Computer Science & IT Books | IGI Global. (2019). <https://www.igi-global.com/book/handbook-research-internet-things-applications/221875>. (Accessed November 22, 2020).

Hu, Z., Bai, Z., Yang, Y., Zheng, Z., Bian, K., & Song, L. (2019). UAV aided aerial-ground IoT for air quality sensing in smart city: Architecture, technologies, and implementation. *IEEE Network*, *33*, 14–22. Available from https://doi.org/10.1109/MNET.2019.1800214.

Hu, Z., Cong, S., Song, T., Bian, K., & Song, L. (2020). AirScope: Mobile robots-assisted ccooperative indoor air quality sensing by distributed deep reinforcement learning. *IEEE Internet of Things Journal*, *7*, 9189–9200.

Jin, M., Liu, S., Schiavon, S., & Spanos, C. (2018). Automated mobile sensing: Towards high-granularity agile indoor environmental quality monitoring. *Building and Environment, 127*, 268–276.

Jo, J., Jo, B., Kim, J., Kim, S., & Han, W. (2020). Development of an iot-based indoor air quality monitoring platform. *Journal of Sensors, 2020*.

Joshi, R., & Ahmed, S. (2016). Status and challenges of municipal solid waste management in India: A review. *Cogent Environmental Science, 2*, 1139434.

Kiruthika R., & Umamakeswari, A. (2018). Low cost pollution control and air quality monitoring system using Raspberry Pi for Internet of Things. In *2017 International conference on energy, communication, data analytics and soft computing (ICECDS)* (pp. 2319–2326). IEEE. Available from https://doi.org/10.1109/ICECDS.2017.8389867.

Kumar P., Verma, P., Singh, R., & Patel, R.K. (2016). Proceeding of international conference on intelligent communication, control and devices (pp. 979–989). Available from https://doi.org/10.1007/978-981-10-1708-7.

Kumar, S., Smith, S. R., Fowler, G., Velis, C., Kumar, S. J., Arya, S., … Cheeseman, C. (2017). Challenges and opportunities associated with waste management in India. *Royal Society Open Science, 4*. Available from https://doi.org/10.1098/rsos.160764.

LoRA and IoT Networks for Applications in Industry 4.0 - Nova Science Publishers, (n.d.). <https://novapublishers.com/shop/lora-and-iot-networks-for-applications-in-industry-4-0/>. (Accessed November 22, 2020).

Marques, G., Pires, I. M., Miranda, N., & Pitarma, R. (2019). Air quality monitoring using assistive robots for ambient assisted living and enhanced living environments through internet of things. *Electronics, 8*, 1375.

Marques, G., & Pitarma, R. (2019). A cost-effective air quality supervision solution for enhanced living environments through the internet of things. *Electronics, 8*. Available from https://doi.org/10.3390/electronics8020170.

Mekki, K., Bajic, E., Chaxel, F., & Meyer, F. (2019). A comparative study of LPWAN technologies for large-scale IoT deployment. *ICT Express, 5*, 1–7. Available from https://doi.org/10.1016/j.icte.2017.12.005.

Monroy, J., & Gonzalez-Jimenez, J. (2019). Towards odor-sensitive mobile robots. In *Rapid automation: Concepts, methodologies, tools, and applications* (pp. 1491–1510). IGI Global.

Nikoukar, A., Raza, S., Poole, A., Gunes, M., & Dezfouli, B. (2018). Low-power wireless for the internet of things: Standards and applications. *IEEE Access, 6*, 67893–67926. Available from https://doi.org/10.1109/ACCESS.2018.2879189.

Parcheta, C. E., Pavlov, C. A., Wiltsie, N., Carpenter, K. C., Nash, J., Parness, A., & Mitchell, K. L. (2016). A robotic approach to mapping post-eruptive volcanic fissure conduits. *Journal of Volcanology and Geothermal Research, 320*, 19–28.

Popa, C. L., Carutasu, G., Cotet, C. E., Carutasu, N. L., & Dobrescu, T. (2017). Smart city platform development for an automated waste collection system. *Sustainability, 9*, 1–15. Available from https://doi.org/10.3390/su9112064.

Rathour, N., Gehlot, A., & Singh, R. (2019). Spruce-A intelligent surveillance device for monitoring of dustbins using image processing and raspberry PI. *International Journal of Recent Technology and Engineering, 8*, 1570–1574. Available from https://doi.org/10.35940/ijrte.B1106.0882S819.

Saini, J., Dutta, M., & Marques, G. (2020). Indoor air quality monitoring systems based on internet of things: A systematic review. *International Journal of Environmental Research and Public Health, 17*, 4942. Available from https://doi.org/10.3390/ijerph17144942.

Salman, H., Rahman, M. S., Tarek, M. A. Y., Wang, J. (2019). The design and implementation of GPS controlled environment monitoring robotic system based on IoT and ARM. In *2019 4th International conference on control and robotics engineering (ICCRE 2019)* (pp. 93–98). IEEE.

Samkaria, R., Singh, R., Gehlot, A., Pachauri, R., Kumar, A., Singh, N. K., & Rawat, K. (2018). IOT and XBee triggered based adaptive intrusion detection using geophone and quick response by UAV. *International Journal of Engineering & Technology*, *7*, 12–18.

Sethi, J. K., & Mittal, M. (2019a). Ambient air quality estimation using supervised learning techniques. *EAI Endorsed Transactions on Scalable Information Systems*, *6*.

Sethi, J. K., & Mittal, M. (2019b). A new feature selection method based on machine learning technique for air quality dataset. *Journal of Statistics and Management Systems*, *22*, 697–705. Available from https://doi.org/10.1080/09720510.2019.1609726.

Sethi J. K., & Mittal, M. (2020). Analysis of air quality using univariate and multivariate time series models. In *Proceedings of the confluence 2020: 10th international conference on cloud computing, data science & engineering* (pp. 823–827). IEEE. https://doi.org/10.1109/Confluence47617.2020.9058303.

Sethi, J. K., & Mittal, M. (2021). Prediction of air quality index using hybrid machine learning algorithm. *Lecture notes in networks and systems* (pp. 439–449). Springer. Available from https://doi.org/10.1007/978-981-15-5421-6_44.

Singh, R., Anita., Khilrani, J., Sah, A., Parmar, M., & Kumar, S. (2019). Path learning algorithm for fully indigenous multimedia and guide robot. *International Journal of Mechatronics and Applied Mechanics*, *1*, 7–16.

Singh, R., Gehlot, A., Jain, V., & Malik, P. K. (Eds.), (2020). *Handbook of research on the internet of things applications in robotics and automation*. IGI Global. Available from https://doi.org/10.4018/978-1-5225-9574-8.

Singh, R., Gehlot, A., Khilrani, J. K., & Mittal, M. (2020). Internet of things–triggered and power-efficient smart pedometer algorithm for intelligent wearable devices. In *Wearable and implantable medical devices* (pp. 1–23). Elsevier. Available from https://doi.org/10.1016/b978-0-12-815369-7.00001-x.

Singh, R., Gehlot, A., Mittal, M., Samkaria, R., & Choudhury, S. (2017). Application of icloud and wireless sensor network in environmental parameter analysis. *International Journal of Sensors, Wireless Communications and Control*, *7*, 170–177.

Singh, R., Gehlot, A., Samkaria, R., & Choudhury, S. (2020). An intelligent and multi-terrain navigational environment monitoring robotic platform with wireless sensor network and internet of robotic things. *International Journal of Mechatronics and Automation*, *7*, 32–42. Available from https://doi.org/10.1504/IJMA.2020.10030350.

Singh, R., Gehlot, A., Thakur, A. K., & Mohan, P. (2019). Outdoor localization of robot with RSSI for IoT connected smart devices. *International Journal of Recent Technology and Engineering*, *8*, 2050–2054. Available from https://doi.org/10.35940/ijrte.C4538.098319.

Singh, R., Gehlot, A., Thakur, A. K., & Prudhvi. (2019). Indoor localization of robot in biofuel lab with web of things. *International Journal of Scientific & Technology Research*, *8*, 3697–3701.

Singh, R., Samkaria, R., Gehlot, A., & Choudhary, S. (2018). Design and development of IoT enabled multi robot system for search and rescue mission. *International Journal of Web Applications*, *10*, 51. Available from https://doi.org/10.6025/ijwa/2018/10/2/51-63.

Thu M. Y., Htun, W., Aung, Y. L., Shwe, P. E. E., & Tun, N. M. (2019). Smart air quality monitoring system with LoRaWAN. In *Proceedings of the 2018 IEEE international conference on internet of things and intelligence system (IOTAIS 2018)* (pp. 10–15). IEEE. Available from https://doi.org/10.1109/IOTAIS.2018.8600904.

Zheng, K., Zhao, S., Yang, Z., Xiong, X., & Xiang, W. (2016). Design and implementation of LPWA-based air quality monitoring system. *IEEE Access*, *4*, 3238–3245. Available from https://doi.org/10.1109/ACCESS.2016.2582153.

Chapter 16

Artificial Intelligence and Internet of Things readiness: inclination for hotels to support a sustainable environment

Mudita Sinha[1], Leena N. Fukey[1] and Ashutosh Sinha[2]
[1]*Christ University, Bengaluru, India,* [2]*Robonomics AI India Pvt Ltd., Bangalore, India*

Introduction

Around 50 billion interconnections are expected to be in place by 2020 with the continuous increase in internet usage and is expected to grow further (Guarda et al., 2017). Artificial Intelligence (AI) is a chant of success and growth in all the sectors in current scenario. AI is not new, the concept was conceived in 1956 by John McCarthy. Since then AI had a very bumpy road and has come a long way through progressing in different areas and becoming an unavoidable part of our daily life (Atzori, Iera, & Morabito, 2010) and making it more comfortable (Shah & Yaqoob, 2016). Like other field AI has a lot of contribution towards application and implementation Internet of things (IoT). The word IoT was originated by Kelvin Ashton in the year 1999 which is becoming smarter day by day. IoT aims at saturated nation installed with smart gadgets, often referred to as "smart objects" interconnected via the Internet or any other means of communication (Hassan & Madani, 2017 and Fortino & Trunfio, 2014). IoT systems capture a lot of data from sensors around us which can be further utilized for improved customized services to companies, citizens and Government (Atzori et al., 2010). IoT plays a major role in formation of smart cities, smart agriculture, health services and many more. Combination of AI and IoT will help the industries to create a more automated and customized services.

Smart cities take a unique approach handling problems and growing challenges cities face and provide better services for its residents (Alshekhly, 2012). A smart city's principal objective is to link existing infrastructure, information and communication technology (ICT) infrastructure, social

Cognitive Computing for Human-Robot Interaction. DOI: https://doi.org/10.1016/B978-0-323-85769-7.00015-X

infrastructure and business infrastructure (Naphade, Banavar, Harrison, Paraszczak, & Morris, 2011). Smart cities are very closely related to the concept of smart infrastructure, smart transportation, smart power management smart health, environmentally-aided living, crime prevention, community protection, governance, facilities status management and maintenance, crisis and disaster management, smart buildings and smart tourism (Aguilar & Mendes, 2017). Economy of any country can be boosted at considerable rate through tourism as it largest sector across boundaries. Tourism is being supported by Government as it is blue eyed sector for revenue generation for any country, as it not only contributes individually but also collectively to the economic well-being of tourism cities. Tourism contributes to uplifting the economy, individually as well as collectively of tourist places, which in turn forms spine of the hotel industry. Tourists needs and wants are fulfilled by hotels by providing different cuisines and beverage services along with entertainment, recreation, vacation sports and several other amenities (Miočić, Korona, & Matešić, 2012).

Approximately 17.5 million hotel rooms are anticipated around the globe (Mesirow & Blumenthal, 2019) which brings it to the top five sectors of viable energy consumption apart from its economic contribution. As hotels form important subsector of hospitality sector therefore it is accountable for substantial quantity of the entire carbon footprints and harmful emissions produced by the whole hospitality sector. It's depicted by several research that there is possibility of resource management in hospitality sector as there is high level of energy wastage. In the light of the growing understanding of the world and the climate conservation, the hotels are under pressure to take sustainable practices. Nevertheless, only since the 1990s has the hotel industry followed and implemented many green practices. Environment and hotel partnership have both positives and negatives associated with it. The effect of the hotel industry on the environment has positive impact on the local community through usage of promotion of sustainable goods, recycling and "green" initiatives (Ahn & Pearce, 2013). The issues like extreme resource utilization, generation of enormous quantity of different kinds of waste, air, water and noise pollution, land degradation and deforestation are the important issues which needs immediate attention of the policy makers and industry experts (Pearon & Parambil, 2019). Hotel sectors are causing irreversible damage to the climate, that is why green philosophy is followed by handful of hotels as the part of their merchandise assortment. The main objective of green philosophy is to streamline the use of natural resources, minimize water generation, keeping a check on contamination of the environment (Table 16.1).

"Green Hotels" are associated with incorporating environment friendly practices via efficient consumption of water, energy and other ono renewable resources while providing high value service to the guests. Green hotels focus on recycling and reusing space, reducing power usage along with

TABLE 16.1 Green practices by few high-end hotels for promoting sustainability.

Category	Practices	Benefits
Indoor environment quality	Good quality lighting and focus on daylight saving. Ample amount of air purification, low volatile organic compounds usage. Mold protection, Improved acoustic output	Reliable and constructive spaces indoor. Providing ideal indoor atmosphere for enhanced safety and well-being of occupants
Building operations and maintenance	Smart maintenance goods, prevention and management of indoor pests, reduction and recycling of waste, conservation of water and energy, Smart floor care, contacting and educating guest electronically or through written documents	Lower energy consumption, well managed waste, reusing and economical utilization of equipment.
Materials and resources	Green products and materials, Building waste management Recyclable goods, Local goods, locally sourced Rapidly recycled materials	Resource saving and Reduced environmental impacts
Energy efficiency	Solar orientation, High performance envelopes (efficient windows and good insulation value), High quality heating, ventilation and air conditioning, automated buildings, efficient lighting and daylight saving. Sources of on-site renewable energy sources (photovoltaics)	Fuel savings Reduction in greenhouse gasses and Lower production costs
Efficient water usage	Water saving fittings and equipment, rainwater harvesting systems	Lower operating costs
Sustainable land usage	Sustainable landscaping, public transport with solar design, Cyclone water protection	Reduce environmental impacts and Quality of site use

Source: Adapted from Ahn, Y., & Pearce, A. (2013). Green luxury: A case study of two green hotels. *Journal of Green Building, 8*(1), 90–119.

minimization of solid waste. It has advantages like cost reduction and burdens, better returns and low point-risk ventures, improved professionalism which are observed and witnessed by hotels. The recognition of such gains and opportunities has allowed the reputation of green hotels to increase.

Unwanted wastage of water, enormous water generation exorbitant usage of nonrenewable natural resources, inappropriate usage of resources inside the hotels are few of the most highlighted problem area of hotels. If individually hotels are considered there is not such environmental damage, but the damage comes to picture when entire hotel industry is been taken into account primarily due to over consumption of resources which is accounted for additional running costs.

Recently IoT has been added to smart hotels arouse a lot of attention (Les Roches, 2018) after Hilton announced an IoT solution in 2017 post which MarrIoT also made a similar announcement. One effective way in which energy utilization can be optimized is IoT which can be identified as "A network of physical objects that contain embedded technology to communicate and sense or interact with their internal states or the external environment."

Thus this research article aims to find out the strategic positions of a hotel in terms of sustainability, AI and IoT technology. Components that will be considered by Hotels for the strategic intention of adopting AI and IoT for environmental sustainability. Different development and modification needed to be taken if management wants high sustainability readiness and/or IoT readiness.

Methodology

This conceptual paper tries to focus on current knowledge of hotel industry readiness on AI and IoT enabled sustainable practices. According to Tranfield et al. following objectives should to be fulfilled by literature reviews: Firstly a review must consolidated all the findings of the research by studying and fusing various pieces of literature of particular research area to find the research gap and act as a catalyst for upcoming research. Secondly large amount of data can be gathered by reviewing appropriate literature which will be reliable, precise and can be retroflexed as per the requirement. Whenever sustainability and hotel industry come into picture plant, people and profit are embossed as emerging research thrust area for future. Current research looks to answer the following:

1. The key concepts of sustainability operational in Hotel Industry
2. What are the AI and IoT driven sustainability practices?
3. Readiness for AI and IoT driven sustainability in hotel industry?

Key concept of sustainability in hotel industry

Sustainability can be best defined as fulfilling the needs of current generation without affecting that of future (Ali, Murphy, & Nadkarni, 2016; Higgins-Desbiolles, Moskwa, & Wijesinghe, 2017). In short sustainability deals with

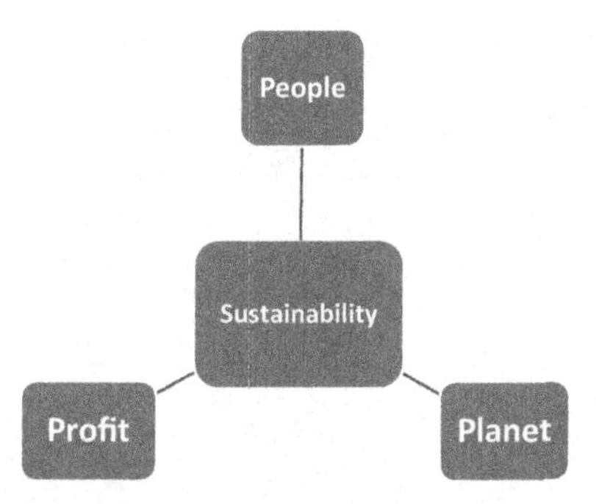

FIGURE 16.1 Triple bottom line of sustainability. *Adapted from Elkington, J. (1997).* Cannibals with forks—Triple bottom line of 21st century business. *Stoney Creek, CT: New Society Publishers.*

the three Ps' of triple bottom line concept-People, Planet and Profit (Fig. 16.1). This model also aims at increasing the economic activity without negatively affecting the environment (Higgins-Desbiolles et al., 2017). Keeping future in consideration it is unavoidable to consider recent changes in the global environment, as has magnified the attention needed towards sustainability (Kim, Lee, & Fairhurst, 2017). The topic of sustainability in hotel industry has become a major concern from past few years due the amount of wate generated (Davids, 2016). As a result to which different Government bodies made guidelines for different hotel chains to reduce the carbon footprint for safeguarding the environment.

Incorporation of sustainable practices is an added advantage for the stakeholders of the hotel. However, the essence of the industry, that strives to create unforgettable guest experiences, frequently results in administrative apprehension to introduce such measures. The challenge of calculating the viability of environmental projects systematically impacts the desire of the upper management to engage in these initiatives (Rosenbaum & Wong, 2015).

Public pressure in addition to recent changes has obligated the undivided attention of the hospitality professionals towards sustainable practices. However, as per the Industry experts it has been observed that though hotels are expected to follow sustainable practices yet it should not be visible to the guests (Withiam, 2011). It has been brought into light by prior researches that the adoption of sustainability practices in Hotels by the stakeholders depends primarily on 3 parameters: Cost of Implementation, Support from the employee and above all support from the customer (Fig. 16.2) (Cvelbar & Dwyer, 2013; Deraman, Ismail, Arifin, & Mostafa, 2017). Although it is known fact by stakeholders of hotels that implementing sustainable practices are virtuous for the business in long run but they all are hesitant to make the

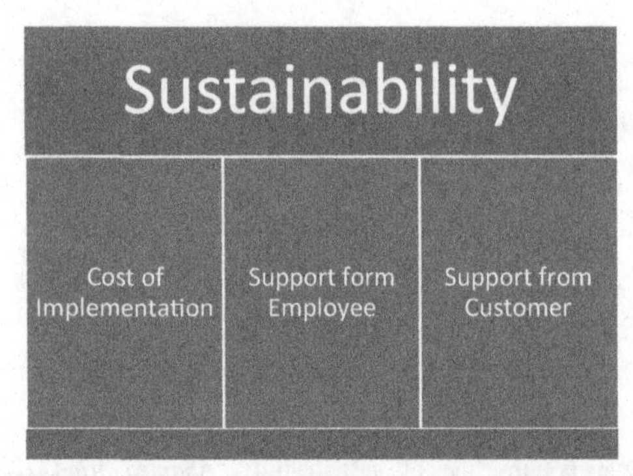

FIGURE 16.2 Important parameters for sustainably practices in hotels. *Adapted from Cvelbar, L., & Dwyer, L. (2013). An importance—Performance analysis of sustainability factors for long-term strategy planning in Slovenian hotels.* Journal of Sustainable Tourism, 21*(3), 487–504; Deraman, F., Ismail, N., Arifin, A., & Mostafa, M. (2017). Green practices in hotel industry: Factors influencing the implementation.* Journal of Tourism, Hospitality and Culinary Arts, 9*(2), 305–316.*

initial investment that the properties require to implement such practices (Deraman et al., 2017).

Regulatory policies are crucial for driving the significant reduction of the hotel carbon footprint. Luxury hotels are often hesitant to adopt sustainable practices because of a misconception that sustainability is still assumed to be incompatible with the status quo that luxury hotels need to retain to attract their target markets (Sourvinou & Filimonau, 2017). The idea of a green hotel begins at the stage of construction, a hotel with features and facilities that help to conserve energy, water and minimize waste can only be built at the stage of the project. The argument is that the strategies and techniques of hotels must be modified focusing on the environment preservation as a prime concern. This message must be circulated among core and peripheral customers by the management. The hotels which are already functional can also convert the hotel into an environment affable, as needed by maintaining a management commitment to the environment. With respect to sustainability, documenting current and chronological use, setting baselines, goals and objectives, prioritizing the course of action, observe and quantify that appropriate information and inspiration to the employees to be initiated (Goldstein & Primlani, 2012a).

There are different kinds of impacts that hotel sector imposes on the environment that depends on the level of its operation. Significant amount of solid waste and effluent is generated due to the constant consumption of energy and water by the guests (Goldstein & Primlani, 2012). Thus there are initiatives taken by hotels to practice and to get certified based on the certifications available to enter sustainability title which is presented in the following table (Amandeep, 2017) (Table 16.2).

TABLE 16.2 Major certification for hotels in India.

S. No.	Certification name	Organization	Description	Source
1.	Leadership in Energy and Environmental Design (LEED)	United States Green Building Council	Certification from LEED validates that the building has been constructed keeping green standards in mind, for example, conservation of energy. The LEED certification scheme, which was established by the United States Green Building Council in 2000, provides industrial properties with a scoring system to meet requirements in fields like area and transportation, resources and energy, and water quality, among others	http://www.usgbc.org
2.	Green Key Global	Hotel Association of Canada, LRA Worldwide, Inc.	Environmental hotel certification scheme. Provides professional direction. Depending on the adherence to requirements, participating facilities are awarded between 1 and 5 green keys	http://www.greenglobe.com
3.	ECOTEL certification	HVS	The ECOTEL certification stays one of it's groundbreaking programs which has integrated sustainability principles with an	http://www.ecotelhotels.com

(*Continued*)

TABLE 16.2 (Continued)

S. No.	Certification name	Organization	Description	Source
			emphasis on environmental protection. The Employee Knowledge and community support horizon predicted an environmental management solution via the establishment of a green squad and employee members to motivate them to activate the sustainability agenda. Overtime, the ECOTEL system moved its focus towards India	
4.	ISO 14000 certification	International Standards Organization	The International Organization for Standards (1996) is basically a blueprint for organiations looking to adopt a structured sustainable management program. After the strategy is introduced within an entity, ISO sends regulators for approval to evaluate the entire system and to monitor energy use, recycling activities, etc.	http://www.iso.org

(*Continued*)

TABLE 16.2 (Continued)

S. No.	Certification name	Organization	Description	Source
5.	The Indian Green Building Council (IGBC)	The Confederation of Indian Industry and the Godrej Green Business Center	IGBC continually strives for the broad application of sustainable building principles within the Indian economy. More than 687 ventures have been documented or accredited in the last 10 years in compliance with the green building guidelines established by IGBC in India. They also empower and certify hotels which are using and filling in their hotels with the required environmentally sound processes	http://www.igbc.in
6.	Sustainable Tourism Eco-Certification Program (STEP)	Sustainable Travel International	Tour operators, hotels, attractions, transportation and the cruise industry have an environmental certification program. Gives recommendations, self-assessment method and an energy saving-logo ranking system of 2–5 stars. Separate Luxury Accommodation Certification offered	http://www.sustainabletravel.org

Source: Adapted from Goldstein K. A., & Primlani, V. R. (2012). *Current trends and opportunities in hotel sustainability.* HVS Sustainability Services.

When it comes to customers' expectation towards still there are disagreement of various researchers and the stakeholders. Rodríguez-Antón, del Mar Alonso-Almeida, Celemín, and Rubio (2012) identified that customers on relaxation (B2C) focused more on the sustainability practices when compared to B2B that consists of business clients. On the contrary Tend et al. in a recent study emphasized that B2B clients are more considerate and have more information about the caron footprints of the hotel. In addition, to this it has been observed that inclination of guests of luxury hotel is more towards the safe hotels and are ready to pay premium for the same as well in comparison to the hotel guests in the economy (Chen et al., 2019 and Kang, Stein, Heo, & Lee, 2012).

AI and IOT driven sustainability

World has come a climate crisis (Leahy, 2019) so now the days to have discussion about global warming is over. What we need today is action to work towards preserving our environment (The Carrington, 2019). AI technologies deliver three major benefits. Firstly, AI allows important yet monotonous and laborious tasks to be automated thus enabling people to concentrate on high-skilled jobs. Secondly, AI can change perceptions that are still stuck inside vast volumes of unorganized data which once involved high level of human supervision and evaluation, such as video-spawned data, images, transcribed records, commercial records, posts form social internet communities or electronic-mails. Thirdly, to resolve extremely complex problems, AI and IoT will combine thousands of computers and other tools for better results.

In this view, we suggest that AI and IoT can help in reducing the environmental and energy needs of human activities within different organizational structure. AI and IoT's true value will not focus on how people and community reduce their energy, water and land-use intensities instead the accurate meaning of AI and IoT at later stage will be understood in the way it simplifies and encourages successful control of the environment sustainability. The official and casual environmental regulations that facilitates human actions in taking appropriate decision which influences how society will look at preserving environmental reserves (Linkov, Trump, Poinsatte-Jones, & Florin, 2018). The environmental, economic and social aspects are known as the three interlinked components that relate to sustainable development (Davidson, 2010; McKenzie, 2004; Morelli, 2011; Olawumi & Chan, 2018)

Environment sustainability can be understood best as meeting the requirements of existing and upcoming generation with no effect on the ecosystem balance (Morelli, 2011). Features of environmental sustainability can include low-impact transport, sustainable agriculture, environmental resources management, sustainable land-use, waste management and emissions prevention (Callicott & Mumford, 1998).

AI and IoT enabled systems in the management of water resources have received much academic focus in past few decades as continuous decreasing levels of water is a major concern for the coming years. Different research on biodiversity identified different array of practices for providing assessments of land-related maintenance stipulations. New methods using AI and IoT can deliver better probability measures of time and space apparatuses of biodiversity. During the process of writing this research article we came across more than 260 article from the 2015 to 1019 which focused mainly on the AI and IoT application for energy conservation, distribution and production of energy of which the major focus was on Solar energy. AI powered computer vision (CV) and IoT backed decision support technologies were found very effective in the areas of managing traffic, traffic safekeeping, municipal transport and inner-city mobility (Liyanage, Dia, Abduljabbar, & Bagloee, 2019). Implementation of CV can be used for marking of roads, detection of incidences and identification of independent vehicle detection (Aymen & Mahmoudi, 2019).

Environmental sustainability thorough AI gives a bigger picture on conservation of natural resources and pollution management. Climate change can be scenarios very well predicted by implementation of Evolutionary Computation (EC), Fuzzy logic (FC) and Expert systems Machine Learning (ML) and several other models. AI has not yet reached cognitive ability at the human level but has grown broadly to be included in planning at strategic level and subsequent follow up action plans (Ghallab, Nau, & Traverso, 2016) and for self-improvement and learning (Omohundro, 2007).

Readiness for AI and IOT driven sustainability in hotel industry

In adopting emerging technological developments, the hospitality industry has traditionally lagged (Pizam, 2017) but continuous expectation from the guests to provide technology enabled touchpoints has compelled the industry to move towards that direction. There are several applications that focus on how energy consumption can be reduced (Kratzert et al., 2020) same is applicable for hotel industry as well (Chamarti, 2016). Energy management tops the chart when it comes to enabling Ai and IoT to sustainability (Naimat, 2017) which interest hotel industry to a great extent due to the high level of energy consumption with limited resources. Water consumption is another gigantic question to be addressed quickly by hotel industry due to the already existing stress on water levels. The overall consumption of water for a day of an overnight guest staying in an a luxury hotel is somewhere in between 200 and 1000 litters which is way more that the water consumption of local population of that region (Becken, 2014).

AI and IoT has been implemented as a potential resolution for the energy consumption concerns in hotels. AI and IoT can be used to promote an eco-friendly hotel business. Although there are lots of application of IoT and AI in different functional departments of hotels which may or may not be sustainability related. In this research article researcher will be focusing only on how different functional units of Hotels can become more sustainable as the hotel industry burdens the environment the emission of Greenhouse Gases (GHG), huge water consumption, electricity usage, waste generation and land usage.

"SMART" paperless hotels

One of the most essential commodities for any office use is paper. Traditional majority of departments rely on paper for the simplest tasks to be accomplished. Despite of digitization at its peak in current times the idea of a paperless desk is far from reality. Humans are used to a convenient and comfortable method of reporting and note-taking on paper. AI enabled document technology can completely convert the desk paperless. The challenges associated with handling large quantities of data written on paper can be effortlessly resolved using these technologies (Namee, 2020).

Due to exorbitant usage of paper in the Hotels is in one of the major changes that need to be though upon by Hotels for sustainability. It is not by choice that any hotel should do paperless, but it is the need of the situation where lakhs of forests are slaughter to fulfill the needs of paper which is depleting the natural resource. Hotel industry is a complex industry unlike other industries that work only for a specific time this industry operates for 24 hours, like hospitals. In hospitals the patients have prescribed needs, but hotel guests have the wants and demand of to be fulfilled at any given point of time because of which there is a constant resource consumption that compels stakeholders to think towards possible use of modern technology towards sustainable measures.

Imagine the stress of stepping in the queue for long formalities to check-in what if it can be replaced by a mobile check-in, where you can encounter the lavish and relaxing stay by heading paperless with seamless customer service. Over the years, technology advancement and increasing demand for industry 4.0 has transformed the hospitality industry and helped to give guests enriching experience along with enhancing energy efficiency and asset management. Today, a guest would be able to easily use their gadgets for work obligations, entertainment or on-the-go contact, rather than the free streaming of movies or room service on request. Technology accessibility in digital room automation is becoming a top priority for guests to experience luxury living through trouble-free check-in and keyless locks in the room. Paperless check-in check outs are now days practiced by several hotel in booking communication and check-in/check-out information is been collected with the help of data mining program. Every booking for correspondence on an average consumes around to five sheets of paper if the process is not automated. One identification proof used for one lodger in every accommodation and a hotel bill copy need to be protected during check-in/check-out saving thousands of papers in a year (Eric, Fevzi, & Wilco, 2017).

Use of electronic handy tablets incorporated as menu option in the food and beverage outlet of hotels and as guest directory and a survey tool can reduce considerable amount of paper usage by saving printing costs and makes it easier to update menu items more effectively and hassle free. Smart gadgets so extremely popular among people now days because of convenience to use that incorporating it in menu section will help provide better guest satisfaction. Through AI and IoT enabled smart devices the order once placed by the guest can be recorded along with the feedback which can later be used to provide customized service. Hotel staff can schedule room assignment, plan VIP Lists for floor managers, check and update room status without wastage of paper with the use of AI enabled e -housekeeping program (Eric et al., 2017).

A minor switch, like including digital compendiums to rooms which include electronic guestroom tablets, will replace paper waste and make your hotel more environmentally friendly. It is estimated that around two pounds of waste per night is generated by a hotel guest; half that waste is accounted for by paper, plastic and cardboard. Printed menus, flyers, and in-room directories make a positive contribution to an enormous amount of waste in many other hotels, so switching to a digital tablet for guest rooms can have an instant effect by cutting down printing costs and waste generation which on other hand can provide an excellent channel to upsell the additional product and services to boost the revenue for instance hotels can improve revenue from ancillary products like spa by posting well-timed notifications and marketing through their digital guest directory. Going green with digital directories is a win-win scenario for hotels.

Reusable tags

Hotel Industry handles humongous amount of guest each day which bring with them a lot of luggage along with them. One of the most promising technology is RFID (radio frequency identification) which can be incorporated for automatic object recognition and surveillance in an IoT grid (López, Ranasinghe, Harrison, & McFarlane, 2012). RFID tags can be of great help in tracking and identification of luggage in hotel industry. DeVries (2008) reflected upon the value of improving RFID infrastructure to achieve efficient luggage monitoring and traceability, he indicated that expense involved with RFID is certainly one of the major downsides to RFID execution. RFID can make luggage surveillance more efficient against conventional bar code technology. RFID has better capability, but a high cost involved when equated to the usage of bar code in the supervising of luggage makes it a second choice. When using RFID in baggage tracking, mobile devices can be used to display the tracking data. Mobile monitoring helps users to track and map luggage location anywhere at any time. Smartphone usage for tracking baggage links baggage location and guest details. One of the recent mobile monitoring research paper discusses on the use of in build RFID bracelet which is to be worn by the guest which tracks the check-in stage of the guests. The RFID signals are processed by the program database and messages are send to the bracelet when there is movement of luggage form one place to other (Sennou, Berrada, Salih-Alj, & Assem, 2013). Check-in processes of hotels can be enhanced in providing updated luggage services for the consumers.

The benefits of RFID in the handling of luggage are not only limited to mishandlings and providing greater prominence but also updated monitoring along with reduced usage of paper glue tags. Using the recyclable tag or installing the RFID tags in your baggage will reduce every year dumping of trillions of gum-based single use luggage tags. This has a constructive effect on the atmosphere by reducing wastage. Usages of recycled tags will be more environment friendly as it will save massive quantities of printed adhesive luggage tags. The frequent travelers can get benefited by the RFID tags as they can continuously track the luggage position during multiple stays. The continuous tracking of the luggage can be done through mobile devices once the "on board" signal of luggage is active and messages like 'in transit to the hotel' and 'at the hotel' status to make it convenient for the guest without worrying for the luggage collection at various check points.

Water management

Water is such a critical tool for the growth of tourism, a core feature of the various tourism sectors activities important to the scenic beauty of the landscape and required for the promotion of accommodation (Gössling, Hall, & Scott, 2015). World Tourism Organization (World Tourism Organization, 2017) states that the number has risen with respect to International visitors and is likely to continue. Requiring additional tourism-support and initiatives services for example, hotels, eateries, health spa, and several other leisure services that needs high water consumption which will eventually exert pressure on the supply of water (Hawkins & Bohdanowicz, 2012). Potential for improved water supply, demand would worsen and any existing issues with water shortages and eventually lead to increased conflict over access to existing water supplies. Therefore water quality and recycling, wastewater management and drinking water quality are the top three critical requirements (World Tourism Organization, 2004).

According to recent research by SIWI, almost 20% of the world's residents is living in areas of physical water scarcity. The water scant area is one where the production of water supplies is "approaching or exceeding sustainable limits" and "more than 75% of river flows are diverted for irrigation, industry and domestic purposes." By 2030, the planet will face 40% of global demand/supply. The private sector is a big water user and is also totally dependent on water for development and delivery of services.

Hospitality is one of those sectors where water plays a key role in daily operations and future development (Tupen, 2013).

To focus on the efficient usage of energy and to stimulate sustainability a lot of technologies are used by smart building. An innovative water management technology is used by the hotels where the collected rainwater is recycled into two parts the brown water is sued for toilet and the recycled ware is used for other purposes. Hotel guests must be educated about reusing towels and toilet paper must be replace d by showers (Walmsley, 2011). Low cost IoT enabled meres can help track the usage water. Water usage in hotels can be further traced by installation of smart bathrooms that consists of smart flow-controlled toilets, smart sinks and smart showers (Kansakar, Munir, & Shabani, 2019). IoT enabled automated system to tract water usage and wastage through wireless technology which will collect the data through Wi-Fi or LAN and inform the user through cell phone can be good choice by hotels (Saseendran & Nithya, 2016).

The smart technology system called Aguardio, which gathers data on the length of the shower and the actions of the guests. Developed by a Danish company specially for hospitality market Aguardio incorporates machine learning algorithms for showering activities classification through the Watson AI IBM platform (Pereira & Romero, 2017).

In Hotel and lodging industry water use is a crucial element it and adds considerably to carbon emissions. Are hotels realizing this and are they ready to spend and practice! Well, industry has taken a turn towards implementing these initiatives to reach greater heights in coming years. Policy makers also play crucial role as they are making it mandatory for organizations to initiate and implement to take a step towards sustainable future.

Rainwater harvesting

Clean water is a precious asset of the 21st century. Therefore rainwater harvesting is becoming such an important issue. This can be broadly defined as the collection and storage of rainwater for a variety of uses—including irrigation, livestock, washing, indoor heating, etc. If one has the proper technological resources, the harvested water can even be converted into drinking water.

Entire world is dealing with the problem of water shortage. It is high time tourism industry must start focusing on effective and efficient water usage at different degrees tourism happens in every corner of the world and establishes itself as one of the industry with swift growth in the World (World Travel & Tourism Council, 2011) which accounts for high impact on economy and environment. Environment management practices is adopted by many big hotels that have four or five start rating (Holcomb, Upchurch, & Okumus, 2007; Hsieh, 2012). These hotels have potential to administer such measures. By implementing such positive changes these hotels can achieve a different status in the market (Farsari, 2012).

The method of rainwater harvesting collects water from the roof and pumps it to the feeder tank from an underground tank. Water can also be used for washing machines and gardening from the system of rainwater. To distinguish from the main water supply rainwater pipe must be marked clearly. Cooling towers can be replaced with cooled water or air system which are energy efficient. IoT enabled timers can be used for irrigation and planting as it needs less water to water gardens and to wash vehicles hand-held trigger nozzles can be used (Sustainable hotels web).

Building management systems

All the major growing industries are taking special interest in sustainable products and services due to the increased concern of the environmental damages and people becoming educated on the same. To gain strategic differential advantages hotel operations are taking special interest to reduce the environmental impact by reengineering their operational strategies because of increased interests of the people in creating a sustainable environment by efficient and effective utilization of natural resources (Balaji, Jiang, & Jha, 2019; Molina-Azorin, Tari, Pereira-Moliner, Lopez-Gamero, & Pertusa-Ortega, 2015).

To obtain a sustainable operational practice the very first thing which is of prime importance is the Heating, Ventilation and Air Conditioning (HVAC) systems of a hotel, as guests spend more time inside the hotel and hence maximum consumption energy is accounted by it. Heating, cooling and lighting of different passages and places inside the hotel forms a major energy consumption point of the building. Other factors include designs of buildings and materials, which have a significant effect on the energy consumed, as well as on the selection of the set End-use facilities (IEA, 2004). Decreasing the consumption of energy or reducing the wastage of energy without hindering the quality of service and comfort to the guest is becoming a tedious task (Vvakloroaya, 2014). It has been highlighted by different literatures that 40% of world energy is consumed in building management systems (BMS) (Harish & Kumar, 2016; Shaikh, Nor, Nallagownden, Elamvazuthi, & Ibrahim, 2014 and Ahmad, Mourshed, Mundow, Sisinni, & Rezgui, 2016) and accounts for 30% of total carbon do oxide emission (Shaikh et al., 2014). BMS is associated with different kinds of consumer centric activities which has varied usage of energy. With automated technologies implemented in a building there are possibilities of reduction of wastage of energy as well as reduction in carbo di oxide emission. These automated systems will enhance the energy savings by 30% (Shaikh et al., 2014). By incorporation of BMS which are computer aided centralized hardware and software networks monitoring and controlling different facilities of building; energy consumption of HVAC system can be enhanced by 20%–60%, lighting which is another source of maximum energy consumption can be reduced by 20%–50%, other electronics, refrigeration and water heating can be brought down to 20%–70% (Harish & Kumar, 2016).

Hospitality sector is looking forward to providing tailor fit services to consumer which can provide better experience to the guests, IoT enabled BMS can pitch in to this scenario along with focused sustainability practices. Exchange of data becomes extremely easy with the usage of IoT enabled devices, through which general spaces can be customized by guest to smart rooms (Eskerod, Hollensen, Morales-Contreras, & Arteaga-Ortiz, 2019). Apart from providing tailor fit services IoT enriched BMS will be a catalyst to smart devices and energy efficient services. Hotel an minimize the expenditure on energy costs by 20%–50% by imbedding smart devices for lighting and temperature control. Occupancy sensors installed uses machine learning algorithms on continuous basis for smart HVAC systems which will enable hotels to manage the light usage based on the trend followed in the occupancy trend. Sensor enabled BMS are a good medium to save on energy consumption and reduces maintenance cost as well. Energy usage per fixture can be monitored and considered from the control center (Eskerod et al., 2019). Building Management systems can be a helping hand to the policy makers and stakeholders to understand and monitor how the building energy consumption can be modified for better and efficient performance. Not only monitoring is important, but analysis of the collected data is equally important for appropriate decision to be taken for efficient usage of energy and reducing emission of harmful gases.

Waste management

To begin with the kind of initiatives hotel industry generally practices, let's look into it. Management of generated water in hotels is becoming a challenging job and disposing of waste is a matter of concern. If a product or services ids manufactured efficiently waste management becomes easy and money can be saved by saving on the resources. Apart from that some money can be saved and generated by selling and recycling old resources. This will ultimately result in minimizing the cost of waste management as waste generation will be kept under check.

This can start at a very small level of using reusable soap dispensers, instead of small toiletry bottles reusable dispensers, washable fabric and serving dishes as an alternative to not reusable goods, installing water filters in place of packaged plastic water bottles. Using water filters instead of plastic bottles, minimizing and reusing packing supplies, minimizing number of paper items and a switch to LED fixtures (Lawson, 2018).

There is especially important role played by hotels in the generation of waste which accounts for around 28.5 pounds of trash per day and is increasing rapidly during to expansion and growth of the industry. Industry must implement different measure to incorporate sustainable practices with measurable steps which will include benchmarking of processes and fiscal accessing of the decision taken and implemented for the measurement of research. Creating, organizational preparation, is impenetrable despite of that Some considerations, such as increased control and increased utility costs (Goldstein & Primlani, 2012b).

A research conducted in Hue City prominently showed very high rates of reprocessing the recycling capabilities reflected to very small economy class hotel to very high-end hotels, because of their waste generation and water management ratio. Waste Management Practice (WMP) is referred as a factor affecting the process of recycling with the added advantage (Son, Matsui, Trang, & Thanh, 2018).

Several research highlight on specific category of waste produced in hotels. For example, in some reports, aluminum, plastics, glass, steel, carton and cooking waste are listed as the key elements of water generated by hotels. According to other study (Zein, Wazner, & Meylen, 2008), the elements of hotel generated waste with their sources are presented in Table 16.3 which indicates whether they are hazardous or not.

Looking at the types of waste in the above table IOT implementation with respect to waste management remains a big question to answer by hoteliers. Though there are few studies and countries who have initiated and adopted the practices towards managing waste through IoT

At the large economic level, with the rapid development of technology as an excess waste has become a problem for the environment and community, building on Porter's value chain model, Garido-Hidalgo, Olivares, Ramirez, and Roda-Sanchez (2019) suggested using the Reverse Supply Chain (RSC) method. Garido-Hidalgo et al. further clarified that the performance of the RSC model relied on the effective use of radio frequency identification (RFID) in the Broad IoT (LIoT) network to communicate across the supply chain for efficient decision taking. Although the initial goal of Garido-Hidalgo et al.'s research was to use LIoT and RFID to handle waste inside the SCM framework, the significant outcome of Garido-Hidalgo et al.'s research was the applicability of Porter's value chain model in the context of the fourth industrial revolution. Al-Aomar and Hussain (2018) have confirmed that hoteliers who have adopted lean methodology such as Just in Time (JIT), increased cost-efficiency, decreased waste and increased revenue. Unfortunately, hotel industry is yet to practice these methodologies in terms of waste recycling through IoT and this remains a challenge.

Information technology and front of the house

In current scenario guests coming to hotels easily adjust to technical shift. They are also happy with everyday life which gives a smooth electronic connectivity. Brewer, Kim, Schrier, and Farrish (2008) in his research stated that the latest use of hospitality technology is very lucrative to guests.

TABLE 16.3 Nonhazardous waste generated by hotel industry (Zein et al., 2008) IoT application.

Nonhazardous waste type	Components	Source -hotel departments & IoT application
Wastes	Cooking related waste, waste, paper and packaging material, plastic packaging material or bags, composted packages	Different departments of hotel. IoT: Order tracking and optimizing supplies, tampering sensors and temp monitoring
Cardboard	Packing	Hotel's purchasing and other departments. IoT: Digital labels, Temp sensors, reuse
Paper	Printed documents, brochures, menus, maps, magazines, newspaper	Administration, reception, guests room, restaurants IoT: Use of Tabs, Browsing displays, Mobile devices, In-house navigation
Plastic	Bags, bottles (that did not contain hazardous substance), domestic goods, separate portion wrappings for different products	Kitchen, restaurants, bars, guest room, Administration IoT: Guest preference on Go Green can be used to change services using packaging to 100% recycled plastic at a cost
Metal	Metal container, jar covers, juice cans, food cans, mayonnaise, mustard and tomato puree tubes, aluminium packaging	Kitchen, restaurants, bars, guest room. IoT: tracking of the consumption and preference to optimize the use
Glass	Bottles, jars, flacks	Kitchen, restaurants, bars, guests' rooms. IoT: Monitor usage and in-house recycling
Cloth	Tablecloth, bed-linen, napkins, clothes, rags	Kitchen, restaurants, bars, bathrooms, guest rooms. IoT: Digital tracking and optimizing
Wood	Wooden packing pallets	Purchasing department. IoT: Reuse and prevention of damages
Organic waste	Fruit and vegetables peelings, flowers and plants, twigs, leaves, grass	Pantry, eateries, bars, guests' rooms, gardens. IoT: Tracking through temperature to avoid perishing

Future developments of the industry forecasts strong consensus that Information technology has a critical role in enhancing the productivity of employees (79.9%) rising the number of customers' satisfaction (82.4%) and sales production (71.3%). The respondents, too, accepted that consumers are most interested in using Wi-Fi networks (82%). with that, a lot of business passengers along with those who wants to surf the net or test it out private e-mail accounts is becoming a frequent element in many hotels / resorts.

Another evolving technology that is becoming common in hospitality industry is machine (ML) learning (Berthelsen, 2017) which is a computer aided technology which mimics human intelligence by learning from human intelligence on sustainability issues (El Naqa & Murphy, 2015). As ML has the capacity to understand exactly forecast requirement hotels can implement it in forecasting nearly the precise amount of demand expected.

Hotel guest acknowledged the value addition by technology advances, although the introducing industry is discussing if more is essential. Technically innovative facilities like automatic self-checking and check-out, will eliminate face-to-face contact with Clients (O'Neill, 2012).

ICT popularly known as Information and Communication Technology is a revolution in progression of goods and services from construction lines to vibrant multiinvestor structures blend computer hardware, antennas, data storage, microprocessors, applications, networking and supply. A fresh generation of intelligent technology that reengineers preeminent practices and pushes the providers of service to dynamically maximize the output.

The daily routine of people has been modified and changed by wireless internet services and mobile service on daily basis. Hotel sector paying attention to specialized technological advancements to match up with the customer's needs and increased requirements throughout the reservation process. Many guests enjoy and understand the significance of loyalty services, customer support, and location comfort enriched by technological advancement.

Intelligent systems use environmental technologies (sensors, telecommunications networks, IoT and AI) providing environmental reserve value and additional insights into complex data process for companies and the clients (Salguero & Espinilla, 2018). IoT is a brand-new technical concept that links everything and everyone at any time anywhere, to give rise to innovative new technologies and support (Lu, Papagiannidis, & Alamanos, 2018).

Hotels are constantly assessing and assessing the intelligent conditions in the guest rooms to help them to handle their rooms better environment and cocreation of business developments (Sheivachman, 2018). Hilton and Marriott are exploring how environmental technology can be efficiently combined with sensors.

The performance of the company is determined by two separate factors reasons for this (Capra, Francalanci, & Slaughter, 2012) are the quality of expertise-based tools a lot of additional sophistication than conventional equipment like the control panel of room, which switches the light on when the guest arrives the hotel room and switches it off automatically when the guest exits. In fact, because these apps are technology-based guests are continually persuaded to update and optimize functionality and allow them to do so bring the new technology with you. Such principles are proving results of Ali and Frew's research (2014) that implies that sustainable hospitality can be extra successful in the case of advanced technology products or procedures incorporated by hoteliers.

Knowing guests' understanding of sustainable technology should facilitate hoteliers in creating and introducing sustainable technology. Environmental development activities to meet the needs and desires of visitors because guests can make a contribution. This should be the responsibility of hoteliers and development entrepreneurs to protect the environment reduce or even reduce the costs of acceptance. A realistic approach that could minimize the size of the cost of green technologies could be a strategic benefit for the hotel. For example, offering exclusive deals or rewards to regular users may be an enticing technique that is not the only one encourages visitors to remain dedicated to their ecological duty, but still holds visitors faithful to the name of the hotel. Hoteliers and product designers need to streamline sustainable technology implementations as much as feasible. These can be made user responsive. Guests ought to know about the interface and how it functions. Installation or activity will influence the climate. Perhaps it is possible to receive inputs from guests who have used green technologies and that is the only realistic way to learn how to use the technologies which will add value towards sustainable development. Hospitality firms, however, are spending excessively and paying attention towards technological needs which intern enhances internal processes and service quality.

Conclusion

Mankind was always concerned about the health of people and their well-being. Now particularly with the increasing awareness and issues known to public, environment sustainable development has been in the prime focus. Choices are clearly suggesting preference towards Eco-friendly brands. In the hospitality industry economics and environmental friendliness go together. As normally seen, the client always seeks an authentic experience and sustainable practices across all departments of the hotel not just for tools that conserve electricity but also for eco-design. The biggest investment in setting up the eco boutique hotel is related to the sustainability agenda it aims to follow well-built facilities and adaptability to quickly implement emerging technology in guest rooms and open areas means that in upcoming developments the eco hotel can remain in synchronization with the modern hotel. Green marketing is a strong tool for achieving a strategic edge over conventional marketing. The hotels which attract the most visitor are the ones who encourage green practices in terms of sustainability policies and boosts business performance. Therefore the primary goal of hotels is to guide the client as well as staff towards adopting a green system.

Hoteliers ought to do better than the absolute minimum to truly create a difference with the community and achieve a strategic advantage. Sustainability needs to be more than simply box-ticking ecological concerns-your principles and strategies must reflect a sincere dedication to the cause. Although certain hoteliers claim they cannot afford to be successful or their clients do not care for their actions, the reverse is real.

Travelers are constantly opting for eco-friendly travel choices, and some are also willing to pay extra to help the sustainability plan of a hotel. Sixty eight percentage of global travelers expect to live at an Eco-accommodation in 2018 (Smartvatten, 2019). A selection of technology that companies already have at their fingertips would be required to implement technical structures that would specifically enhance business growth and environmental performance.

The adoption will be part of a cultural shift that will entail a radical transformation in organization in the hospitality sector. Given the potential of these technologies to tackle climate change, we need policies to enable the hospitality sector to embrace them. Policymakers need to implement approaches for the hospitality industry's management techniques and services to practice sustainability through IoT usage.

This work can be seen as a qualitative appraisal of IoT's position in sustainability in the hospitality industry, focused on a range of comments. Investigators intend to perform systematic analysis with a view to validating the results and expanding it to various communities to enable more testing. The article lays forth more detailed methodological recommendations.

References

Aguilar, J.F.A., & Mendes, L. (2017). Conceptual theoretical approach about smart cities. In 2017 IEEE first summer school on smart cities (S3C) (pp. 132–136). IEEE.

Ahmad, M. W., Mourshed, M., Mundow, D., Sisinni, M., & Rezgui, Y. (2016). Building energy metering and environmental monitoring–A state-of-the-art review and directions for future research. *Energy and Buildings, 120*, 85–102.

Ahn, Y., & Pearce, A. (2013). Green luxury: A case study of two green hotels. *Journal of Green Building, 8*(1), 90–119.

Al-Aomar, R., & Hussain, M. (2018). An assessment of adopting lean techniques in the construct of hotel supply chain. *Tourism Management, 69*, 553–565. Available from https://doi.org/10.1016/j.tourman.2018.06.030.

Ali, A., & Frew, A. J. (2014). Technology innovation and applications in sustainable destination development. *Information Technology & Tourism, 14*(4), 265–290. Available from https://doi.org/10.1007/s40558-014-0015-7.

Ali, A., Murphy, H., & Nadkarni, S. (2016). Hospitality employers' perceptions of technology for sustainable development: the implications for graduate employability. *Tourism and Hospitality Research, 18*(2), 131–142.

Alshekhly, I. F. (2012). Smart cities: Survey. *Journal of Advanced Computer Science and Technology Research*, *2*(2), 79–90.

Amandeep, A. (2017). Green hotels and sustainable hotel operations in India. *International Journal of Management and Social Sciences Research*, *6*(2), 13–16.

Atzori, L., Iera, A., & Morabito, G. (2010). The internet of things: A survey. *Computer Networks*, *54*(15), 2787–2805.

Aymen, F., & Mahmoudi, C. (2019). A novel energy optimization approach for electrical vehicles in a smart city. *Energies*, *12*(5), 929.

Balaji, M. S., Jiang, Y., & Jha, S. (2019). Green hotel adoption: A personal choice or social pressure? *International Journal of Contemporary Hospitality Management*, *31*, 3287–3305.

Becken, S. (2014). Water equity—Contrasting tourism water use with that of the local community. *Water Resources and Industry*, *7/8*, 9–22.

Berthelsen, C. (2017). Machine learning can make hospitality businesses more predictable. *Fourth.* Retrieved from https://www.fourth.com/en-gb/blog/machine-learning-can-make-hospitality-businesses-morepredictable

Brewer, P., Kim, J., Schrier, T. R., & Farrish, J. (2008). *Current and future technology use in the hospitality industry.* Las Vegas: American Hotel and Lodging, Association & University of Nevada.

Callicott, J. B., & Mumford, K. (1998). *Ecological sustainability as a conservation concept. Ecological sustainability and integrity: Concepts and approaches* (pp. 31–45). Dordrecht: Springer.

Capra, E., Francalanci, C., & Slaughter, S. A. (2012). Is the software "green"? Application development environments and energy efficiency in open source applications. *Information and Software Technology*, *54*(1), 60–71. Available from https://doi.org/10.1016/j.infsof.2011.07.005.

Carrington, D. (2019). Why the Guardian is changing the language it uses about the environment. https://www.theguardian.com/environment/. Accessed 12 July 2019.

Chamarti, A. (2016). Internet of things is making hotels smarter, Hotel Online. Available at: https://goo.gl/a16AAV. Accessed 9 July 2020.

Chen, L. F. (2019). Hotel chain affiliation as an environmental performance strategy for luxury hotels. *International Journal of Hospitality Management*, *77*, 1–6.

Goldstein, K.A., & Primlani, V.R. (2012a). *Current trends and opportunities in hotel sustainability. HVS Sustainability Services.*

Cvelbar, L., & Dwyer, L. (2013). An importance—Performance analysis of sustainability factors for long-term strategy planning in Slovenian hotels. *Journal of Sustainable Tourism*, *21*(3), 487–504.

Davids, G. (2016). UAE hotels urged to open door to sustainability. *Middle East Construction News.* Available at: http://meconstructionnews.com/19278/uae-hotels-urged-to-open-door-tosustainability. Accessed 29 June 2020.

Davidson, M. (2010). Social sustainability and the city. *Geography Compass*, *4*(7), 872–880.

Deraman, F., Ismail, N., Arifin, A., & Mostafa, M. (2017). Green practices in hotel industry: Factors influencing the implementation. *Journal of Tourism, Hospitality and Culinary Arts*, *9*(2), 305–316.

DeVries, P. D. (2008). The state of RFID for effective baggage tracking in the airline industry. *International Journal of Mobile Communications*, *6*(2), 151–164.

El Naqa, I., & Murphy, M. J. (2015). *What is machine learning? Machine learning in radiation oncology* (pp. 3–11). Cham: Springer. Available from http://doi.org/10.1007/978-3-319-18305-3_1.

Elkington, J. (1997). *Cannibals with forks—Triple bottom line of 21st century business.* Stoney Creek, CT: New Society Publishers.

Eric, S. W. C., Fevzi, Okumus, & Wilco, Chan (2017). The applications of environmental technologies in hotels. *Journal of Hospitality Marketing & Management, 26*(1), 23–47. Available from https://doi.org/10.1080/19368623.2016.1176975.

Eskerod, P., Hollensen, S., Morales-Contreras, M. F., & Arteaga-Ortiz, J. (2019). Drivers for pursuing sustainability through IoT technology within high-end hotels-An exploratory study. Sustainability (Switzerland). https://doi.org/10.3390/su11195372

Farsari, I. (2012). The development of a conceptual model to support sustainable tourism policy in north Mediterranean destinations. *Journal of Hospitality Marketing & Management, 21* (7), 710–738.

Fortino, G., & Trunfio, P. (Eds.), (2014). *Internet of things based on smart objects: Technology, middleware and applications.* Springer Science & Business Media.

Garido-Hidalgo, C., Olivares, T., Ramirez, J., & Roda-Sanchez, L. (2019). An end-to-end internet of things solution for reverse supply chain management in Industry 4.0. *Computers in Industry, 112*, 1–13. Available from https://doi.org/10.1016/j.compind.2019.103127.

Ghallab, M., Nau, D., & Traverso, P. (2016). *Automated planning and acting.* Cambridge University Press.

Goldstein, K. A., & Primlani, R. V. (2012b). *Current trends and opportunities in hotels sustainability. HVS Global.*

Gössling, S., Hall, C. M., & Scott, D. (2015). *Tourism and water.* Bristol, UK: Channel View Publications.

Guarda, T., Leon, M., Augusto, M., Haz, L., Cruz, M., Orozco, W., & Alvarez, J. (2017). Internet of things challenges. In *12th Iberian conference on information systems and technologies* (pp. 1–4). Lisbon, Portugal.

Harish, V. S. K. V., & Kumar, A. (2016). A review on modeling and simulation of building energy systems. *Renewable and Sustainable Energy Reviews, 56*, 1272–1292.

Hassan, Q. F., & Madani, S. A. (Eds.), (2017). *Internet of things: Challenges, advances, and applications.* CRC Press.

Hawkins, R.; Bohdanowicz, P. (2012). Responsible hospitality: Theory and practice; goodfellow: Oxford, UK.

Higgins-Desbiolles, F., Moskwa, E., & Wijesinghe, G. (2017). How sustainable is sustainable hospitality research? A review of sustainable restaurant literature from 1991 to 2015. *Current Issues in Tourism, 20*(10), 1–30.

Holcomb, J., Upchurch, R., & Okumus, F. (2007). Corporate social responsibility: What are top hotel companies reporting? *International Journal of Contemporary Hospitality Management, 19*(6), 461–475.

Hsieh, J. (2012). Hotel companies' environmental policies and practices: A content analysis of their web pages. *International Journal of Contemporary Hospitality Management, 24*(1), 97–121.

IEA. (2004). Energy balances for OECD countries and energy balances for non-OECD countries; Energy statistics for OECD countries and energy statistics for non-OECD countries, Paris.

Kang, K. H., Stein, L., Heo, C. Y., & Lee, S. (2012). Consumers' willingness to pay for green initiatives of the hotel industry. *International Journal of Hospitality Management, 31*(2), 564–572.

Kansakar, P., Munir, A., & Shabani, N. (2019). *Technology in the hospitality industry: Prospects and challenges. IEEE Consumer Electronics Magazine.*

Kim, S., Lee, K., & Fairhurst, A. (2017). The review of "green" research in hospitality, 2000–2014: Current trends and future research directions. *International Journal of Contemporary Hospitality Management, 29*(1), 226–247.

Kratzert, T. Collignon, H. Broquist, M. and Vincent, J. (2016), The internet of things- A new path to European prosperity, available at: https://goo.gl/7iaDm7. Accessed 12 July 2020.

Lawson, E. (2018). https://www.hotel-online.com/press_releases/release/effective-ways-of-waste-management-in-the-hotel-industry-and-its-importance/.

Leahy, S. (2019). Climate study warns of vanishing safety window—Here's why. https://www.nationalgeographic.com/environment/2019/03/climatechange-model-warns-of-difficult-future/. Accessed 1 July 2019.

Les Roches, (2018). Smart hotels and the internet of things (IoT): Myths uncovered, Les Roches [Online]. Available: https://www.lesroches.edu/blog/smart-hotelsand-the-internet-of-things-IoT-mythsuncovered/. Accessed 3 July 2020.

Linkov, I., Trump, B. D., Poinsatte-Jones, K., & Florin, M. V. (2018). Governance strategies for a sustainable digital world. *Sustainability*, *10*(2), 440.

Liyanage, S., Dia, H., Abduljabbar, R., & Bagloee, S. A. (2019). Flexible mobility on demand: An environmental scan. *Sustainability*, *11*(5), 1262.

López, T., Ranasinghe, D., Harrison, M., & McFarlane, D. (2012). Adding sense to the internet of things as architecture framework for smart object systems. *Pers Ubiquit Comput*, *16*, 291–308.

Lu, Y., Papagiannidis, S., & Alamanos, E. (2018). Internet of things: A systematic review of the business literature from the user and organisational perspectives. *Technological Forecasting and Social Change*, *136*, 285–297.

McKenzie, S. (2004). Social sustainability: Towards some definitions.

Mesirow, R., & Blumenthal, J. (2019, April 18). Why IoT is the "Smart" Solution for the Hospitality Industry. *Phocus Wire*. https://www.phocuswire.com/PwC-opinion-IoT-for-hospitality. Accessed 3 June 2020.

Miočić, B. K., Korona, L. Z., & Matešić, M. (2012). Adoption of smart technology in Croatian hotels. *Proceedings of the 35th international convention MIPRO* (. 1440–1445). IEEE.

Molina-Azorin, J., Tari, J., Pereira-Moliner, J., Lopez-Gamero, M., & Pertusa-Ortega, E. (2015). The effects of quality and environmental management on competitive advantage. A mixed methods study in the hotel industry. *Tourism Management*, *50*, 41–54, [CrossRef].

Morelli, J. (2011). Environmental sustainability: A definition for environmental professionals. *Journal of Environmental Sustainability*, *1*(1), 2.

Naimat, A. (2017). *The internet of things market: A data-driven analysis of companies developing and adopting IoT technology*. Canada: O'Reilly Media, Inc. Available at: https://goo.gl/PEvjoe. Accessed 9 July 2020.

Namee, J. (2020). Achieving paperless operations and document automation with AI and ML. readwrite.com. Accessed 9 July 2020.

Naphade, M., Banavar, G., Harrison, C., Paraszczak, J., & Morris, R. (2011). Smarter cities and their innovation challenges. *Computer*, *44*(6), 32–39.

O'Neill, J. (2012). Face time in the hotel industry. *Journal of Hospitality & Tourism Research*, *36*(4), 478–494.

Olawumi, T. O., & Chan, D. D. W. (2018). A scientometric review of global research on sustainability and sustainable development. *Journal of Cleaner Production*, *183*, 231–250.

Omohundro, S. M. (2007). The nature of self-improving artificial intelligence. *Singularity Summit*, 2008.

Pearon, S., & Parambil, M. (2019). Hotel industry and environmental impact [Online]. Available from: http://bit.ly/2OGYPoE

Pereira, A. C., & Romero, F. (2017). A review of the meanings and the implications of the Industry 4.0 concept. *Procedia Manufacturing*, *13*, 1206–1214.

Pizam, A. (2017). The internet of things (IoT): The next challenge to the hospitality industry. *International Journal of Hospitality Management*, *62*, 132–133.

Rodríguez-Antón, J. M., del Mar Alonso-Almeida, M., Celemín, M. S., & Rubio, L. (2012). Use of different sustainability management systems in the hospitality industry. The case of Spanish hotels. *Journal of Cleaner Production*, *22*(1), 76–84.

Rosenbaum, M., & Wong, I. (2015). Green marketing programs as strategic initiatives in hospitality. *Journal of Services Marketing*, *29*(2), 81–92.

Salguero, A. G., & Espinilla, M. (2018). Ontology-based feature generation to improve accuracy of activity recognition in smart environments. *Computers and Electrical Engineering*, *68*, 1–13.

Saseendran, S., & Nithya, V. (2016). Automated water usage monitoring system. In *International conference on communication and signal processing, ICCSP*. https://doi.org/10.1109/ICCSP.2016.7754501.

Sennou, A. S., Berrada, A., Salih-Alj, Y., & Assem, N. (2013). An interactive RFID-based bracelet for airport luggage tracking system. *4th international conference on intelligent systems, modelling and simulation* (. 40–44). IEEE.

Shah, S. H., & Yaqoob, I. (2016). A survey: Internet of things (IoT) technologies, applications and challenges. In *IEEE smart energy grid engineering (SEGE)* (. 381–385). IEEE.

Shaikh, P. H., Nor, N. B. M., Nallagownden, P., Elamvazuthi, I., & Ibrahim, T. (2014). A review on optimized control systems for building energy and comfort management of smart sustainable buildings. *Renewable and Sustainable Energy Reviews*, *34*, 409–429.

Sheivachman, A. (2018). Smart hotel guest rooms are almost here. *Skift*. Available at: https://skift.com/2018/02/13/smart-hotel-guest-rooms-are-almost-here.0

Smartvatten. (2019). The ultimate guide to sustainable hotel practices [Online]. Available from: http://bit.ly/37wY2iA. Accessed 25 Nov 2019.

Son, L. H., Matsui, Y., Trang, D. T., & Thanh, N. P. (2018). Estimation of the solid waste generation and recycling potential of the hotel sector: A case study in Hue City, Vietnam. *Journal of Environmental Protection*, *09*(07), 751–769. Available from https://doi.org/10.4236/jep.2018.97047.

Tupen, H. (2013). https://www.greenhotelier.org/know-how-guides/water-management-and-responsibility-in-hotels/. Accessed 25 July 2020.

Vvakloroaya, V. (2014). A review of different strategies for HVAC energy saving. Energy conversion and Management, Western Sydney University.

Walmsley, A. (2011). Climate change mitigation and adaptation in the hospitality industry. In. In R. Conrady, & M. Buck (Eds.), *Trends and issues in global tourism 2011* (pp. 315–321). Berlin/Heidelberg: Springer. Available from https://doi.org/10.1007/978-3-642-17767-5_9.

World Tourism Organisation. (2004). Indicators of sustainable development for tourism destinations: A guidebook. Madrid, Spain: The World Tourism Organisation.

World Tourism Organisation. (2017). International tourism results: The highest in seven years. Available online: http://media.unwto.org/press-release/2018-01-15/2017-international-tourism-results-highest-sevenyears

World Travel and Tourism Council. (2011). Travel and tourism 2011. Retrieved October 26, 2011, from World Travel and Tourism Council website: http://www.wttc.org/site_media/uploads/downloads/traveltourism2011.pdf

Zein, K., Wazner, M. S., & Meylen, G. (2008). Best environmental practices for the hotel.

Further reading

Ashton, K. (2009). That 'internet of things' thing. *RFID Journal*, *22*(7), 97–114.

Bandyopadhyay, D., & Sen, J. (2011). Internet of things: Applications and challenges in technology and standardization. *Wireless Personal Communications*, *58*, 49–69.

Buhalis, D., Harwood, T., Bogicevic, V., Viglia, G., Beldona, S., & Hofacker, C. (2019). Technological disruptions in services: Lessons from tourism and hospitality". *Journal of Service Management*, *30*(4), 484–506. Available from https://doi.org/10.1108/JOSM-12-2018-0398.

https://hoteltechreport.com/news/sustainable-hospitality. Accessed 4 July 2020.

https://www.hotelierindia.com/business/10763-from-social-distancing-to-contactless-check-in-and-dining-interactive-technology-will-be-key-in-dealing-with. Accessed 4 July 2020.

India Hospitality Review. (2012). Indian hospitality going the green way by SanaMirza.

Kasim, A., Gursoy, D., Okumus, F., & Wong, A. (2014). The importance of water management in hotels: A framework for sustainability through innovation. *Journal of Sustainable Tourism*, *22*(7), 1090–1107. Available from https://doi.org/10.1080/09669582.2013.873444.

Malcheva, M. (2019). Green boutique hotels-marketing and economic benefits. Лзвестия на Съюза на учените-Варна. Серия Лкономически науки, *8*(3), 179–187.

Omidiani, A., & Hashemi Hezaveh, S. (2016). Waste management in hotel industry in India: A review. *International Journal of Scientific and Research Publications*, *6*(9), 670–680.

Pham Phu, S. T., Fujiwara, T., Hoang, M. G., Pham, V. D., & Tran, M. T. (2018). Waste separation at source and recycling potential of the hotel industry in Hoi A city, Vietnam. *Journal of Material Cycles and Waste Management*. Available from https://doi.org/10.1007/s10163-018-0807-5.

Sagahyroon, A., Al-Ali, A., Sajwani, F., Al-Muhairi, A., & Shahenn, E. (2007). Assessing the feasibility of using rfid technology in airports. In *1st annual RFID Eurasia* (. 1–5). IEEE.

Seuring, S., & Gold, S. (2012). Conducting content-analysis based literature reviews in supply chain management. *Supply Chain Management: An International Journal*, *17*, 544e555.

Subbiah, K., & Kannan, S. (2011). The eco-friendly management of hotel industry. In *International conference on green technology and environmental conservation (GTEC-2011)*. https://doi.org/10.1109/gtec.2011.6167681

Sustainable Hotel website [Online]. Available: http://www.sustainablehotel.co.uk/Microgen%20Rainwater%20Harvesting.html

United States Energy Information Administration. (2013). Heating and cooling no longer majority of United States home energy use [cited 2017 1.8.]. Available from: http://www.eia.gov/todayinenergy/detail.php?id = 10271&src = %E2%80%B9%20Consumption%20%20%20%20%20%20Residential%20Energy%20Consumption%20Survey%20(RECS)-b1.

Vakiloroaya, V., Samali, B., & Madadnia, J. (2011). Component-wise optimisation for commercial central cooling plant. In *The 37th international conference of the IEEE industrial electronics (IECON)* (pp. 2686–2691), Melbourne, Australia.

Chapter 17

Design and fabrication of an automatic classifying smart trash bin based on Internet of Things

Bo Sun[1], Quanjin Ma[1,2], Guangxu Zhu[1], Hao Yao[1], Zidong Yang[1], Da Peng[1] and M.R.M. Rejab[1,2]

[1]*School of Mechanical Engineering, Ningxia University, Yinchuan, China,* [2]*Structural Performance Materials Engineering (SUPREME) Focus Group, Faculty of Mechanical & Automotive Engineering Technology, Universiti Malaysia Pahang, Pekan, Malaysia*

Introduction

Municipal solid waste (MSW) refers to the materials discarded, including commercial wastes, collected and disposed of by the municipalities (Guerrero, Maas, & Hogland, 2013). With rapid economic growth and massive urbanization, municipal solid waste is the major challenge to recover and recycle as alternative renewable energy (Cheng & Hu, 2010). According to Allied Market Research, Portland, Oregon, waste management worldwide is expected to grow at an annual rate of 6.2% by 2023, with more significant growth in the emerging Asia Pacific region. In Europe, this sector grew by more than 30% in 2016 and growth is expected to continue to accelerate due to the presence of advanced infrastructure and the high demand of several interested sectors (Pardini et al., 2019). Moreover, the quantity of MSW generation in 214 large and medium-sized Chinese cities was 188.51 million tons in 2016 (Yang et al., 2018). With the increasing amount of wasted material, there is an urgent demand for trash garbage with classification and recycling functions (Dahlén & Lagerkvist, 2010). Due to the lack of relevant infrastructure and facilities, garbage classification and recycling have large potential application fields.

The conventional trash bins adopt the concept of foot stepping, hand lifting or opening, as illustrated in Fig. 17.1. The garbage could not be effectively be classified, and does not have several functions. The garbage becomes moldy and stinky, such as various bacteria and flies. These not only lead to it not being convenient and opening the lids with hands, but also

Cognitive Computing for Human-Robot Interaction. DOI: https://doi.org/10.1016/B978-0-323-85769-7.00013-6

FIGURE 17.1 Image of traditional trash bins.

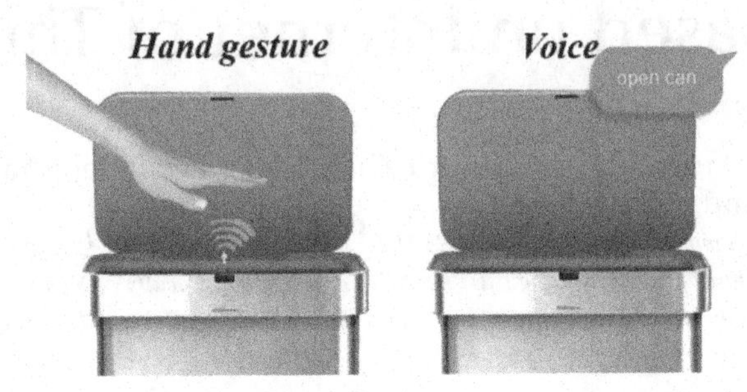

FIGURE 17.2 Schematic illustration of smart trash can from simple human company. *https://www.simplehuman.com/.*

polluting the surrounding environment, particularly in the shopping and restaurants. However, people are trying to depend on high technology to handle such burdensome issues (Hannan et al., 2016).

There is a smart trash bin with mature technology on the market, which can automatically open its cover (Al Mamun, Hannan, & Hussain, 2014; Saranya et al., 2020). Fig. 17.2 indicates the smart trash bin in the current market. As the function of this garbage can is still relatively single, it does not actualize the automatic garbage classification and overflow reminding, which either results in the extra workforce in later classification and regular inspection. Therefore there are still many areas for improvement to solve the environmental issues we confront. However, everything is being based on internet of things (IoT), and our lives have become increasingly digitalized (Elminaam et al., 2009; Mehmood et al., 2017). Fast evolution of cloud computing and IoT make all the devices used in our life are interconnected with IoT (Gaur et al., 2015). Besides, modern public places can provide a stable power supply, extensive coverage of WiFi, a mature management system, and a sufficient flow of people (Lambrechts & Sinha, 2016). These advances above are suitable for the use of smart trash bins. It is expected that they will be seen soon, especially in some international financial districts.

This paper aims to describe the hardware and software design ideas of trash bins in the implementation process, and the mechanism of garbage classification and the application of the IoT in garbage recycling are emphatically studied. The device is tested and evaluated according to real conditions, which offers good response performance. The results highlighted that the system could run stably and achieve accurate classification within 2 seconds, which reduces the waste of resources, changes the traditional garbage management mode, improves management efficiency, and realizes recycling of garbage.

General scheme design

The automatic classifying smart trash bin under a WiFi environment mainly includes hardware section and software section (Kumar et al., 2017). The hardware comprises a master control module, WiFi module, sensor module, steering engine module, motor drive module and trash bin body. Furthermore, the software part is connected to the trash bin through a mobile phone APP and WiFi, and the smart trash bin is controlled by a computer program to realize its functions. The smart trash bin's primary functions are garbage automatic classification, garbage compression, garbage overflow reminder, automatic opening and closing of trash bin flip lid, and real-time garbage monitoring. The general scheme design is shown in Fig. 17.3.

System hardware design

The control system is selected to use Arduino as the master microcontroller, which refers to several portable devices (Pradipta, Hendrawan, & Quanjin, 2020; Quanjin et al., 2017, 2018, 2019a; 2019b). The hardware of the system is mainly composed of a master control module, WiFi module, sensor module, motor drive module, steering engine module, and trash bin body, all of

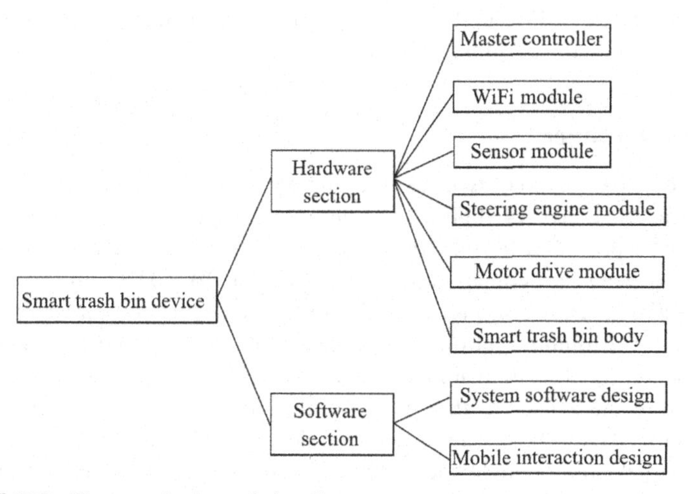

FIGURE 17.3 The general scheme design diagram.

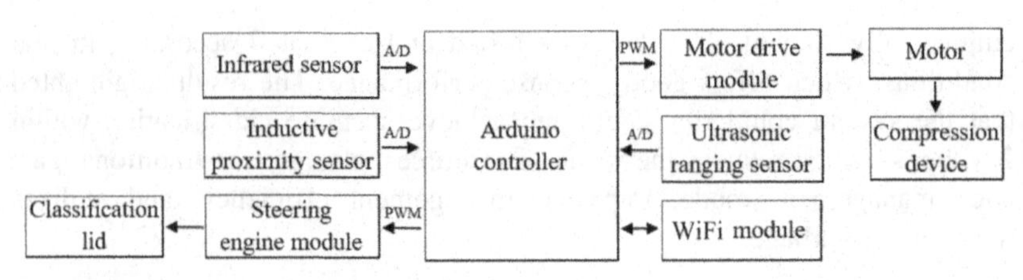

FIGURE 17.4 Hardware module design diagram.

TABLE 17.1 Main specifications of components used in this study.

Specification	Parameter
Arduino uno microcontroller	5 V/Digital I/O pins: 14
ESP8266-12F WiFi module	3.0 V ~ 3.6 V/SPI Flash: 32 Mbit/I/O: 9
Infrared sensor	DC 5 V/Range: 3–80 cm
Inductive proximity sensor	3 wire NPN/DC 6 ~ 36 V/Range: 4 mm
Ultrasonic ranging sensor	Angle < 15 degrees/Range: 2 ~ 400 cm[fg\fg]
SG90 steering engine	Torque: 1.6 kg/cm/Rotation: 90–180 degrees
Led lamp	5 mm 2 Pins/2 ~ 2.2 V/20 mA

which are connected, and Arduino controller is used as the central control module among them to adjust the operation of each module. An ESP8266 WiFi module shows the communication between people and the smart trash bin in the WiFi environment (Ling et al., 2020). The hardware module design is shown in Fig. 17.4, and the relevant specifications and several parameters of modules are summarized in Table 17.1.

Master controller

Arduino UNO R3 is used as the master controller, and its model is ATmega328. It has 14 digital I/O pins, six analog input terminals, a USB connection port, and a 16 Mhz crystal oscillator. Due to the bootloader, it can download programs directly through USB. The image of the ATmega328 is shown in Fig. 17.5. It is designed to transmit the detected signals to the Arduino controller to process, and the Arduino uno controller sends those signals out again (Kumar et al., 2016). The processed signal initially from the sensor module is transmitted to the motor drive module to perform accurate classification, compression, and flip-lid opening and closing; The detection module and the alarm module are used to receive the processed ranging signals.

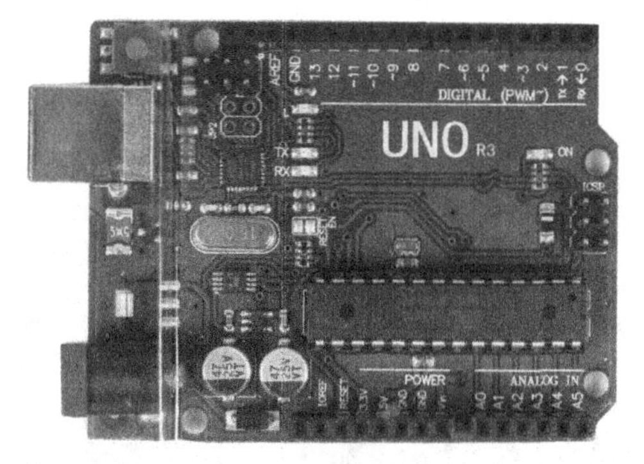

FIGURE 17.5 Image of ATmega328.

FIGURE 17.6 Image of the WiFi module.

WiFi module

It is the ESP8266-12F WiFi module that is used to establish a connection between users and devices. Once WiFi emitted from the system is connected to a usable router, and corresponding modules in the system transmit corresponding instructions through WiFi (Al Mamun et al., 2013). Furthermore, the trash bin's system hardware enables to send and receive data through the WiFi connection and Arduino interaction. Hence, the users can send action instructions to the Arduino controller to make corresponding control, such as driving the trash bin to open and close through the steering engine module (Zhang et al., 2020). By the connection, the trash bin sends real-time data to the user terminal (mobile phone APP) as well as realize remote monitoring. The image of the WiFi module is shown in Fig. 17.6.

Sensor module and alarm module

The sensor module comprises an E18-D80NK-N infrared sensor, LJ18A3-8-ZBX inductive proximity sensor, and ultrasonic ranging sensor. The E18-D80NK-N infrared sensor is a photoelectric sensor integrating transmitting and receiving. The emitted light is modulated and sent out, and the receiving

FIGURE 17.7 Image of an infrared sensor and the inductive proximity sensor.

FIGURE 17.8 Image of the ultrasonic ranging module.

end demodulates and outputs the reflected light, thus resulting in the interference of visible light is effectively avoided. Besides, with a lens, the sensor can detect objects within 80 cm as far as possible. An oscillator is mounted inside the LJ18A3-8-ZBX inductive proximity sensor. When the target objects with different dielectric constants and volumes approach, the oscillator starts to vibrate, which is processed by the circuit and converted into electrical signals, thus realizing the purpose of detecting the object types. The image of the infrared sensor and the inductive proximity sensor is shown in Fig. 17.7.

United States-100 ultrasonic ranging module used in this system owns the noncontact ranging of 2 cm–4.5 m and has its temperature sensor to correct the ranking results. It has various communication modes such as GPIO and serial port, and there is a watchdog inside. The sensor is used to detect the garbage stacking height in the trash bin. The LED lamp is used to flash and remind when the trash bin overflows, thus alarming the users. The image of the ultrasonic ranging module is shown in Fig. 17.8.

Motor drive module

The deceleration motor mainly does the compression of the trash bin. In order to drive the deceleration motor to work, the L298N chip is selected in the system, which has excellent driving ability, convenient operation, good

FIGURE 17.9 Image of the L298N chip.

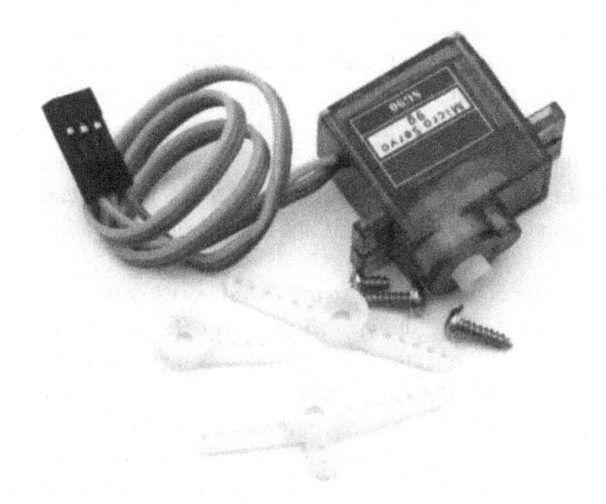

FIGURE 17.10 Image of steering engine module.

stability, and excellent performance. The image of the chip is shown in Fig. 17.9.

Steering engine module

The steering engine used in the trash bin provides a PWM interface, its power supply voltage range is DC6V-8.4 V, and the positioning accuracy is 0.2 degrees. It is mainly responsible for opening and closing the trash bin's flip lid in the classification process. The image of the steering engine module is shown in Fig. 17.10.

FIGURE 17.11 The finalized design structure.

Smart trash bin body

The smart trash bin's finalized design structure is shown in Fig. 17.11, and it is divided into three sections, including the actual bin, cover, and flip lid. The actual bin is designed as two storage boxes to collect recyclable and unrecyclable garbage, respectively. The device's main framework has the most massive shape to support and steady the upper sections. The cover is used to provide sufficient room to place all modules demanded and support the flip lid. On the one hand, the lid tilts with the help of a steering engine attached to diametrically opposite ends of the cover achieve segregation functions (Saranya et al., 2020). On the other hand, the lid can seal the trash bin's opening to keep the inner smell not release out due to incorporating the cover.

The 3D prototype is designed using Solidworks software to present the design concept in detail, as shown in Fig. 17.12. Firstly, the inductive proximity sensor is fixed beneath the flip lid's center to detect the garbage quickly. Next, the infrared sensor and the steering engine are placed on both sides of the cover in order, the led lamp is placed in the front part of the cover to show the working status, and two ranging ultrasonic sensors are mounted on both sides of the cover, perpendicular to the position where the infrared sensor is. Ultimately, the Arduino controller, WiFi module, and motor are placed on the necessary bin's bottom. The necessary bin and the cover are connected using the nested structure, which is assembled and fabricated.

System software design

The smart trash bin system's software design mainly includes Arduino controller software design, sensor-based garbage classification, steering engine control, motor control, WiFi communication establishment, and mobile interaction design.

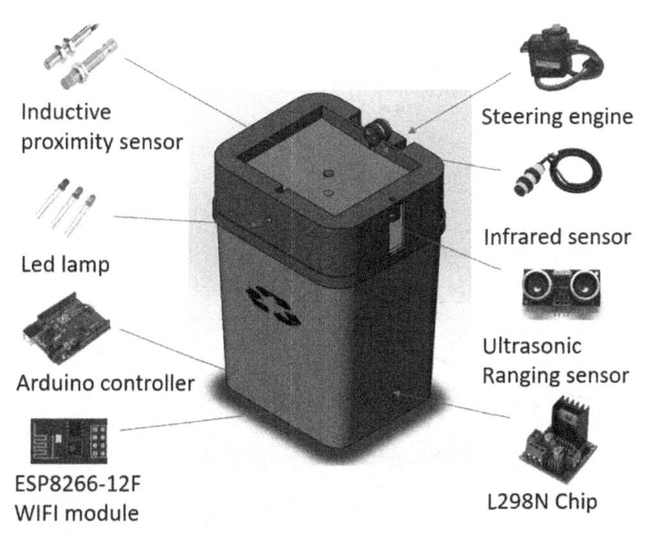

FIGURE 17.12 The design concept in detail.

Software design of the Arduino Uno controller

The general control flow chart of the Arduino controller system is shown in Fig. 17.13. After the system is powered on, the Arduino controller and other external devices are initialized first, and WiFi is automatically connected. Then the detection is divided into two parts: (1) Determine whether there is an object on the trash bin classification plate: If yes, detect the type of the object through the inductive proximity sensor, control the steering gear to rotate, and put the garbage into the corresponding trash bin to achieve the classification; If it is judged as no, it returns. (2) Judge whether the garbage stacking height in the trash bin is beyond threshold value; If no, it returns; If the judgment is yes, the motor is controlled to rotate forward and backward to compress the garbage container in the trash bin. The LED lights and the classification system stops working to remind the user to check the user through APP.

Sensor-based garbage classification control

For the procedure of classification, Fig. 17.14 indicates that it is mainly involved object type identification system that is made up of inductive proximity sensor and infrared sensor, and transmission system which consists of the flip lid and steering engine. Arduino controller collects data of garbage detected by the infrared sensor and inductive proximity sensor and compares the detected data by setting parameters to obtain whether the garbage belongs to metal or nonmetal. The Arduino controller sends an action

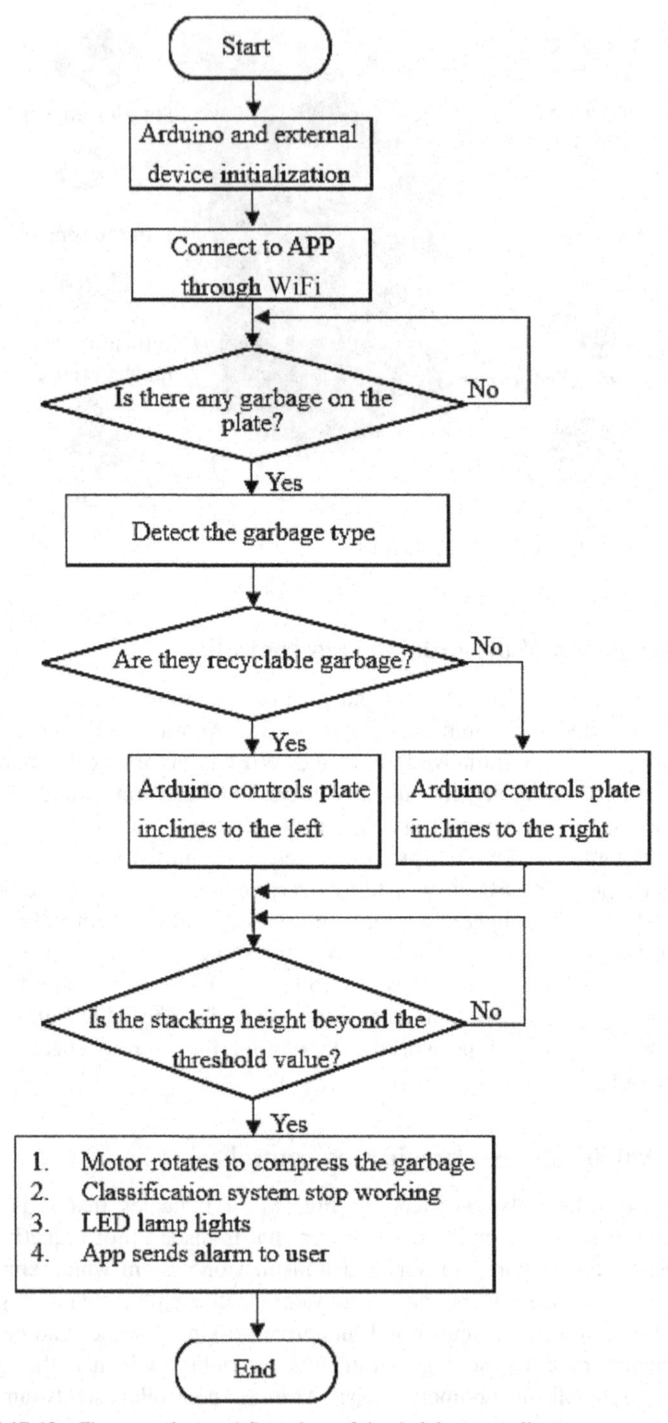

FIGURE 17.13 The general control flow chart of the Arduino controller system.

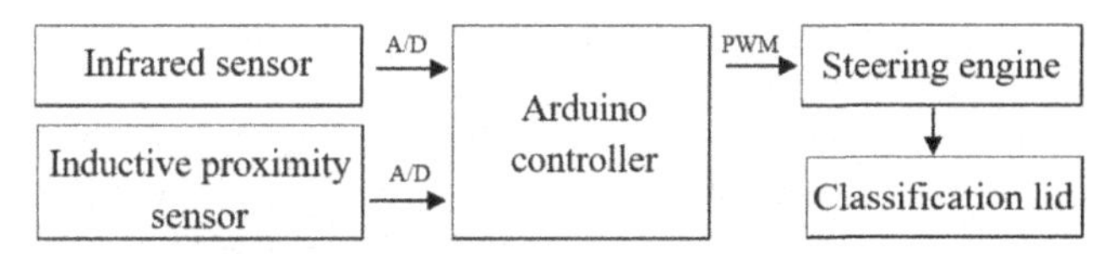

FIGURE 17.14 Sensor-based garbage classification control diagram.

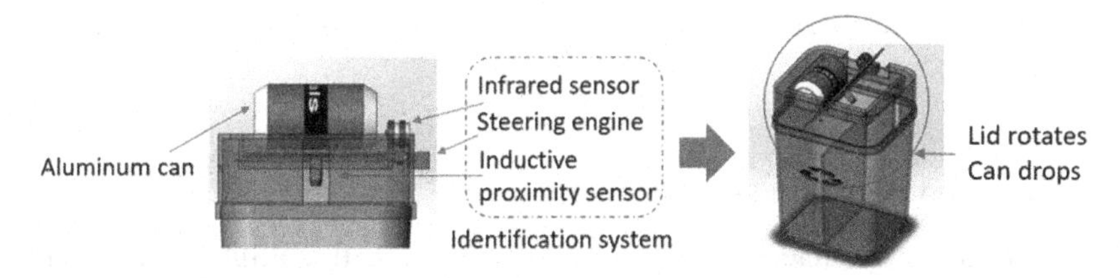

FIGURE 17.15 The identification process of aluminum can.

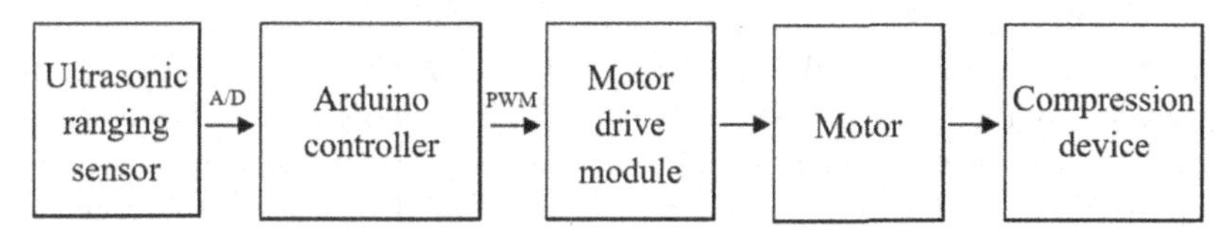

FIGURE 17.16 The hardware design of motor control.

instruction to the steering engine to let it rotate 90 degrees, so the lid will tilt from the horizontal to the vertical to let the garbage drop into one of the bins simultaneously. On the contrary, if detecting it is nonmetallic waste, such as plastic, paper. The steering engine will rotate opposite 90 degrees, and the garbage will drop into another bin. The identification process in the software section is shown in Fig. 17.15.

Steering engine control

The steering engine is mainly used to drive the rotation of the classification lid. The Arduino controller is connected with the sensors. When the sensors detect the type of garbage, the sensors send a signal to the Arduino controller, and then it controls the steering engine to rotate the classification lid.

Motor control

The motor is mainly applied to compress garbage in the system. By controlling the motor positive and negative rotation, the connecting rod is driven to rise and fall and finally compresses the trash bin's garbage. The hardware design is shown in the following Fig. 17.16. Here is its mechanism, when the ultrasonic ranging sensor detects that the distance between the highest position of garbage stacking in the trash bin and the sensor is less than 8 cm, the sensor will send a digital signal to the Arduino controller, and then it instructs the motor drive module to control the moving of the motor (Zhao, 2019). What calls for special attention is that one end of the motor drive module is used to receive the PWM signal, and the other end needs to be connected with the motor (Ma et al., 2020). Ultimately, this motor is connected with the compression device to realize the compression. The control diagram of compression is represented in Fig. 17.15.

WiFi communication establishment

This system is connected to the Internet through the ESP8266-12FWiFi module, and it is built a mobile APP with the help of the Gizwits platform to monitor the real-time data of this system. The Gizwits platform is now one of the essential development platforms for the world's IoT. Developers can quickly create an APP by the application programming interface (API) provided by the Gizwits platform for the Android mobile APP. Besides, on this basis, what equipment is required to connect to the system, and then the system generates corresponding communication protocols to complete data mutual transmission for further remote viewing of the trash bin's data. To connect the Gizwits platform through the ESP8266-12FWiFi module, it is necessary to transplant the communication firmware library into the Arduino controller. Then it is required to burn GAgent communication firmware of the Gizwits into the controller so that the data obtained by the cloud can be converted into serial port data to realize cloud communication (Zhao, 2019), the flow chart of connecting to the Internet is illustrated in Fig. 17.17.

Mobile software design

A mobile APP is designed to make it easier for users to remotely control and operate the trash bin and monitor the trash bin's real-time status. When opening the APP home page, first detect mobile phone WiFi and establish a socket connection under the available WiFi connection. If the former step is successful, keep on to the next step. However, if it fails, remind the users immediately; The next step is to enter the APP home page to send the instruction to control the trash bin and simultaneously monitor WiFi's connection status and the data uploaded by the monitoring device. If under normal circumstances, control the trash bin to make corresponding instructions. Otherwise, remind the user. In the end, when all is finished, close the link and preserve the user's information. The flow chart of mobile software design is demonstrated in Fig. 17.18 (Zhang et al., 2020).

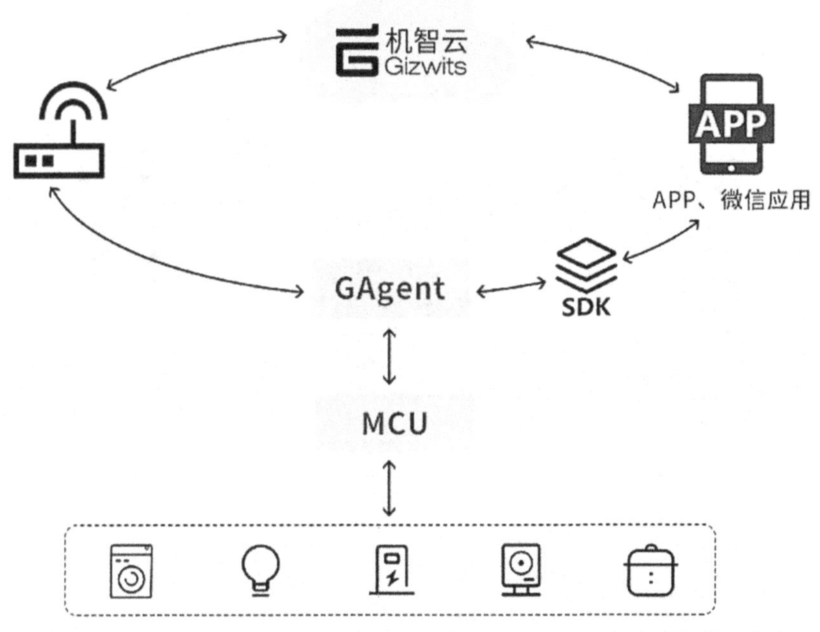

FIGURE 17.17 The flow chart of connecting to the Internet through the Gizwits platform. *https://www.gizwits.com/*.

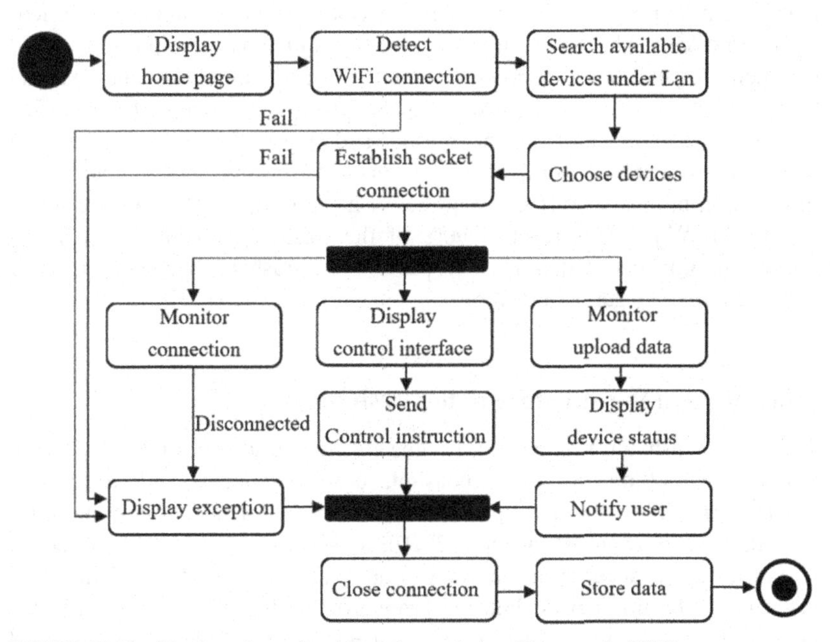

FIGURE 17.18 The flow chart of mobile software design (Zhang et al., 2020).

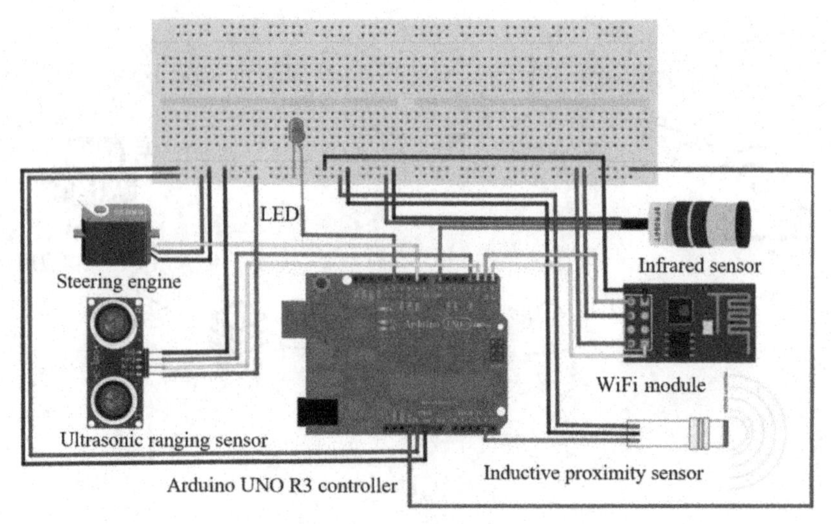

FIGURE 17.19 Wiring diagram of the smart trash bin.

Fabrication

Electric circuit connection

Fig. 17.19 indicates the smart trash bin's wiring diagram, which involves an Arduino Uno controller, WiFi module, an ultrasonic ranging sensor, an inductive proximity sensor, an infrared sensor, a steering engine, and a ledger lamp circuit. The sensitivities of three types of sensors could be adjusted via their potentiometers, depending on the usage conditions and external environment. In particular, since the three types of garbage to be classified and identified are mainly metal, plastic, and paper, the concept adopts an inductive proximity sensor as its metal sensor. Ultrasonic sensor modules are installed on both sides of the top in the trash bin to detect the garbage stacking height. When it is reached 90% of the trash bin's volume, the lid is opened automatically. It is used to prevent the trash bin from overflowing due to continuous garbage feeding.

Fabrication of the structure of the trash bin

The smart trash bin's image is designed using Solidwork software, which refers to structural design, dimension, design concept, and component arrangement, the product specification and the details of dimensions (Susanto, Hendrawan, & Quanjin, 2020), as shown in Fig. 17.20. Based on its fabrication procedure, as shown in Fig. 17.21, the cover is designed and fabricated. The lid is fixed into the groove of the top of the cover. The led lamp and the transmission system are respectively arranged at the front and

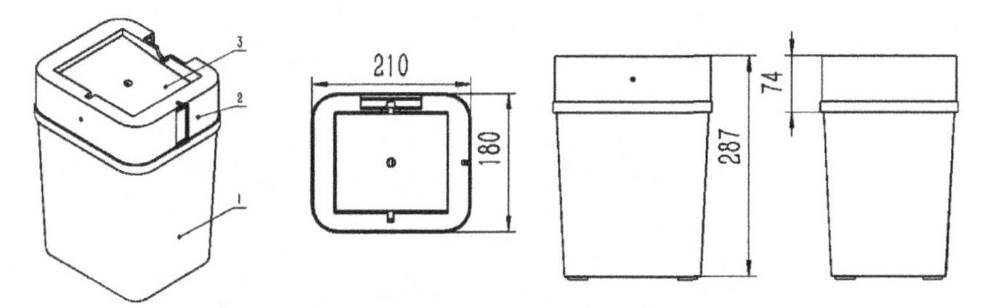

FIGURE 17.20 The structural dimensions of the smart trash bin.

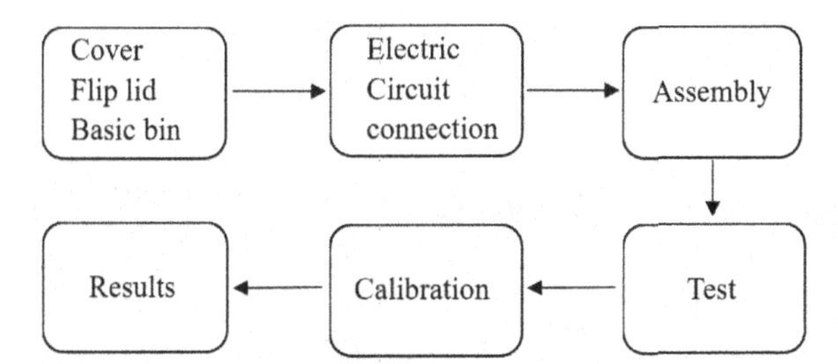

FIGURE 17.21 Fabrication procedure of the smart trash bin.

backside of the cover. The identification system is connected with a lid fixed on the cover, and the stacking height detection system is put on both longitudinal sides of the cover. The various modules are placed in the control system zone at the bottom of the trash bin, and the framework is assembled. Furthermore, the smart trash bins are fabricated with high-density polypropylene, and its cover and lid are manufactured by PLA using the 3D printing technique. The calibration procedure is performed to test the device response performance according to its design function.

Results and discussion

The 3D prototype structure is presented in Fig. 17.22A, and the prototype device is shown in Fig. 17.22B. It is concluded that the smart trash bin has been successfully designed and fabricated. According to the testing results, the smart trash bin has succeeded in automatically detecting three different garbage types and sorting them out. Besides, it has a significant effect in compressing the garbage and overflow alarm when the garbage stacking height is over the threshold value, making fair use of the storage room of trash bin volume and preventing the trash bin from overflowing continuous garbage feeding. With the WiFi module, the smart trash bin connects to the Internet and sends real-time data to the mobile APP. It perfectly develops communication between people and the smart trash bin.

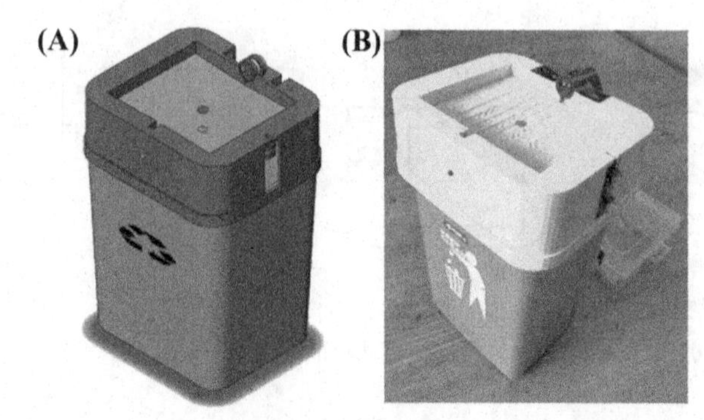

FIGURE 17.22 Schematic structure of the smart trash bin: (A) 3D prototype structure; (B) prototype device.

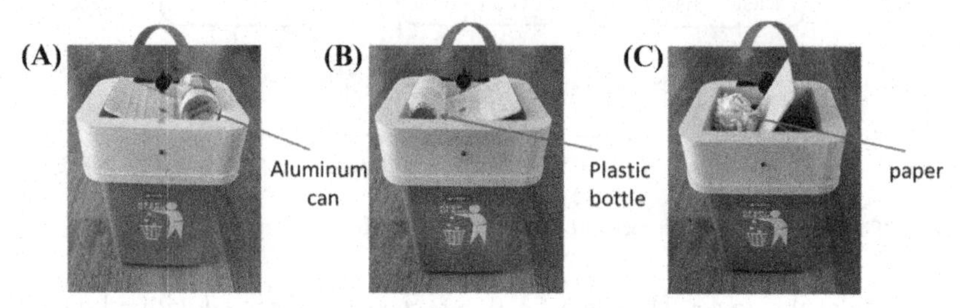

FIGURE 17.23 The classification performance of the smart trash bin: (A) metal waste-aluminum can; (B) nonmetallic waste-plastic bottle; (C) nonmetallic waste-scraps of paper.

When any exceptions from the system occur or the trash bin volume is almost full, the smart trash bin can send notifications to users at once. Besides, the users can have remote control or monitoring either. It offers much acceptable and handy method to improve garbage recycling compared to the conventional trash bin, and it reduces the chance to touch the trash bin directly and subverts people's impression of traditional trash bin as "dirty, messy, and smelly" to make our living environment more comfortable.

The smart trash bin's response performance is studied and tested, which involves garbage classification and overflow reminder. The garbage classification is as shown in Fig. 17.23. It is found that the smart trash bin can automatically sort garbage out with high accuracy when the new garbage is put on the flip lid. For metal garbage, the aluminum can on the flip lid is accurately detected by the classification system that is fixed inside the cover. For plastic garbage, it is tested using the plastic bottle to put on the flip lid. As for paper, it is detected by just putting the paper on the lid using a hand. It is tested that the response time is approximately 0–2 seconds to finish classification on three different conditions fully.

Fig. 17.24 presents the overflow reminder procedure, which is used to prevent the trash bin from overflowing after continuous garbage

FIGURE 17.24 The overflow reminder procedure.

TABLE 17.2 The test results of the smart trash bin.

Tested parameters	Value
Boot time (s)	2.0–3.6
Metal detection and segregation time (s)	0.6–1.7
Nonmetallic detection and segregation time (s)	0.9–1.9
Overflow remind reaction time (s)	0.5–0.8
Rotation angle (degrees)	0–90

compression. When the measured value collected by the ultrasonic sensor is more than the threshold value, indicating that one of the storage box volumes reaches 90% of the volume of the trash bin, the garbage lid will be controlled not to be opened, and the led lamp simultaneously lights in red to warn that situation. Based on current results, it is concluded that the smart trash bin illustrates a remarkable response performance to precisely classify the metal and nonmetallic waste and serves as a reminder of the overflow situation in a mighty short time. The testing parameters are indicated in Table 17.2.

In some public places like shopping malls, it is mainly required to keep the environment clean and tidy, and even provide several intelligent functions. There have been several similar products on the field launched into the market. To better emphasize the features and advantages between the smart trash bin and others on sale, a detailed investigation of products similar to the developed one has been carried out and shown in the comparison results in Table 17.3. Based on Table 17.3, electronic technology has been broadly adopted in the household hardware connection and a mature industrial production approach through the elaborate contrast.

Although the Micro Control Unit and its corresponding programmable software dominate the software section, Arduino Uno either gradually plays

TABLE 17.3 Comparison results of smart trash bins in the current commercial market.

Machine view	Hardware Section	Software section	Connection method	Main function	Costs	Sources
	Electronic	Micro Control Unit/ Programmable software	Wire connection	Automatic induction opening Odor control system	$55	http://www.amazon.com
	Electronic	Micro Control Unit/ Programmable software	Wire connection	Automatic induction opening Touch switch (Kick, Knock)	$32	http://www.detail.tmall.com
	Electronic	Micro Control Unit/ Programmable software	Wire connection	Automatic induction opening Touch switch Classification bins	$13	http://www.detail.tmall.com
	Electronic	Arduino Uno platform	Wire Connection/ Wireless connection	Automatic induction opening automatic classification filling reminder system Information gathering Wireless Remote control	$34	This paper

a prominent role in software control on smart home appliances because of its powerful open-source platform. Due to its convenient and flexible modular hardware connection and the easy-use integrated development environment, it is preferable to standard programmable software for designers, engineers, and anyone interested in creating interactive objects or environments.

Besides, many of the proposed researches only specialize in the ideas of the automatic induction opening system. Just a few add excess features to the products, such as an odor control system. The lack of competitive creations is coupled with innovation and practicability on the smart trash bin system. Consequently, this device represents an advanced trash bin system that provides various functions to meet more various requirements than the new homogeneous products, and it is highlighted that it is more promising to the future smart city. Depending on the future stable WiFi in the buildings, the smart trash bin can send the data to the corresponding platform, so it is significant to help managers achieve the remote control and monitoring of trash bins and further effectively and timely administer and arrange the garbage processing.

Conclusion

The automatic classifying smart trash bin is successfully designed, fabricated, and tested according to real conditions, simulating possible real usage situations. Based on current results, it is concluded that the smart trash bin shows a remarkable response performance to precisely classify, compress, and monitor the garbage in a mighty short time. Compared with a conventional trash bin, the smart trash bin effectively saves the time needed for classifying garbage, and at the same time, it dramatically reduces the workload of cleaning staff. Furthermore, the humanized product and useful functions beautify the public environment, especially in some international cities' financial districts, improve garbage recycling efficiency and reduce labor costs. The smart trash bin provides a novel solution in garbage recycling, a practical idea for future life.

Declaration of conflicting interest

The author(s) declared no potential conflicts of interest for the research, authorship and publication of this article.

Credit authorship contribution statement

Bo Sun: Writing—original draft, Formal analysis, Validation. Quanjin Ma: Validation, Investigation, Methodology. Guangxu Zhu: Resources, Conceptualization. Hao Yao: Date curation, Validation. Zidong Yang: Date curation. Da Peng: Conceptualization. M.R.M. Rejab: Supervision, Funding acquisition, Methodology.

Acknowledgment

The authors are grateful to the Ministry of Education Malaysia: FRGS/1/2019/TK03/UMP/02/10 and University Malaysia Pahang: PGRS180319 for funding this research. This research work is strongly supported by the *Structural Performance Materials Engineering (SUPREME) Focus Group* and the *Human Engineering Focus Group (HEG)*, which provided the research materials and equipment.

References

Al Mamun, M. A., et al. (2013). Wireless sensor network prototype for solid waste bin monitoring with energy efficient sensing algorithm. In *IEEE 16th international conference on computational science and engineering*. IEEE.

Al Mamun, M. A., Hannan, M., & Hussain, A. (2014). Real time solid waste bin monitoring system framework using wireless sensor network. In *International conference on electronics, information and communications (ICEIC)*. IEEE.

Cheng, H., & Hu, Y. (2010). Municipal solid waste (MSW) as a renewable source of energy: Current and future practices in China. *Bioresource Technology*, *101*(11), 3816–3824.

Dahlén, L., & Lagerkvist, A. (2010). Evaluation of recycling programmes in household waste collection systems. *Waste Management & Research*, *28*(7), 577–586.

Elminaam, D. S. A., et al. (2009). Energy efficiency of encryption schemes for wireless devices. *International Journal of Computer Theory and Engineering*, *1*(3), 302.

Gaur, A., et al. (2015). Smart city architecture and its applications based on IoT. *Procedia Computer Science*, *52*, 1089–1094.

Guerrero, L. A., Maas, G., & Hogland, W. J. W. (2013). Solid waste management challenges for cities in developing countries. *Waste Management*, *33*(1), 220–232.

Hannan, M., et al. (2016). Content-based image retrieval system for solid waste bin level detection and performance evaluation. *Waste Management*, *50*, 10–19.

Kumar, N. S., et al. (2016). IOT based smart garbage alert system using Arduino UNO. In *IEEE region 10 conference (TENCON)*. IEEE.

Kumar, S. V., et al. (2017). Smart garbage monitoring and clearance system using internet of things. In *IEEE international conference on smart technologies and management for computing, communication, controls, energy and materials (ICSTM)*. IEEE.

Lambrechts, J., & Sinha, S. (2016). *Population growth in developing countries and smart city fundamentals. The internet-of-things and wireless sensor networks. Microsensing networks for sustainable cities* (pp. 29–62). Springer.

Ling, J., et al. (2020). Intelligent trash can supervision system based on the internet of things. *Scientific and Technological Innovation*, *23*, 94–95.

Ma, Q., et al. (2020). Design and fabrication of the portable cultural relic display cabinet for public exhibition. *Journal of Advanced Research in Applied Mechanics*, *7*(1), 1–10.

Mehmood, Y., et al. (2017). Internet-of-things-based smart cities: Recent advances and challenges. *IEEE Communications Magazine*, *55*(9), 16–24.

Pardini, K., et al. (2019). IoT-based solid waste management solutions: A survey. *Journal of Sensor and Actuator Networks*, *8*(1), 5.

Pradipta, A. W., Hendrawan, A., & Quanjin, M. (2020). *Designing a portable LPG gas leak detection and fire protection device. Engineering, information and agricultural technology in the global digital revolution* (pp. 107–111). CRC Press.

Quanjin, M., et al. (2017). *Design and optimize of 3-axis filament winding machine. IOP conference series: Materials science and engineering*. IOP Publishing.

Quanjin, M., et al. (2018). Design of portable 3-axis filament winding machine with inexpensive control system. *Journal of Mechanical Engineering and Sciences*, *12*, 3479–3493.

Quanjin, M., et al. (2019a). *Wireless technology applied in 3-axis filament winding machine control system using MIT app inventor. IOP conference series: Materials science and engineering*. IOP Publishing.

Quanjin, M., et al. (2019b). Design and fabrication of the portable artwork preservative device (PAPD). *Journal of Modern Manufacturing Systems and Technology*, *3*, 31–38.

Saranya, K., et al. (2020). *Smart bin with automated metal segregation and optimal distribution of the bins. Emerging technologies for agriculture and environment* (pp. 115–125). Springer.

Susanto, A. W. P., Hendrawan, A., & Quanjin, M. (2020). Designing a portable LPG gas leak detection and fire protection device. In *Engineering, information and agricultural technology in the global digital revolution: Proceedings of the 1st international conference on civil engineering, electrical engineering, information systems, information technology, and agricultural technology (SCIS 2019)*, July 10, 2019. Semarang, Indonesia: CRC Press.

Yang, Q., et al. (2018). Evaluating the efficiency of municipal Solid Waste Management in China. *International Journal of Environmental Research and Public Health*, *15*(11), 2448.

Zhang, K., et al. (2020). Design of smart trash bin based on WiFi environment. *Information Research*, *46*(03), 74–78.

Zhao, G., et al. (2019). Research and design of smart trash bin based on NB-IoT. *Information and Communication*, *12*, 113–114.

Index

Note: Page numbers followed by "*f*", "*t*," and "*b*" refer to figures, tables, and boxes, respectively.

D

E

M

N

O

P

Q

R

S

T

Printed in the United States
by Baker & Taylor Publisher Services